T.C. SMOUT AND MAIRI STEWART

THE FIRTH OF FORTH

AN ENVIRONMENTAL HISTORY

BIRLINN

For

Penny

Andy and Sophie

John and Tom

First published in 2012 by
Birlinn Limited
West Newington House
10 Newington Road
Edinburgh
EH9 1QS

www.birlinn.co.uk

Copyright © T. C. Smout and Mairi Stewart 2012

The moral right of T.C. Smout and Mairi Stewart to be identified as the authors of this work has been asserted by them in accordance with the Copyright, Designs and Patents Act 1988.

All rights reserved. No part of this publication may be reproduced, stored or transmitted in any form without the express written permission of the publisher.

ISBN: 978 1 78027 064 7

British Library Cataloguing-in-Publication Data
A catalogue record for this book is available from the British Library

Typeset by Brinnoven, Livingston
Printed and bound by Gutenberg Press, Malta

CONTENTS

List of Figures	iv
List of Colour Plates	vii
List of Tables	viii
Preface and Acknowledgements	ix
Picture Credits	xii
Introduction	1
1 Hunting and Gathering	9
2 Fishing in a Firth of Plenty	24
3 Oyster Wars	46
4 Herring Boom and Herring Bust, 1820–1950	74
5 Lines and Trawls	96
6 Traps and Nets in the Estuary	125
7 Pollution	144
8 Land Claim from the Sea	175
9 The Bass and its Gannets	201
10 The Isle of May and the Other Seabird Colonies	224
11 Seals: The Bone of Contention	251
Conclusion	269
Notes	278
Some Further Reading	294
Index	297

LIST OF FIGURES

0.1 The Modern Forth
0.2 Forth Rail Bridge, 1888
0.3 Longannet *c*.1980
0.4 Rosyth Bay, 1909
0.5 Pittenweem fish-market, 1930s
1.1 Archaeological sites
1.2 Prehistoric rubbish tip, Polmont, 1983
1.3 Seaweed use
1.4 Seaweed cart, Pittenweem, *c*.1910
2.1 Crail burgh seal, 1357
2.2 Kilrenny burgh seal, 1578
2.3 Fishing communities 1500–1800
2.4 Crail harbour, 1880
2.5 Pittenweem quay, early nineteenth century
2.6 Fisherrow fishwife, *c*.1950
3.1 Firth of Forth oyster beds
3.2 Newhaven fisherman, *c*.1845
3.3 Newhaven fishwives, *c*.1847
3.4 Oyster dredge, 1960
3.5 Oyster boats fishing, *c*.1870
4.1 Fishing inshore Dunbar, early nineteenth century
4.2 Mr and Mrs Livingstone, Fisherrow, *c*.1900
4.3 Herring fishing, 1830–1950
4.4 Crans of herring landed in Anstruther Fishery District, 1854–1954
4.5 Methods of fishing
4.6 Women gutting herring, Dunbar, *c*.1905
4.7 A fishing fleet waiting for the tide, *c*.1880
4.8 Union Harbour, Anstruther, 1890

4.9   Boats leaving Anstruther, 1934
4.10   Fisherrow herring catch, 1955
5.1   Women baiting lines, St Andrews, *c.*1847
5.2   Baiting lines in North Berwick, *c.*1920
5.3   Mussel cart, Musselburgh, *c.*1920
5.4   Professor Thomas Huxley
5.5   Professor W.C. M'Intosh
5.6   The fishing fleet in 1912
5.7   Golden Haddock Award, 1973
6.1   Fishing in the upper firth
6.2   Fish trap near Culross
6.3   Herring and sprats – the difference
6.4   Newhaven sprat boats, *c.*1950
6.5   Salmon nets, 1983
6.6   Salmon nets at Kinghorn
7.1   The Forth Catchment Area
7.2   Craigentinny Meadows, 1847
7.3   Main polluted rivers, *c.*1870
7.4   A kettle from the River Almond, 1959
7.5   Edinburgh sewage disposal
7.6   Building the new sewers, Edinburgh, 1972
7.7   Oiled scaup from the Firth of Forth
8.1   Grangemouth from the air
8.2   Sea dykes at Queenshaugh, Stirling
8.3   Roy's map of the shore between the Carron and the Avon
8.4   Location map: Upper Firth
8.5   The shoreline at Airth
8.6   Land reclamation on the Forth Estuary
8.7   The site of Longannet power station in 1965
8.8   Luftwaffe photograph, Longannet, 1939
8.9   Property of Scottish Oils, 1923
8.10   Kinneil shoreline in 1929
9.1   Shooting on the Bass, 1876
9.2   The Bass around 1903
9.3   The Bass from the air, 1969
9.4   The Bass around 2009
9.5   The Bass lighthouse keepers, 1909
10.1   Islands of the Firth of Forth

10.2 Isle of May, 1979
10.3 Creatures of the Forth, 1684
10.4 Isle of May Low Light
10.5 Breeding populations of certain seabirds in the Firth of Forth
10.6 Shags on the Isle of May
11.1 Seals at the Isle of May, 1979
11.2 Seals on Fair Isle, *c.*1920
12.1 Whales ashore near Dunbar, 1940
12.2 Families on Kinghorn beach, 1969
12.3 Oiled birds at Musselburgh, 1978
12.4 Fire at Grangemouth, 1987

LIST OF COLOUR PLATES

1 Matthew Paris's map of Britain, *c*.1250.
2 Laurence Nowell's map of *c*.1562.
3 The inner Firth from the air in 2005.
4 The Bass Rock in the outer Firth, seen from the air around 1990.
5 A young grey seal swimming above the dense underwater kelp forests of laminaria.
6 The blue whale washed ashore near Longniddry in 1869, drawn by Sam Bough.
7 Newhaven shore in the later nineteenth century, by Keeley Halswelle.
8 Steam drifters in Anstruther in 1920, by James Gilmour.
9 The *Reaper*, the only surviving Fifie.
10 The banner of the Cockenzie fishermen from *c*.1900.
11 Skinflats as it was, a reclamation from the sea for agriculture.
12 Skinflats today, with the tide and the habitat for birds restored.
13 The Bass Rock in the late seventeenth century, by an unknown artist.
14 The underwater fauna of the Firth: wolf fish and Devonshire cup coral.
15 Diving gannets from the Bass.
16 Puffins on the Isle of May.

LIST OF TABLES

5.1 The fishing fleet in 1912
9.1 Bass gannet prices
9.2 Bass valuations
9.3 Gannet nest counts
10.1 Breeding seabirds of the Firth of Forth, *c.*1935–*c.*2000

PREFACE AND ACKNOWLEDGEMENTS

This is a book born of looking out of the window. The house in which Anne-Marie and I have lived in Anstruther for three decades sits 30 yards back from the Firth of Forth and faces the Isle of May, six miles away: every hour of every day the view is different, as the weather races its changes across the sky, and the sea below responds. As we have looked at it, and watched and counted the birds every day, we have seen both continuity and change. The fishing boats no longer return to the spacious harbour with cod and haddock, let alone with the herring on which the prosperity of the Victorian town was built – the fish come by road from Peterhead, and the harbour is used only for little lobster and crab boats, and as a tourist marina. In the bay, the goldeneye in winter have fallen from a score in the 1980s to ones and twos, and the scaup duck have gone altogether. The distant Bass Rock grew whiter each summer as the returning gannets nested on more and more of the surface, until now it is completely covered, in the sunshine as if with icing sugar. We were also aware there were more puffins and more seals breeding on the Isle of May and the Berwickshire cliffs and coves, and sometimes now files of bottle-nosed dolphins pass the house. How could there be fewer fish caught, but more fish-eating birds and mammals about? The waves of the sea looked the same, raged the same and were calm in the same way. The environment of the Firth of Forth has changed, but you do not fully see that by looking at it.

Towards the end of my career as a Scottish social and economic historian at the University of St Andrews, I became an environmental historian, one who studies the changing relationship between the people and nature, from the perspective both of the people and of the other organisms with which we share this earthly space. Initially my speciality was woodland history, but when I had completed, with colleagues, our *History of the Native*

Woodlands of Scotland, 1500–1920, I cast around for a new challenge, and staring across to the Isle of May, discovered it. I was inspired to understand that environmental history did not have to be terrestrial by Marc Cioc, *The Rhine: an Eco-Biography* (Seattle, 2006), and tuned to the excitements of the sea by the Danish historian Poul Holm. But ultimately it was curiosity about what the view before me concealed that drove me to seek new explorations.

This ambition was realised through the generosity of the Leverhulme Trust, who awarded me an Emeritus Fellowship in 2009, for which I am deeply grateful. The nub of the problem to be investigated was the relationship in the Firth of Forth between human social and economic history on the one hand and natural habitat and biodiversity on the other: what had man done to nature? That the effect on the environment had been profound was clear, but that it could include outcomes that most conservationists and members of the public would consider as good ones was less so: that it can result in degradation was also clear, but what are the circumstances in which it could result in restoration, which has also happened?

Among other things, the Fellowship enabled me to engage the help of Mairi Stewart of the University of the Highlands and Islands, who co-authors the book with me. I wrote the actual text apart from Chapter 8, but we did the research together and read and criticised each other's texts, so that it became in every sense a joint venture. Margaret Richards' secretarial help was as unfailingly efficient as ever, and she drew the maps. Anne-Marie read much of the text, made many helpful suggestions and assisted me in my electronic illiteracy.

Every chapter brought its own problems and friends and colleagues willing to offer help in the many areas where we were ignorant and badly needed it. The archaeology was assisted by advice from Caroline Wickham-Jones and John Coles; the fishing chapters by the resources and staff of the Scottish Fisheries Museum, Anstruther; the study of pollution particularly by Tom Leatherland; that of reclamation by John Harrison, Richard Oram and the staff of the RSPB; the chapters on seabirds by the Scottish Seabird Centre at North Berwick, by members of staff of Scottish Natural Heritage and also by Bryan Nelson, Ron Morris, Sarah Wanless, Robert Furness and Francis Daunt; that on seals was greatly helped by John Harwood, Ian Boyd, Jeremy Greenwood and Sophie Smout of the Scottish Oceans Institute at St Andrews. David Clugston and Archie Rennie loaned books, and the staff of Special Collections at St Andrews University Library helped endlessly with finding nineteenth-century publications and again with photographic material. The National Records of Scotland were their usual efficient and helpful selves, as were the Royal Commission on the Ancient and Historical Monuments of Scotland which runs the wonderful photographic database, SCRAN. The Carnegie Trust for the Universities of Scotland generously made a grant towards the colour plates.

That does not even begin to exhaust the list of those to whom we are indebted. We would especially like to mention the following: John Anderson, Elizabeth Ashton, Geoff Bailey, John Ballantyne, Stephen Bastiman, John Cairns, Roger Crofts, Alastair Dawson, Tom Dawson, Alan Drever, Mike Elliott, Alexander Fenton, Kirsty Golding, Nancy Gordon, the late William Halcrow, Sir Hew Hamilton-Dalrymple, Alistair Jump, Rob Lambert, Jim McCarthy, Norman Macdougall, Donald McLusky, Colin Martin, Stuart Murray, Alan Phillips, Robert Prescott, Harry Scott, Richard Smout, Colin Taylor, Richard Tipping, Alex Woolf, John Young, Bernie Zonfrillo, and the members and staff of the Forth Forum. There are bound to be omissions, for which we apologise. All mistakes in the book are our own.

Finally, we would like to express our appreciation to Hugh Andrew, Andrew Simmons and other staff of Birlinn Publishers for their support, patience and skill in seeing this venture through from text to book.

Christopher Smout

Picture Credits

Figure 0.1 Drawn by Margaret Richards
Figure 0.2 © Getty Images
Figure 0.3 © Scottish Mining Museum (Licensor www.scran.ac.uk)
Figure 0.4 © St Andrews University Library Special Collections, RMA-H69
Figure 0.5 © St Andrews University Library Special Collections, GMC-22/38
Figure 1.1 Drawn by Margaret Richards
Figure 1.2 © The Scotsman Publications Ltd (Licensor www.scran.ac.uk)
Figure 1.3 Drawn by Margaret Richards
Figure 1.4 © National Museums Scotland (Licensor www.scran.ac.uk)
Figure 2.1 From A.H. Miller, *Proceedings of the Society of Antiquaries of Scotland* (1903)
Figure 2.2 From S. Lewis, *Topographical Dictionary of Scotland* (1846)
Figure 2.3 Drawn by Margaret Richards
Figure 2.4 © St Andrews University Library Special Collections, JV-19610
Figure 2.5 © Royal Commission on the Ancient and Historical Monuments of Scotland (Licensor www.scran.ac.uk)
Figure 2.6 © East Lothian Library Service (Licensor www.scran.ac.uk)
Figure 3.1 Drawn by Margaret Richards
Figure 3.2 © St Andrews University Library Special Collections, ALB24/63
Figure 3.3 © St Andrews University Library Special Collections, ALB6/92
Figure 3.4 © National Museums Scotland (Licensor www.scran.ac.uk)
Figure 3.5 © City of Edinburgh Museums and Galleries (NC) (Licensor www.scran.ac.uk)

PICTURE CREDITS

Figure 4.1 © Gordon C. Easingwood (Licensor www.scran.ac.uk)
Figure 4.2 © Fisherrow Old Men's Club (Licensor www.scran.ac.uk)
Figure 4.3 Drawn by Margaret Richards
Figure 4.4 Derived from P. Smith, *The Lammas Drave and the Winter Herrin'. A History of the Herring Fishing from East Fife* (Edinburgh, 1985)
Figure 4.5. From James R. Coull, *Sea Fisheries of Scotland* (1996)
Figure 4.6 © Gordon C. Easingwood (Licensor www.scran.ac.uk)
Figure 4.7 © Dundee University Archives Services: Heddle Collection
Figure 4.8 © Royal Commission on the Ancient and Historic Monuments of Scotland; F/1887 (Licensor www.scran.ac.uk)
Figure 4.9 © St Andrews University Library Special Collections, JV-A83
Figure 4.10 © Fisherrow Old Men's Club (Licensor www.scran.ac.uk)
Figure 5.1 © St Andrews University Library Special Collections, ALB6-91
Figure 5.2 © East Lothian Museums Service (Licensor www.scran.ac.uk)
Figure 5.3 © East Lothian Library Service (Licensor www.scran.ac.uk)
Figure 5.4 © National Portrait Gallery, London
Figure 5.5 © St Andrews University Library Special Collections, JHW 9
Figure 5.6 Drawn by Margaret Richards
Figure 5.7 © The Scotsman Publications Ltd (Licensor www.scran.ac.uk)
Figure 6.1 Drawn by Margaret Richards
Figure 6.2 © Tom Dawson
Figure 6.3 From D. Herbert (ed.) *Fish and Fisheries*, (Edinburgh, 1883) p. 38
Figure 6.4 © City of Edinburgh Museums and Galleries (Licensor www.scran.ac.uk)
Figure 6.5 © Scottish Fisheries Museum (Licensor www.scran.ac.uk)
Figure 6.6 © Scottish Fisheries Museum
Figure 7.1 Drawn by Margaret Richards
Figure 7.2 © National Library of Scotland (Licensor www.scran.ac.uk)
Figure 7.3 Drawn by Margaret Richards
Figure 7.4 © The Scotsman Publications Ltd (Licensor www.scran.ac.uk)
Figure 7.5 Drawn by Margaret Richards
Figure 7.6 © The Scotsman Publications Ltd (Licensor www.scran.ac.uk)
Figure 7.7 © The Scotsman Publications Ltd (Licensor www.scran.ac.uk)
Figure 8.1 © Royal Commission on the Ancient and Historical Monuments of Scotland

Figure 8.2 © John Harrison
Figure 8.3 © The British Library (Licensor www.scran.ac.uk)
Figure 8.4 Drawn by Margaret Richards
Figure 8.5 © Geoff Bailey
Figure 8.6 Drawn by Margaret Richards
Figure 8.7 © The Scotsman Publications Ltd (Licensor www. scran.ac.uk)
Figure 8.8 © Royal Commission on the Ancient and Historical Monuments of Scotland (Licensor www.scran.ac.uk)
Figure 8.9 © Falkirk Museums (Licensor www.scran.ac.uk)
Figure 8.10 © NCVAP/aerial.rcahms.gov.uk
Figure 9.1 From the *Graphic* magazine, 9 September 1876: thanks to Ron Morris.
Figure 9.2 © St Andrews University Library Special Collections, JV-40993[B]
Figure 9.3. © RAF/Operation Seafarer
Figure 9.4 © S. Murray
Figure 9.5 © St Andrews University Library Special Collections, RMA-Q456
Figure 10.1 Drawn by Margaret Richards
Figure 10.2 © The Scotsman Publications Ltd (Licensor www.scran.ac.uk)
Figure 10.3 © National Library of Scotland (Licensor www.scran.ac.uk)
Figure 10.4 © St Andrews University Library Special Collections, GAL-31
Figure 10.5 From R. Murray et al., *The Breeding Birds of South-East Scotland; A Tetrad Atlas* (Scottish Ornithologists Club, 1998)
Figure 10.6 © St Andrews University Library Special collections, GAL-37
Figure 11.1 © The Scotsman Publications Ltd (Licensor www.scran.ac.uk)
Figure 11.2 © George Waterston Memorial Centre (Licensor www.scran.ac.uk)
Figure 12.1 © The Scotsman Publications Ltd (Licensor www.scran.ac.uk)
Figure 12.2 © The Scotsman Publications Ltd (Licensor www.scran.ac.uk)
Figure 12.3 © The Scotsman Publications Ltd (Licensor www.scran.ac.uk)
Figure 12.4 © The Scotsman Publications Ltd (Licensor www.scran.ac.uk)

INTRODUCTION

The Firth of Forth is the great gaping mouth of Scotland. It is about 20 miles wide from Fife Ness to Dunbar, and about 60 miles long from that opening to Craigforth above Stirling, where the River Forth ceases to be tidal. For millennia it was the obvious way through which boats, people, goods, books and ideas entered the country. Edinburgh, now the national capital, seat of Parliament, law and kirk, for security's sake was originally set back from the edge but reaches the sea through Leith, its satellite port. The early towns of Linlithgow, Haddington and Dunfermline also lie well back. Stirling, the strategic key to the medieval kingdom, sits at the head of the Firth, but 12 miles up a twisting river. All this shows a certain nervousness about who might arrive by sea to take a town by surprise, yet even in the Middle Ages, and increasingly so in the sixteenth and seventeenth centuries, the shores of the Forth became themselves directly lined on both sides with trading, fishing and manufacturing settlements, making their living from the junction of the sea and the land. That land, especially in Lothian and Fife, is among the most fertile and the driest in the country. Kings built their palaces, monks their abbeys, lords their castles, merchants their burghs, peasants their farmtouns, in due course attracting foreigners, rivals and reformers to trade with them, loot them and sometimes to build again in their place.

English medieval maps were vague about the Firth of Forth but noticed a few essential features. Mathew Paris around 1250, in four similarly ill-informed maps, drew Scotland as wasp-waisted, only Stirling Bridge joining the two halves, the northern part called Scocia Ultramarina – Scotland beyond the sea (Plate 1). This at least confirms that strangers thought the Firth of Forth a great divider of land mass. The Gough map of about 1360, after a period of war between Scotland and England but also of increasing trade between Scotland and Europe, had a clearer idea of the relative position of places but still joined the east coast to the west at Stirling. However, it did show that you had to pass two islands, correctly named as the May and the Bass, to enter the Firth to approach Edinburgh. By the sixteenth century, familiarity and better methodologies had bred much greater accuracy. The map made by Laurence Nowell early in Elizabeth's reign in England, probably utilising work by Alexander Lyndsay, a Scottish surveyor of the coast in the 1540s, had everything nearly as right and recognisable in the Firth of Forth as they are on modern maps. Significantly, Nowell called it 'the Scottish Frith' (Plate 2). A century later, the great *Atlas Novus* published by Joan Blaeu in Amsterdam in 1654, had maps by the Scottish cartographers Timothy Pont and Robert Gordon of Straloch, with good outlines of the Firth of Forth and properly detailed depictions of Lothian, Fife and the upper estuary. The Firth was now geographically well described for the outside world.[1]

Already before the end of the sixteenth century the industrial use of the Firth was well established. Salt-making had been practised since at least the twelfth century on monastic and lay estates, by boiling and evaporating sea water or salty mud (the latter by a process known as sleeching) in large iron pans. As coal mining itself expanded after around 1550, the manufacture and export of salt became increasingly important, and travellers in the mid seventeenth century commented upon the scores of pans along the shores almost up to the head of the estuary, especially near Prestonpans and Culross, but also in Fife from Inverkeithing east to Weemys and later to Pittenweem. 'The works . . . are placed all along the shore, at least thirty English mile', said Sir William Brereton in 1636; 'all the seacoast here for certaine miles is smoking with salt pans' said another English visitor, William Blundell, in 1657.[2] Just as with the power stations in the twentieth century, what was critical to their location was not the quality of the water, but the close availability of fuel on land. They were not usually situated where the sea was most salty, but where the mineral outcrops came closest to the shore, or (in the upper estuary in the Middle Ages) where first wood and then peat had been plentiful.

Dynastic politics of the seventeenth and eighteenth centuries ultimately united Scotland and England, in the two main steps of the Union of the Crowns in 1603 and the Union of Parliaments in 1707. Between the two, Edinburgh had already grown to become among the biggest towns in Britain, at 40,000 vying with Norwich and Bristol for second place after London. In the eighteenth century that intellectual flowering that we now

INTRODUCTION

call the Scottish Enlightenment brought the capital European fame, and the city grew in size, wealth and cultural reputation. The 'new town' was constructed. Publishing activity along with the law courts created a big demand for paper from the late seventeenth century onwards, on a scale that ultimately led to a host of mills and thus to the first serious river pollution. The modern industrial revolution began in Scotland not in the west but in the east, between 1750 and 1775, signalled by the long-lived iron works at Carron near Falkirk, the chemical factory at Prestonpans, and the short-lived cotton mill at Penicuik on a stream that fed into the Firth. The observant young French tourist Alexandre de La Rochefoucauld said of the central belt of Scotland as early as 1786 that 'the people . . . have proved themselves. They've cleared their lands, working daily to improve them, and have reached the point of taking on manufactures . . . in these richer parts, one has a sense of real opulence'.[3]

The nineteenth century saw a tremendous growth of population and industry around the Firth of Forth. The first railway ferry in the world was opened between Granton and Burntisland in 1849 and the great rail bridge at Queensferry, completed in 1890 (Fig. 0.2), became an icon of Victorian engineering brilliance. The railways made it easy to move perishable fish around quickly, and the fisheries grew especially in the small towns of the East Neuk of Fife. The coal mines, also using the railways, enormously expanded in Lothian, Fife and Stirlingshire, now for the first

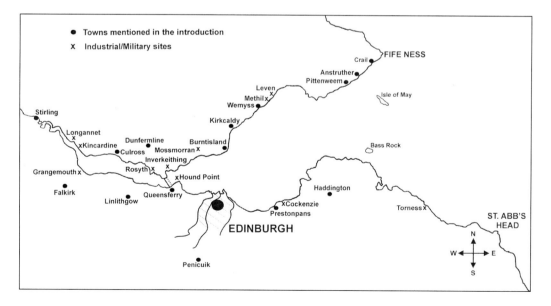

Fig. 0.1 – The modern Firth of Forth

Fig. 0.2 – Icon under construction: Forth Rail Bridge, 1888.

time operating well inland as well as along the shores. They employed tens of thousands of miners in devastating conditions underground and often housed them vilely, but cheap steam power had economically liberating consequences. Edinburgh became a city of nearly half a million and the largest manufacturing town in Scotland after Glasgow.

The twentieth century, with its tearaway attitudes to growth and energy consumption, built the road bridges, the first at Kincardine in 1936 (Plate 3) and the next at Queensferry nearly 70 years later, plants to refine oil and gas at Grangemouth and Mossmorran, an off-loading base for tankers at Hound Point, an oil-rig construction and repair base at Methil, an array of power stations, coal-fired at Longannet (Fig. 0.3), Kincardine and Leven (both now demolished), and Cockenzie, nuclear at Torness, and also a naval base at Rosyth (Fig. 0.4). By around 1980, within the 1,800 square miles that comprise the catchment of all the streams that feed into the Firth of

INTRODUCTION

Fig. 0.3 – Electricity and pollution from coal: Longannet power station and its opencast mine in 1997.

Forth, there lived about 1.3 million people, a quarter of the population of Scotland, strong in the service industries but weaker than before in mining and manufacturing. And until that time almost all their sewage still entered the Firth of Forth raw, virtually untreated, but well mingled with industrial waste.

The theme of this book is the interrelationships between people and biodiversity. The Firth of Forth has been, and in many respects, despite all this human exploitation, remains, a wonderful place for wild nature. At one time there were oyster beds off Edinburgh that covered 50 square miles; they may have been the largest in Britain. In 1860, on the opening day of the salmon fishing on the River Forth, one set of nets near Stirling caught 56 fish weighing 900 lb, fresh from the sea. When scientists in 1862 discovered for the first time how the herring spawned off the Isle of May, they described them as lying on the bottom 'in tiers covering several

Fig. 0.4 – Rosyth Bay before Rosyth Naval Base: a pastoral idyll in 1909.

square miles'. In 1889, the *Garland*, the Scottish Fisheries Board's first research vessel, found a shoal of young whiting that 'extended like a sheet from the Oxcars lighthouse to some eight miles beyond the Isle of May, a distance of thirty-six miles'. Today the oysters are virtually gone, there is no commercial fishing for either demersal species like whiting or pelagic ones like herring, and the salmon are diminished. On the other hand, within the twentieth century, there occurred an extraordinary expansion of seabird numbers, puffins on the Isle of May, for example, increasing a thousandfold from six pairs in 1924 to 68,000 pairs in 2002, and gannets on the Bass Rock from 3,000 pairs around 1900 to 48,000 pairs a century later (Plate 4 and Figs 9.3–9.5). Grey seals did not breed in the Firth in the nineteenth century, though they had done so earlier. They started to do so again in 1956, and in 2000 some 2,133 pups were born on the Isle of May, with others on the Berwickshire coast. These are now among the largest grey seal colonies on the east coast of Britain. There are also ancient continuities, like the relatively small areas of the seabed where reefs escaped disturbance by trawlers, off the Isle of May and St Abb's Head, where scuba divers can go down today and find among the underwater kelp beds and rock surfaces an astonishing beauty of marine life (Plates 5 and 14). This is a distinctive assemblage fed by cold northern waters, including bizarre wonders like the wolf fish, which has been known in the Forth since the days of the earliest naturalists in the seventeenth century. There is a risk that it might not survive climate change that warms the seas.

This book is about the juxtaposition of these facts, that is, about the history of the relationship between people and nature in the Firth of Forth. What have we lost through our activities and why? What have we gained, either through our deliberate interventions in nature conservation or as an accidental consequence of activities designed for some other, entirely anthropocentric, purpose? By polluting the sea with organic human and industrial waste we made the Forth the best place in Britain to see concentrations of sea duck in winter, flocks of scaup, pochard and goldeneye off the sewer outfalls at Leith in the early 1970s amounting at times to tens of thousands of birds: by reversing our previous actions a few years later, we also removed the duck. By putting domestic and other types of waste into uncovered landfill tips we helped to cause a population explosion in gulls; by covering the tips again we decimated them. In 2008 the herring gull came onto the Royal Society for the Protection of Birds' red list of endangered species because it had suffered a population decline in the UK of more than 50 per cent since their records began in 1969.

Conservationists, though, would profit from more history, which is one of the reasons for this book. Had the RSPB taken a different baseline, and gone back to 1900 or earlier, they could hardly have reached the same conclusion without making a qualification. When Gray in 1871 wrote of the herring gull in Scotland he said they were never commoner in early spring than in Fife, where on Leven sands they 'assemble in companies numbering thirty or forty birds', and when Baxter and Rintoul first went to the Isle of May in 1907 they found one breeding pair. By 1994, flocks of up to 5,000 were feeding on the Fife tips and there were 2,122 nests on the Isle of May. But in 2008 the largest flock in Fife was near a reservoir, of 1,200 birds, though in fact the number of nests on the May was much as before. No doubt nationally there had been a big decrease since 1969, but surely not back to levels of a century ago.

In Pittenweem, as late as the interwar years, fishermen auctioned their cod and haddock in the open air (Fig. 0.5), and there are still Roman numbers incised in the harbour stones to show where each man had his stance. No doubt there would already have been a few birds hovering hopefully in the background when our picture was taken, but to have an open-air fish-market there now would be impossible. It would be an invitation to the feast for the hundreds of gulls that wait in the Firth for the boats coming into Pittenweem to throw out their by-catch on the way back from the prawn beds. Unfortunately there never could be a fish-market for cod and haddock in Pittenweem today in any case, as the demersal and pelagic fish

Fig. 0.5 – Pittenweem fish-market in the 1930s: cod and haddock could still be laid out for sale without risk of being snatched by gulls.

in the Firth have been commercially fished out, leaving little to sell except crabs and lobsters, a few scallops, and especially the nethrops, the prawns that earlier generations regarded as unsaleable for human consumption, though suitable breakfast for a hungry crew member.

This is one of the values of environmental history – it provides another perspective. Environmental history shows us a world of conundrums and a morass of unintended consequences. It shows us that whatever we do is not neutral for the innumerable other organisms with which we share our planet. Scientists have begun to term the period since 1800 in which we now live 'the Anthropocene', the age when our actions are the main determinant of the fortunes of all creatures that on earth do dwell. This book chooses one theatre, small in global terms, to tell the story of how we have impacted on nature. If usually it has been destructive, sometimes it has been for better. Environmental history in an economically successful modern Western society, such as Scotland is today, is not always a story of unmitigated doom and gloom for biodiversity, though often it is. We will tell our tale by a series of studies taking us into the history of hunting, gathering and fishing, into industry and pollution, into recreation and conservation, and into the lives and fortunes of many sensate and insensate beings other than ourselves.

CHAPTER 1
HUNTING AND GATHERING

1. The Prehistoric Environment

In 1996, the Fife Regional archaeologist, keeping a watching brief on the construction of the thirteenth green of a new golf course at Fife Ness at the eastern extremity of the Firth of Forth, noticed a discolouration in the earth as the turf was stripped away, and a scatter of worked flints and burnt hazelnut shells. Subsequent excavation revealed something remarkable, traces of a camp set up by Scotland's first peoples about 7,500 years ago.[1] The Mesolithic hunter-gatherers had entered the country after the glaciers of the last Ice Age had finally melted. Archaeologists consider the Mesolithic in Scotland to stretch from nearly 11,000 years ago to around

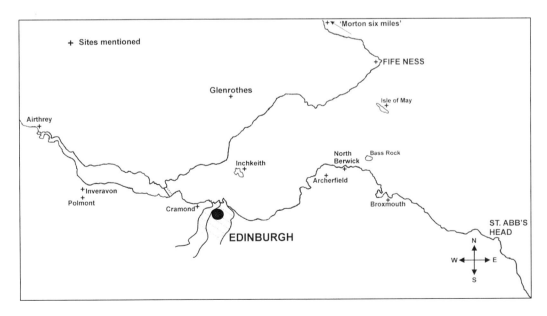

Fig. 1.1 – Archaeological sites

6,000 years ago, when it was succeeded by the Neolithic, so it lasted for between two and three times as long as the time since the birth of Christ. The very earliest known scatter of worked stone fragments in the Firth of Forth, (one of the two earliest in Scotland) is also associated with a temporary camp and remains of hazelnuts, at Cramond on the edge of Edinburgh, dated at about 10,500 years old.[2] These are unlikely to have been the very first, as the country had possibly been suitable for human occupation for some centuries before that. During the Mesolithic period, early peoples had no knowledge of farming, though they might modify the land by burning the woodland edge to open up a space for grass and berries, to make it more attractive to herbivores that they could then kill. They lived by hunting mammals, birds and fish, and by gathering plants and shellfish.

The camp at Fife Ness appeared to centre on a hearth protected from the north-east by some kind of shelter for which only the postholes remained. There were other pits, and 1,518 fragments of worked flint. Most of these were debris, but there were 70 blades, including ten retouched to become scrapers, a quantity of sharp microliths suitable for setting into hunting spears or arrows, including some unusual crescent-shaped ones considered to be possibly for fishing equipment, a hammer stone and an anvil stone. All the flints were local material picked up in the vicinity, and some had apparently been manufactured on site. Such, however, was the ratio of finished tools to debris that it seemed that other tools had been brought in, and that the primary function of the camp was not the manufacture of flint into implements.

So what was it? The site lay in a small hollow on the top of what is today a raised beach. On the landward side, there would have been woods dominantly of oak and elm, fringed with hazel as the exposure to the sea thinned out the forest. The sea 7,500 years ago was probably slightly lower than it is today, though it is hard to say by how much at Fife Ness: it could have been as much as 300 yards away, with a wide, flat, intertidal zone for the hunter-gatherers to range over. Some of the lithic debris was discoloured by fire at high temperature. The shelter lay to the north-east of the hearth, though the prevailing wind is from the south-west. In late autumn, though, when the wind turns north-east and the weather deteriorates, clouds of migrant thrushes and blackbirds from Scandinavia arrive at Fife Ness, sometimes in many thousands. The site contained no organic remains apart from the hazelnuts. What had been caught had been carried away. The archaeologists interpreted it as a place where food could be cooked and smoked for winter, probably birds, fish and shellfish, a place for hunting and processing.

Fifteen miles north-west of Fife Ness, another, much larger, Mesolithic site was excavated around 1970 by John Coles at Morton, now on the landward edge of Tentsmuir, then on the edge of the sea near the entrance to the Firth of Tay, probably an island linked to the mainland at low tide: the accretion of sand over millennia has moved the shoreline nearly four kilometres east since then, quite apart from any shifts in sea levels.[3] Morton was occupied, probably seasonally, over a period of several generations somewhere between 8,000 and 6,000 years ago. From the different stone tools used, we know the hunters ranged on land along the southern shores of the Tay, along the southern Eden valley and towards St Andrews, picking up suitable pebbles as they went. From the midden into which they threw their waste we know they ate red deer, roe deer, wild boar and aurochs (also hedgehog and some kind of vole), killed in the woods inland. They also ate 20 species of shellfish, mainly the typical species of this sandy and muddy shore to this day, especially cockles, small marine clams and mussels, as well as crabs. There were a lot of fish bones in the midden, especially cod, with some haddock and salmon or sea trout, and a single turbot and a sturgeon of about 250 kilograms, which would have been a very big catch for the hunters, well above the accepted record catch in a British river. Sometimes, however, sturgeon strand like whales. It is a fish that no longer occurs in Scottish rivers or seas, but did so until comparatively recently and probably at one time bred in both the Tay and the Forth. The quantity of sizeable cod suggests that the settlers at Morton had boats well able to work the inshore waters.

It is the bird list from the midden that is the most interesting, however. It contains no swans, geese, duck, herons or wading birds of any description, possibly because they were difficult to kill and not worth spending a precious arrow on. Or, perhaps, if the camp was only occupied in summer, many of these species may have been absent. The only land birds in the midden are crows and thrushes. But the seabird list is extensive: fulmar, gannet, cormorant, shag, greater black-backed gull, kittiwake, razorbill, guillemot and puffin. Where did they come from? They would all have been hard to catch at sea, and if they had been driven on the shore in winter storms, this would only happen when they were starving and so of no nutritional value. But auks and other seabirds are easier to kill on land at their breeding places, possibly with the hunters climbing down the cliffs as they used to do on St Kilda.

The only localities on the east of Scotland where all these species breed together are the islands in the Firth of Forth. All of them actually nest on the Isle of May at present, except for the gannet, which nests on the Bass Rock but has been known to breed on the May, and the cormorant, which is thought possibly to have bred there at one time and breeds now on other Forth islands. It seems possible that the hunters of Morton undertook expeditions in summer to harvest the birds of the Isle of May and perhaps elsewhere, probably in boats of skin and wicker similar to Irish curraghs. Though no trace of any Mesolithic boats except dugout canoes has ever been found, we know from their camps in the Inner Hebrides and Orkney that they were easily capable of sea voyages of the 20 miles or so required to reach the May. The materials from which their boats would have been made do not readily survive in the archaeological record. On the island itself, archaeological investigation in the 1990s searched, among other things, for flints left behind by early hunter-gatherers: they found evidence of Mesolithic activity that was 'strong but not unequivocal'. More certain were later high quality flints of English origin, that suggested that during the late Neolithic and early Bronze Age, when Fife had an impressive ceremonial landscape of standing stones, notably around Balbirnie and Balfarg near Glenrothes, the Isle of May was perhaps a staging post for trade in materials from the south.[4]

Over the ten or twelve millennia since the end of the last Ice Age, the Firth of Forth changed shape under two immense but conflicting natural forces, both released by the decay of the ice caps.[5] One was the vastly increased quantity of water coming from Britain and Scandinavia, creating a sea-level rise that overwhelmed the dry land between Britain and Continental Europe, obliterating the bridge over which man and many other animals had crossed, and creating the Firth itself. The other was the force of isostatic rebound, the uplift and tilt of the land once the great weight of the ice upon it had been lifted. Lift and tilt varied in extent depending on the weight of the ice, being greatest where the ice had been thickest and heaviest, so in our region greater close to the core of the Highland glaciers at the head of the Forth. These two forces acted in complex and unequal ways over a very long period of time, but especially in the first five or six millennia, leaving a series of fossilised raised beaches, some buried today but most still visible, to mark different points of temporary equilibria along the coast, where the sea had cut platforms into the land, only for the rebound to lift the beach beyond the reach of the waves again.

There are several raised beaches because the two forces acted at different rates at different times, and because of the complexities of the geological tilting. The melting ice at the end of the Ice Age quickly enlarged the sea and flooded the land, but the uplift of the land, accelerating after the ice melted, then began to counteract the sea-level rise induced by the melting itself, and to force the flood back again from parts of the land. Then there was a second mighty inrush of the sea when the ice on the Laurentide Ice Sheet that covered Canada and parts of the northern USA melted, and it was this that created what geologists have termed the Main Post-glacial Shoreline. In the western part of the Forth Valley above Stirling, this is visible today at 49 feet above the modern sea level, but only 20 feet above modern levels at Dunbar. The difference is due to their relative proximity to the Highland ice cap. These figures do not mean that the sea was that much higher than it is now, but that uplift and tilt took them to those heights. But on the other hand the sea did indeed come to stretch much further inland than it does now.

In the Firth of Forth, the creation of a 200,000 hectare tidal lagoon at the head of the present estuary was the most dramatic modification of the environment during the Mesolithic period. Between 10,000 and 8,500 years before the present, land uplift exceeded sea rise: the plain above Stirling, gouged out by glaciers, dried out, rose and became covered by vegetation and peat. But then, from about 8,000 years ago, the opposite occurred as sea-level rise was accelerated by the great American melt, culminating about 6,500 years ago when the final limits of the Main Post-glacial Shoreline were reached. The result was a huge marine inlet covering formerly dry land, stretching ultimately as far as the Lake of Menteith, 12 miles west of Stirling. At this spot, the North Sea was then less than ten miles from the Atlantic, as Loch Lomond had been flooded by the same forces and become a sea loch at the head of the Firth of Clyde. For a man to cross from sea to sea was only a long morning's walk.

This environment did not remain static, however. From the moment of its inception the lagoon began to fill with deposits from the land, ultimately to form the deep carse clays of the land above and below Stirling. At one point about 7,000 years ago there are signs that a tsunami struck the east coast of Scotland when part of a Norwegian mountain collapsed into the sea. In Shetland it is believed to have created a sea surge that ran to some 80 feet above normal sea level: down the rest of the east coast of Scotland the height of the tsunami was more like 15 feet. It is clearly detectable within the clay deposits of the lagoon. We cannot tell what impact it had on man, though one Mesolithic camp near Inverness clearly lies beneath the sandy geological horizon formed by the tsunami. Given the scanty, scattered and mobile human population, the impact might nevertheless have been comparatively small, and its force at least within the lagoon largely expended when it came ashore.

After about 6,000 years ago, the uplift of the land gradually began to exceed the supply of new meltwater, and the whole process gradually began to go into reverse. The carse clays were slowly but surely lifted out of the sea; at Grangemouth the sea floor was raised to 16 feet above average high tide level over the next two millennia. The tidal lagoon very slowly became less salt, then it turned into freshwater marshland with great reed beds, and natural succession transformed these into first, grasses, sedges and mosses, then slowly (on the driest parts) into oak, ash and elm woodland. Throughout this time, the ecosystem must have been rich and varied, a paradise for hunters. Eventually the forests were affected by a change in the climate towards more rain, wind and cold, and increasingly under pressure from the domestic animals of a rising human population that had turned to farming. They ceased to regenerate, collapsed and turned into peat bogs. Where there had once been sea there were in places now large raised bogs, like Flanders Moss. The whole process took millennia: the last woodland on Flanders Moss disappeared only two centuries after the birth of Christ, and thereafter the growth of peat was unremitting until the drainage and reclamation of modern times.

Nevertheless, for several thousand years there was a great marine tidal inlet of the Firth of Forth, stretching well above Stirling, which provided rich pickings for the first peoples. Victorian engineers and others, working to recover land for agriculture or to build housing and transport infrastructure, discovered embedded in the carse clays the skeletons of 16 great whales.[6] They were scattered at various points above Grangemouth, 12 of them above Stirling and one right up at Flanders Moss at the western extremity of the lagoon. Archaeology being in its infancy, little attention was paid to the details, but four of the whales were recorded with associated finds of tools made from red deer antlers, evidently used as mattocks to strip the meat from the bones. The largest skeleton was that of a 72-foot-long blue whale found at Airthrey in 1819, with a tool beside it. Another great whale of uncertain species at Meiklewood by Stirling was found alongside a stag's antler with a hole drilled in it to take a handle. The tool was presented to

Fig. 1.2 – The contrasts of history: a 6,000-year-old pile of oyster shells unearthed during pipeline construction in 1983, Grangemouth on the skyline, archaeologist Derek Sloan in the foreground.

the Society of Antiquaries, from whence it found its way into the modern Museum of Scotland. It has been carbon-dated to around 5,900 years ago.

Such whales as these must have come as an inexpressible wonder to Mesolithic man, representing a gift of food and bone generous to the point of superfluity, a resource beyond what even the entire tribe could use. Presumably the whales stranded of their own accord, as they occasionally still do in the Firth of Forth today, though no doubt the people in their boats strove to ensure they did so – unlike in recent times, when the arrival of a whale in the inner Firth has been greeted by flotillas of boats unsuccessfully trying to turn it round and drive it back to safety. A beached whale in winter 6,000 years ago must have been seen as proof of benevolent and merciful gods.

Hunter-gatherers generally leave few traces of their comings and goings on the ground, so it is hard even to guess how many there were or where they lived. However, their middens of discarded oyster shells occasionally reach monumental proportions, nowhere more so than in these ancient upper reaches of the Firth of Forth, where 14 have been identified between Bo'ness and Falkirk, with more on the opposite shore.[7] Some were so large that even in the eighteenth century they were recognised to be ancient monuments, though no one could guess their true meaning or antiquity. The midden at Polmonthill measures 175 yards long by 25 yards wide and is three or four feet deep. That at Inveravon is of similar size and up to six feet deep. Three of the mounds grouped round the mouth of the Avon have been shown to contain at least 3,000 tons of oyster shells each, and the excavations of Derek Sloan show them to have accumulated over the space of as many years, starting about 6,000 years ago (Fig. 1.2).

They are composed almost entirely of millions upon millions of oysters, with a small proportion of mussels, cockles, periwinkles, whelks, razor-shells and crabs. But they also contain some domestic animal bone and some pottery, and some show signs of having been lived in (Skara Brae was also built in a midden). Sloan suggests they were not temporary camps for casual occupation, but places of permanent occupation by a people whose economy was based on the shore. They may have lived there all the year; if not, they apparently returned to their home base annually. They were hunter-gatherers, but they came to know the Neolithic arts of making pots and taming animals, and in any case they could not have lived entirely from the shellfish. Oysters are at their most nutritious in winter and spring when other foods might be scarce. Unlike the whales, which must have been an occasional unexpected bonanza, oysters were a staff of life for generation after generation after generation.

There is no evidence the inhabitants of these middens ever practised agriculture to raise grains: perhaps they felt they had no need to when the sea was so bountiful. At some point about 1,000 years before Christ, the marine environment seems to have changed to a less saline one in the region of the middens, and the oysters moved downstream. In recent historic times their great stronghold was in the waters off Leith and Prestonpans (see Chapter 3).

There are other middens on the shores of the Firth of Forth, beyond the main concentration of oysters at that time, which seem to show the same signs of late Mesolithic and early Neolithic activity. The Neolithic is defined by archaeologists as the period of early farming between about 6,000 and 4,000 years ago, when it was succeeded by the Bronze Age, and then at about 2,800 years before the present, by the Iron Age, as the use of these metals began to spread and replace the use of purely stone tools. But all these divisions of prehistory are matters of convenience for scholars, concealing how slowly one technology or way of life succeeded another, and masking the overlap and continuities between them.

One such midden of which no trace now remains was on a slipped raised beach on the island of Inchkeith, between Kinghorn and Leith.[8] When it was first noticed around 1870 it contained both shells and animal bones. The shells were those of limpets, periwinkles, whelks, a clam, scallops (all still locally common) and some oysters, and the bones were those of sheep, pig, horse, ox and rabbit (the last must have been a modern intrusion), as well as those of grey seal. The seal, which does not seem so remarkable to us, was exciting to the Victorians, as it had by that time long been extinct in the Firth of Forth. When investigated again in 1898, implements of deer horn and bone, including round-headed chisels similar to those found in late Mesolithic deposits in Argyll, were discovered in the deposit, and since then flints have been found on the beach. The domestic animal bones were split and burnt, and could not be earlier than Neolithic. It has now all been destroyed by military building activity, but it may well be contemporary with the activities higher up the Firth, with the hunters of Inchkeith utilising a different set of ecological opportunities.

Further down the coast in East Lothian, archaeologists located in 1907 three middens that they considered, from the associated pottery, to be late Neolithic or early Bronze Age.[9] Two were sealed beneath a medieval floor in North Berwick, and one was at Archerfield near Dirleton. Here the finds were dominated by many thousands of whelks (plainly the local speciality), with some limpets, but with a dozen other local species of seashell in small quantities, and many crabs' claws. Some of the whelks were described as much larger than any in Scotland in modern times, except those in Orkney and Shetland.

There were also so many shells of the brown-lipped land snail, *Cephaea nemoralis*, which the archaeologists reckoned had also been gathered for food. It is a particularly common species on the coast in this region. There were none of the larger and familiar garden snail, *Cornu aspersum*, recognised as edible at least since Roman times, but these are a southern species reckoned only to have arrived in Britain with Celtic immigrants in the Iron Age.

All this evidence shows how prehistoric man remained a gatherer along the shore long after agriculture had arrived, and no doubt many more middens of this sort remain to be discovered and analysed. However, as cultivation methods became more sophisticated, what was gathered began to change. Shellfish continued as part of their diet on a fairly small scale for those who lived by the sea, but they were a critical element only in times of dearth. An Iron Age fort at Broxmouth near Dunbar, from shortly before the dawn of the Christian era, shows how the local farming population, much given to eating meat in good times, turned completely to common winkles and limpets in a famine; the excavators found that bones suddenly disappeared from the site and were replaced by great piles of shells in the ditches.[10] So extreme was the exploitation that the local limpet population appears to have been depleted for a long time thereafter.

2. 'Wrack and Wair'

Hunting has never died out as a major activity in the Firth of Forth, but after prehistory it gradually came to take the form of fishing by a specialised occupation of fishermen, at first light and subsistence for comparatively local markets, eventually heavy and industrial, focused on distant markets. We shall tell that story in later chapters. But it is convenient here to bring the story of gathering from the sea up to the present time, especially in relation to the emerging skills of agriculture.

As time passed, shells were more likely to be burned for building lime than thrown aside, and middens became associated with the accumulation, not of shells and bones, but of human and animal dung mixed with soot, straw and turf, greatly valued for use as manure on the fields. Books were written by the agricultural improvers of the eighteenth century on how to make a good dung heap and what ingredients should go into it. It was said that a Scottish farmer would tip his hat to a well-made midden.

It is difficult to say when people first discovered that it was useful to add seaweed to the midden, or alternatively to put it straight on the land.[11] It

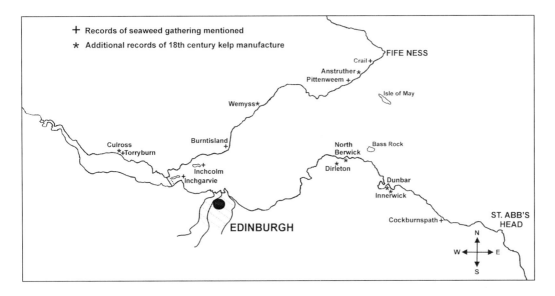

Fig. 1.3 – Seaweed use

leaves no trace in the archaeological record, though the presence of large quantities of flat periwinkle shells, which attach themselves to the fronds in the sea, may denote manuring with seaweed, but equally may denote either that the seaweed itself was being used as food or that the little winkles were eaten. Written sources from as early as the fourth century AD mention manuring with weed in other parts of northern Europe. The first such records in our area, and indeed for Scotland, come from a confirmation by James III of 1479 to the archbishop of St Andrews of his rights to 'wrack and wair' cast up by the sea, and a grant by James IV in 1491 to Sir John Dundas of Dundas of rights to take these materials on Inchgarvie, the island now beneath the Forth rail bridge. Such late references do not, though, necessarily mean that the practice was new in the Firth of Forth, though it may at this time have gained a new emphasis. So poor are the records of Scottish medieval economic history that it could already have been ancient usage in the fifteenth century.

The value of the weed lay in its contribution to the organic structure of thin soils and in its chemical content. Compared to farmyard manure, it was as rich in nitrogen and twice as rich in potassium. It was, however deficient in calcium and phosphates. There is a curious statement by Robert Gordon in Blaeu's *Atlas* of 1654, referring to Fife, about 'the new art of manuring with seaweed, or lime from burnt stone, lightly spread on the surface of the ground, which has been proved by frequent experiment to pour life as it were on thin and sandy soils and to be efficacious beyond belief for fertility.' As it stands that passage makes no sense, as the use of seaweed was already centuries old, though the use of lime as a fertiliser on the land does appear to have been new at this time, at least in the east of Scotland. If, however, 'or' was a slip for 'and', it would make better sense, as, if the two were mixed together, the lack of calcium in seaweed would have been compensated for by the lime, which would unlock other available nutrients, though there was no explicit mention of mixing the two manures. Even if this were done, it would still leave a deficiency, as there was only a third as much phosphorus in seaweed as in the same quantity of farmyard manure. Repeated application of seaweed alone, therefore, may have carried a risk for the soil that the farmer would learn to avoid from experience.

Two main types of seaweed were utilised. One was the species of tangle, *Laminaria*, the long-stalked, wide-fronded weed growing from eight to ten feet in length in underwater kelp forests patrolled by fish and seals, growing especially at the mouth of the Firth, from just below high-water mark to as deep as 20 fathoms half a mile offshore (Plate 5). The weed was thrown ashore in gales, particularly between November and March, and piled itself onto the beaches. Every winter storm was an occasion to suspend other farm work and take the horses down the short roads between the coastal farms and the shore, to bring the sea-ware up before the next high tide carried it to a different place. 'It often happens', said the agricultural reporter for Berwickshire in 1809, 'that one tide leaves enormous quantities on the beach, which are swept away by the tide immediately succeeding'. The weed was taken in carts or in panniers, sometimes in Berwickshire up very steep cliff-edge tracks, as far as two miles inland, but most was used closer to the shore. The other main weed used were species of wracks, *Fucus*, which grew between high tide and low tide on rocky substrata throughout the Firth, where they could be pulled or cut at any time of year. Some, of course, also washed ashore. The wracks were the most valuable, but expensive in time and wages to gather. Other weeds were used as the tides made them available.

The use of seaweed was practised along the shores of the Firth as far up as Culross, with records between 1500 and 1800 from the north shore from Crail, Anstruther, Pittenweem, Burntisland, Torryburn and Culross, as well as many generalised references to Fife, and along the south shore from Dirleton to Dunbar and beyond to Cockburnspath. As might be expected, most of these places lie landward of the *Laminaria* kelp forests in the outer waters of the Firth, where storms hurled the weed ashore, but in the upper reaches where only *Fucus* grew, much of it had to be cut off the rocks.

The rights to gather 'wrack and ware' were highly valued and sometimes contested. For instance, in 1570 at Crail, John Bowsy, presumably a local farmer, was prosecuted by the baillies for trying to stop the inhabitants of the burgh from carrying off seaweed brought ashore 'be violence of the sea'. Two years later the council decreed that, without licence from them, no one living outside Crail was allowed to carry off 'wair or fuilzie' from the sea coast 'within the liberty and privilege' of the burgh, which was a considerably larger area than the bounds of the burgh itself. This would have been difficult to enforce in the courts, as the liberties were only intended to be an area within which the burgh had a legal monopoly of foreign trade and market, not of gathering seaweed. Then there was an incident near Culross in 1663, when the town officer and others came with clubs and staves to stop the servants of one local laird from loading the seaweed claimed by another, forcibly unyoking their horses, seizing their forks and emptying their carts.[12]

The manufacture of kelp ashes by burning seaweed in kilns for industrial use was never such an important matter in the Firth of Forth as it became in the western Highlands and Orkney, but there were a number of places where it was done. *Fucus* weed was cut from the rocks and burned on the spot before sending off for the manufacture of soap and glass, in concerns probably based mainly in Leith. In the *Statistical Account* of the 1790s, this was referred to, for example, on Inchcolm, and in the parishes of East Wemyss, Anstruther Wester, Dirleton, North Berwick, Dunbar and Innerwick. At Wemyss they cut every three years and made 100 tons of ashes from the harvest, but this was exceptional. Presumably it was worth more to do this than to spread the weed on the fields, but it was a noxious, smoky business that could attract strict burgh regulation. In 1694 the baillies of Anstruther Wester stopped an Englishman from burning his kelp except on the western edge of the burgh and when the wind was blowing from the east. In 1721, the baillies of Culross forbade the burning of seaweed within their 'liberties' without the consent of the majority of heritors and occupiers of the burgh acres. Eight years later there was a riot when a manufacturer had his kilns destroyed by the servants of an infuriated local landowner, Colonel Erskine, who suffered from asthma which he claimed was aggravated by the smoke and smell.

A burgh could be fussy in its regulations. In Crail in 1568, the council ordered no one to carry seaweed out of the harbour from Saturday until Monday, in order to preserve the Sabbath (which must have been an irritation after a storm). In 1590, they tried to prevent sharp practice by ordering the inhabitants not to gather weed before 7.00 a.m., or to wade into the sea, but all must stand on dry land and draw the weed together with their forks. They also decreed that heaping of weed, or placing weed in middens, should not take place within the burgh itself. Rotting weed in bulk has a vile smell and attracts flies, so there was a public health concern. Interestingly, scientists investigating fertilising practices on the old burgh soils at Pittenweem have found an enhancement of manganese in the town fields, which could be a marker of using quantities of seaweed, but none in the core of the burgh itself, so maybe it was banned here as well.[13]

The question of how best to use the weed on the land was much discussed by eighteenth-century writers on agricultural improvement, but most thought that it was best to apply it immediately by spreading it on the surface of the ground, rather than to keep it in middens or to plough it in, though some recommended mixing it with manure in the farmyard.

Generally, though, it was only put into middens when there was too much of it to use at once, or if it could not be used immediately because it came ashore at an inconvenient time of year. Most commentators agreed that it was highly beneficial to the ground, especially on light, sandy ground near the sea where potassium was in short supply, and where dung-based manures would clog and not readily bind with the thin soil. Also, unlike farmyard composts, seaweed had the distinct advantage of being free of noxious land weeds. Spreading it was, however, a lot of work. *Laminaria* had to be applied yearly, though the benefits of *Fucus* cut from the rocks could last for three years. Enormous quantities of weed were carted, in the early nineteenth century at the rate of 25–40 loads per acre. In the mid nineteenth century in farms of 100–120 acres in East Lothian this amounted to 3,000 to 3,600 cartloads a season.

Englishmen seldom praised Scottish farming before the nineteenth century, but in 1726 Daniel Defoe, in his *Tour of the Whole Island of Great Britain*, gave an account of 'the application of sea-ware, as they call the weeds which the sea casts up', in East Lothian, and of how 'by laying this continually upon the land, they plow every year without laying their lands fallow as we do: and I found they had as much corn, as our plowmen express it, as could stand upon the ground'. This important point about the application of seaweed allowing unbroken cultivation was made again in an account of practice in Fife in 1757: 'there is no such thing as lay land upon the coast, nor land which has not carried crops since the memory of man'.[14] It was particularly valued on light, sandy land for growing barley. The crop might be relatively light, said some commentators, but it could be sown a week later than inland, and was of good quality, sought by brewers and by inland farmers for seed. Land not otherwise worth ten shillings an acre, it was remarked in 1757, could be rented for up to four times that amount, and there were many subsequent observations about the high value of land with good access to sea-ware.

After the advent of guano imported and purchased in bags from Peru, and eventually of all kinds of mineral and artificial fertilisers, from the middle of the nineteenth century, the attractions of gathering seaweed began to decline. It continued at a lower level of intensity, however, in a few parishes until after the Second World War, since farmers could not resist making use of the free bounty of the sea, should they have time and labour to collect it (Fig. 1.4). In East Fife gathering eventually came to an end in the 1970s, because by then the amount of plastic mixed in with the

HUNTING AND GATHERING

Fig. 1.4 – For centuries, seaweed washed up by storms was collected for fertiliser: this cartload at Pittenweem in the early twentieth century would be taken to fields behind the burgh where the soil retains a distinctive texture from repeated applications.

weed on the shore made it potentially dangerous to animals if it was spread on the fields. It then became, not an asset, but only a problem, as it piled up in the harbours and on the beaches. In St Andrews, the local council in the 1990s cleared the West Sands of any light seaweed deposits in order to achieve 'blue flag' bathing-beach status, despite warnings that this would accelerate the erosion of the dunes in front of the golf course, as indeed happened. The latest European Community guidance on 'blue flag' beaches, however, discourages removal of seaweed unless a threat to public health is involved. In Crail, Cellardyke and Anstruther, more within reach of the heavy deposits of the *Laminaria* beds and liable to trap the drifted weed through the construction of harbour works, the smell of the accumulated weed in summer was said to put off the tourists and damage trade. The council cleared it off, and in Cellardyke removed too much sand with the weed, leaving the bed of the harbour bereft of ragworm and other animals that would naturally have recycled the weed. They took it away to landfill near Pittenweem, where it rotted away in stinky pools, found by rock pipits though nearly a mile from the shore that they usually frequent. When, eventually the landfill site was closed, the problem of what to do with the weed on the shore remained unresolved. In 2010, the contractors were using their tractors and mechanical shovels to drop the weed accumulated on the beach at Anstruther back into the sea there, thus, whether wilfully, or in landlubbers' ignorance of tides, ensuring good short-term prospects of future employment.

CHAPTER 2
FISHING IN A FIRTH OF PLENTY

1. Introduction

Long before the start of heavy exploitation in the nineteenth century, the Firth of Forth and the surrounding waters were seen as a place of immense natural abundance. Even English visitors thought well of it. Daniel Defoe, journalist, travel-writer, novelist and one-time government spy, described in his *Tour* of 1726 a journey from Eyemouth in Berwickshire as far as Kinghorn in Fife. At Dunbar, he admired 'the great herring trade', with a thriving business in smoking kippers, just as in Yarmouth. Later, he commented on the great abundance of white fish on this coast and watched an English boat being loaded with cod for Bilbao in Spain. When he crossed at Queensferry 'in a little Norway yawl', the boys in the boat were catching herring and throwing them aboard with their bare hands. At Buckhaven, he found a community of fishermen, ill-housed but prosperous enough, who lived by selling fish to the Edinburgh market. The sight of a great Dutch fleet sheltering from a storm after fishing for herring outside the Firth reminded him of what opportunities were being missed.

Nor was it only fish. At Kinghorn, he described an enterprise, economically dodgy, that shot porpoises, 'of which very great numbers are seen almost constantly', as well as sometimes 'grampuses and fin fish and several species of small whales'. The grampus was the orca, or killer whale. The oil of all these animals would be used for lamp lighting and industrial lubrication. He also saw a group of eight or nine other whales stranded on the shore that he could not identify, some 20 feet long; possibly they were pilot whales.

Defoe's Scottish contemporary, Sir Robert Sibbald, described how the local fishermen knew dolphins as 'meerswine' and porpoises as 'porpuss': he also mentioned orca and listed a large number of other different whales which he knew to have been stranded at different times in the Firth.[1] Indeed, the mighty blue whale, the biggest animal on earth, was first described by him in 1694 from a casual stranding and consequently became named 'Sibbald's Rorqual'. Another blue whale became stranded near Longniddry in the nineteenth century and was painted by Sam Bough (see Plate 6).

It was not just passing journalists and early scientists who dwelt on the plenty of the Firth. In 1773, its abundance was the subject of the opening verses of Robert Fergusson's splendid celebration of Edinburgh pub life, 'Caller Oysters':

> O' A' the waters that can hobble
> A fishin' yole or sa'mon coble
> An' can reward the fishers' trouble
> Or south or north,
> There's nane sae spacious an' sae noble
> As Frith o' Forth.
>
> In her the skate an' codlin' sail
> The eel, fu' souple wags her tail
> Wi' herrin', fleuk and mackarel
> An' whitens dainty,
> Their spindle-sanks the labsters trail
> Wi' partans plenty.
>
> Auld Reekie's sons blithe faces wear
> September's merry month is near
> That brings in Neptune's caller cheer
> New oysters fresh
> The halesomest and nicest gear
> O' fish or flesh.

In 1825, Fergusson's descriptions of fishy plenty seemed to combine with Defoe's sense of lost opportunity, in an account of a public meeting in Edinburgh reported in *The Scotsman*. On 15 October, a gathering of citizens at the Waterloo Tavern at the head of Leith Walk was called to form a company to improve the supply of fish to the city's market. It was chaired by the Lord Provost and opened by Captain Carnegie, who said that although naturalists had enumerated 200 sorts of fish in our waters, which 'swarm with the abundance of the finest species', we had made no progress in catching them:

> While in every art, our countrymen had been making the most rapid improvements, the art of fishing had been so stationary, that were an inhabitant of Cellardyke or Newhaven, of the sixteenth century, to rise from his grave, he would find himself quite at home: his boat, his tackle, his very bait (which, by the bye, was none of the best) would make him fancy that he had only awakened from a dream.

Mr Dundas of Dundas spoke next. He said that he had been out in his yacht in Aberlady Bay and 'threw out a net three or four times: at each haul he caught a number of fish, such as dabs, plaice and halibut. At one time he caught three turbot and one sole'. If an amateur out for pleasure could do this, he asked, what might a well organised company not do? He had once seen two smacks fishing for cod off Blackness where they had caught 70 score and sold them to London at very high price.

Not everyone was quite so sanguine. The Lord Provost welcomed the idea of a company to improve supply but hoped that nothing would be done to prejudice the livelihoods of the fishermen of Newhaven and Fisherrow. Others expressed concerns about the impact on 'the many Margaret Mucklebackers', the fishwives who carried the catch from these ports to the streets of Edinburgh and were a valued part of the picturesque tradition of the city. Captain Carnegie assured everyone that the interests of the lower classes would be served by such a company, implying that the fishermen and their families would find ample employment, though no one asked if they might want it. In the event the company came to nothing, as a recession immediately followed and no one was willing to invest.

Nevertheless, Captain Carnegie was right both to stress contemporary opportunity and long-term stagnation in methods of exploitation. Exactly what had been the history of fishing in the Firth of Forth up to that point?

2. Early Fishing Enterprise

There can be no doubt of the commercial importance of fishing in the Firth of Forth from an early date. Crail had been founded as a burgh on royal lands in the twelfth century, laid out with lavish provision for markets and attractive to merchants, its surrounding lands granted to the king's Anglo-Norman and Northumbrian servants, as if were intended as an English-speaking Hong Kong at the tip of East Fife, a node of commerce and civility in an otherwise Gaelic-speaking area.[2] The first surviving impressions of its burgh seal date from 1357, and show an undecked boat, sailing under the stars with a single mast, a net and a crew of seven (Fig. 2.1). It is the first depiction in Europe of a herring fishing vessel, and one that would have been recognisable locally as such down to the early nineteenth century. The neighbouring burgh of Kilrenny, by contrast, which includes the fishing town of Cellardyke, has a crest dating to 1578 and shows an open rowing-boat with a crew of five, apparently preparing to drop anchor, with an indication of a line trailing behind. This appears to be the first depiction in Scotland of a boat intended to fish inshore for cod or haddock, though no doubt it would have been more appropriate to have illustrated more clearly the 'sma lines' with numerous hooks (Fig. 2.2). The herring has a soft mouth and was best caught by a suspended net that would entangle its head; in the thirteenth century the gill net became widely adopted in Europe. Cod, haddock and similar white fish, on the other hand, have a hard mouth that can hold a hook. Dunbar, the main early fishing port on the south shore of the Firth, has a charter from 1370, but no special indication of fishing on its coat of arms.

Fig. 2.1 – The seal of the royal burgh of Crail (1357) shows an open boat with a single mast, the sail furled at night, with a herring net and a crew of seven.

Such is the scarcity of sources for all aspects of medieval economic life in Scotland that it is hard to say much in detail of the early history of fishing in the Firth. Most of the scraps of information that we have come from monastic records and relate to questions about the payment and distribution of teind fish, the tithes in kind that were rendered to the church from the catch. Thus in 1222 there was an agreement between the Premonstratensian monks of Dryburgh, who held the teinds of Kilrenny, and the Augustinian canons of St Andrews, to allow the fishermen of their respective communities to use each others' harbours without the necessity of paying tithes except in their native parish. Two years later the papal court in Rome had to settle a dispute between Dryburgh and the Benedictine

Fig. 2.2 – The seal of Kilrenny (1578) shows a rowing boat with a crew of five, apparently line fishing.

monks of the Isle of May, who owned the teinds of Wester Anstruther, over dividing the dues at the anchorage of the Dreel Burn, which was a boundary between their two parishes.[3]

By the close of the fourteenth century, there are some signs that a regular, if modest, export trade was beginning to appear from Scotland, as herring fisheries in England and the Baltic faltered, though records are not very good. The Scots certainly came to play second fiddle to the Dutch, whose highly organised 'Great Fishery', from at least the early sixteenth century, dominated the catch in the North Sea and the valuable international trade that followed. The Dutch fleet, starting to operate in June off Shetland, and moving steadily down the British coast through the late summer and autumn, consisted of hundreds of large, three-masted, partly-decked, herring 'busses' that could each lay up to 45 drift-nets, 1.4 kilometres in length, extending to 22,000 square metres. They were accompanied by other boats packing the catch in salt, and by cargo vessels taking the partly preserved fish back to the Netherlands for repacking and export while the main fleet remained at sea. The Scottish and British governments for centuries tried to encourage imitation of their methods, and also to prevent them from fishing within sight of Scotland's shores. They did not succeed in either objective, though they kept them from fishing within the Firth of Forth itself. Scottish fishing, however, was itself far from unenterprising, though operating on a minor scale compared to the Dutch.[4]

Between 1470 and 1600, the Exchequer kept relatively full records of exports, from which it became evident how a new prosperity and new initiatives began to appear in the Firth of Forth herring fishing fleet, particularly from the late 1530s. Fife fishermen started to use half-decked boats, 'crears' of up to 30 tons, to sail to the north of Scotland in the autumn, after the local 'Lammas Drave' was completed. Sometimes they fished off Orkney and Shetland and sometimes they rounded Cape Wrath and entered the Minch, where they organised the actual fishing from open boats in the sea lochs of Wester Ross and Lewis, returning before Christmas. They carried salt to keep the herring from decay (the fish start to rot within a day or two if not at least partially cured). They normally also

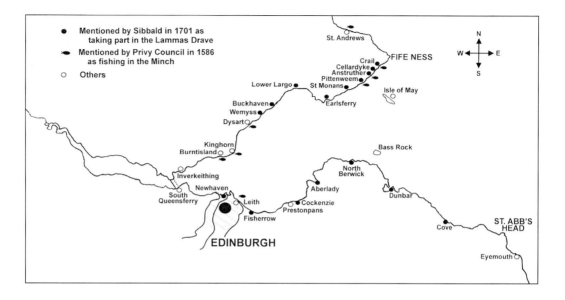

Fig. 2.3 – Fishing communities 1500–1800

carried two small 'yawls' on board, but sometimes they bought new open boats from the Highlanders. By 1587, as many as 100 crears from the south-east of Scotland were to be seen in Aberdeen harbour on their way home from the 'Iyles'.[5]

All this was also accompanied by considerable expansion of fishing within and around the Firth itself, by a surge in exports, especially to the Baltic, and by investment in the fishing burghs by Edinburgh merchants, equipping ships and gear that were considerably more expensive than anything seen before. It also brought an immediate increase in prosperity to the East Neuk of Fife, marked by the erection of more fishing towns into royal burghs with full commercial privileges, Pittenweem in 1541, Anstruther Easter in 1585 and Anstruther Wester in 1587, to the indignation of Crail which had hitherto had a monopoly of foreign trade as far along the coast as Leven. A document of 1586 names all these ports as involved in the fishing in the west, along with St Andrews, Leith, Burntisland, Kinghorn and Dysart, but not at this point, interestingly, Dunbar or any burghs in East Lothian.[6] Exports from the Forth rose from about 76 lasts a year before 1536 (less than the total contents of three average-sized Dutch busses) to nearly 1,000 lasts in 1541, settling down to an average of a little under 500 lasts (1,000 tonnes) in the last half of the century. In 1586, shipments of 'Lochbroom herring' overseas were restrained by the Privy Council to make certain that consumers at home, not least the royal household, got their necessary supplies first. Despite this initiative of sailing all the way round Scotland to catch fish, it seems likely that most of the herring exported from Fife were actually still caught locally. According to customs records, in 1574 and 1575, more than two-thirds, and in 1577, about one-third, of the exports came from the Firth.[7] Internal demand from Edinburgh and the country around would also have been mainly satisfied from within the Firth of Forth.

Eventually the enterprise of the east coast fishermen in the west led to the extraordinary incident of the Fife Adventurers, in which the Crown sponsored an attack on its own Highland subjects.[8] There had been complaints for many years of harassment and lawless extortion by the clans, directed against the Lowland fishermen when they came ashore, albeit, as the latter sourly observed in 1586, that the Lowlanders 'have biggit thair houssis of thair awne tymmer and cover the same with thair schip sailles for saulftie of thair geir fra the rayne'.[9] The Macleods of Lewis in particular were regarded as 'awowed enemies to all lawfull traffique and handling in these bounds': according to the Privy Council, 'the maist profitable and commodious trade of fishing', was, by 'thair barbaritie altogether neglectit and overruin'.[10]

In 1598, the king's cousin, the Duke of Lennox, with royal permission, banded together with 12 of the lairds of Fife, mainly the occupants of tower houses round the coast between Anstruther and Balmerino, including Wormiston and Balcomie, to attack and occupy the Island of Lewis. The intention was to displace the native Macleods, considered rebels as well as barbarians by James VI, and to establish in their place a colony of industrious Fifers, whose fishing was clearly to play a large part in a new prosperity for the western Highlands. The initial onslaught with 500 men, accompanied by Robert Durie, minister of Anstruther to give spiritual comfort, succeeded, but the Macleods counter-attacked and threw the invaders out. Over the next ten years the Adventurers tried on three occasions to capture the island and to establish a burgh at Stornoway, with varying success. At one point the Macleod chieftain was captured, tried in the Lowlands, condemned and executed, his head displayed on a spike in St Andrews. But in the end the Fifers were roundly defeated. The task of subduing the Macleods was handed over to the Mackenzies from the neighbouring mainland. As the author of the Wardlaw Chronicle put it, writing from outside Inverness, the whole episode 'turned to the ruin of diverse of the undertakers, being exhausted in means, not having the language, wanting power to manage such an enterprise in a strange place farr off'. He concluded that 'Highlanders were fittest to grapple with Highlanders, and one divil, as the proverb is, to ding out another'.[11] For the king, this was a forerunner for his plantation of Ulster (where he was continuing the colonialisation begun by Queen Elizabeth). At least one of the Fife Adventurers, Thomas Monypenny of Kinkell, having failed in Lewis, went on to enjoy a successful land grant in County Fermanagh. And Robert Durie, the minister, went on to found a Scottish kirk in the Dutch university city of Leyden, surely a more comfortable outcome for him than shivering it out in Stornoway.[12]

For the Fife fishermen, evidently, this episode was a diversion that did not permanently either favour or harm their efforts in the Minch. They went on operating in the north until the 1640s, and their enterprise only came to an end with the disasters of the civil wars, especially the loss of life at the battle of Kilsyth in 1645 where a large number of local skippers and seamen perished for the Covenanting cause against Montrose, fighting inappropriately in a Lanarkshire bog. The Fife burghs later, in 1651, also suffered sacking and heavy loss of boats at the hands of Cromwell's armies.[13] All this was disastrous in the short term, but by the end of the seventeenth century fishing in the Firth of Forth had managed to make a considerable recovery, although trips to 'the Lewes' were not resumed on any scale, possibly for lack of capital but probably because enough fish could now be caught locally in and close to the Firth itself.

The fullest description of the way the 'Lewes' fishing was organised at the heyday of the expeditions to the west comes from an account of 1733 by Patrick Lindsay, provost of Edinburgh.[14] This was almost 90 years after the fatal battle of Kilsyth, so his testimony is not first hand and one could have doubts, at first sight, about its value. But Edinburgh had been closely involved in financing the whole business of fishing with larger boats, and possibly there was oral recollection in his own family or in the merchant circle about how it was done. The circumstantial detail is at least convincing.

The fishing year, wrote Lindsay, was organised around small decked boats of 15–30 tons which he called 'bushes', apparently the crears of earlier sources. These were described as carrying two large nets and having a single mast, shipped at night and used to dry one net while they shot the other. These boats left in March from the Fife ports to catch cod on the coast of Orkney, using herring caught in the nets as bait for their lines, salting the cod in the hold or in suitable weather drying them on local beaches. They returned in May, washed the salt out of any fish in the hold and completed drying them, selling part for home consumption and part overseas. In early June, the bushes again took their nets, salt and cask, and went after deep water herring in the North Sea. When they had caught what they could, they ran for the shore, obtained new materials and fished (away from Fife) until the end of July. Then they returned 'and fished across the opening of the Firth' for the next month. 'Here' says Lindsay, 'they never failed to fish with success, and gave certain intelligence to the open boats (wherein the same persons were sharers) where to lay their nets for the herring, close by the shore in shallow water.' This last was what became known as the Lammas Drave.

In September, the bushes took in 'a fresh fleet of nets' and sailed to the Lewes, operating in the Minch and off Sutherland and Wester Ross, returning about Christmas. He does not explain what we know from other sources about the open boats used there, or go into further detail, but says the bushes were home before Christmas, after which the fishermen 'went to fishing upon the coast in open boats and were by this constant practice the most expert fishers in Europe'.

In the absence of more contemporary detail, it is hard to be sure if Lindsay's picture is totally accurate, but it would have made sense to find a use for the relatively expensive crears outside the autumn season. Certainly the seamanship of the inhabitants of these small Fife ports became exceptional as they ventured in the Minch and round Cape Wrath and through the Pentland Firth between September and December, at the time of the worst gales of autumn and early winter. It is significant that merchants from Edinburgh and elsewhere were quick to hire, in a carrying trade around Europe, the skippers from Crail to Burntisland, including from Kirkcaldy, a port which did not fish much itself but carried much of Dysart's and Kinghorn's business. Even in the late seventeenth century when the Lewes fishing had declined, the boats of these Fife ports were often among the last coming home through the Danish Sound in November before it iced over, returning from selling a late cargo of herring in Danzig with hemp, flax and sometimes rye.[15]

By the end of the seventeenth century, however, it does seem clear that the bulk of the herring fishing based on Firth of Forth ports now took place locally, and close inshore from open boats much like the one portrayed on the Crail burgh arms. The quantities of fish exported became, however, even more substantial than they had been in the previous century, which perhaps explains why the fishermen no longer troubled to go to the west. The Baltic was taking some 400 lasts a year from the later 1670s, rising between 1700 and 1707 to 870 lasts. Most of the Baltic export trade would have originated from the Firth of Forth. In the year of the Union itself exports touched 1,440 lasts (about 2,900 tonnes). Scotland supplied 49 per cent of all herring to pass the Sound in that decade, partly because the Dutch fleet, operating out to sea, was much more vulnerable to attacks from French privateers than were the inshore Scots. Exports to the Baltic became higher still up to about 1720, after which they went into rapid decline, apparently because the fish moved further offshore in the spawning season.[16]

An interesting account of the Firth of Forth herring fisheries at the opening of the eighteenth century was made in 1701 by Sir Robert Sibbald. The 'Drave', he explained, took place in July and August every year. He listed the ports of the south shore that participated – Newhaven, Fisherrow, Cockenzie, Aberlady, North Berwick, Dunbar, and Cove in Berwickshire. He gave more detail of the north shore, starting with Crail (30 boats, each with eight men); then Cellardyke (12 boats, of seven or eight men); then Anstruther (12 boats), Pittenweem (15 boats), St Monans (15 boats), Earlsferry (three boats), Largo (three boats), East Wemyss (four boats) – all these with seven men; finally Buckhaven (12 boats with six men). The varying number of crew presumably indicates slightly different fishing methods and size of boat; he implied that those in Fife tended to be larger than those in Lothian, and said that 20 more boats came down from further north to the Lammas Drave. The Fife fleet that he listed contained 104 boats and 752 men, but nine years later he revised the list upwards: Crail now had 80 boats (Fig. 2.4), Cellardyke 20, Anstruther 24, and Pittenweem 'more' than before. This would increase the Fife fleet to upwards of 174. The visiting fleet from the north he now put at 200. Patrick Lindsay 20 years later spoke of 600–800 boats engaged in the Firth of Forth Lammas Drave, each employing eight or nine men, but this was not broken down by port and could have been an exaggeration.[17]

The drave boats were, said Sibbald, more than twice the size of a line fishing boat, and each 'ordinarily' contained nine double nets. Lindsay spoke of each boat containing eight nets. An account from Eyemouth of 1785 spoke of ten nets for the 'ground drave' where the nets were anchored and left overnight, each net 40 yards long by 10 yards deep, and ten nets for the 'float drave', where they were hung from a drifting boat, each net 50 yards long and 18 yards deep.[18] Plainly the latter operated in deeper water,

Fig. 2.4 – A harbour for pre-industrial fishing: Crail in 1880, probably little changed since the Lammas Draves of the eighteenth century.

but for much of the eighteenth century the ground drave was the method preferred, at least in Fife.

Sibbald and Lindsay broadly agreed how the boats were owned and operated in the early eighteenth century. Lindsay said that the boats themselves were owned partly by the fishermen who otherwise spent the year catching white fish, but most belonged to shipwrights and others who built them and fitted them out with nets and ropes. He said that two or three full-time fishermen worked with six or seven landsmen to make up the crew for the two months of the drave, these landsmen being either associated with curing, net-making or shipbuilding in the fishing burghs, or from the countryside: 'farmers servants that live near the coast make it a condition with their masters to be allowed the drave to themselves'. This last was quite different from the situation in the later nineteenth century when the economic and social division between fishing and agriculture was absolute.

One of the crew was chosen as keeper of the stock purse. His job was to keep accounts, pay the expenses and divide up the profits at the end of the season; this was done according to a complex formula which Sibbald and Lindsay describe slightly differently. Sibbald says the profits were divided into 20 parts, two of which were allocated to pay the tiends, and seven crew had one part each. Of the remaining 11 parts, the owners of the nine nets each had a part and the owners of the boat the remaining two.

Lindsay's description is similar, but he explains that the nets might be owned either by the fishermen or by those who fitted out the boat. Those who had not been to sea before were entitled to only half of what the others got for their work. This was a method more co-operative than capitalist, with no wage earners as such. As James Miller, historian of Dunbar, put it in 1859, 'all parties are interested in profit and loss'.[19]

From time to time it occurred to government that Scotland would be better served by a joint-stock company that would have the capital to challenge the Dutch at their own game of bush fisheries and offshore fishing. Schemes were dreamt up by Charles I's government in the 1630s, encouraged by an Act of the Scottish Parliament in 1661, more determinedly attempted in 1670, but abandoned by 1690 having achieved nothing, fleetingly considered again in 1700 and in 1720, and, after the 1745 rebellion, actually taking shape on the west coast in the efforts to subsidise the building of larger ships and to found Highland fishing settlements by the British Fisheries Society, as at Tobermory, Ullapool and Wick. As far as the Firth of Forth was concerned, however, the effects of all this attention were minimal. Of more concern here were regulations about what salt might be allowed to be used in the herring cure after the Act of Union, and what subsidies might be paid on herring exported. But above all what most deeply affected the prosperity of the fishing was whether the herring would turn up.[20]

For most of the sixteenth and seventeenth century, and at the start of the eighteenth century, the herring appear to have been relatively reliable in their migrations. John Lamont noted a dearth of herring in 1657, 1658, 1662 and 1668 as both exceptional and catastrophic. In 1658 he wrote in his diary:

> Thir two yeares, ther was few or no herring gotten on the Fyfe side, and not many in Dumbar, so that divers persones beganne to fear yther sould be no dreve hireafter, which was a great prejudice to the poor fisher men, as also to the whole places nire-about, for the like had not beine, as some thinke, for the space of a hundred years before.[21]

The vagaries of Providence were unknowable. Whether the herring came or not probably depended, in reality, on offshore currents which themselves depended on the North Atlantic weather system. The fishermen themselves certainly caught far too few to have any impact on herring numbers from one year to the next.

The Lammas Drave in the Firth of Forth occupied, at most, two months between July and early September, the fishing taking place close inshore in the 'traiths' where the herring bred. Usually it took place at night, but exceptionally by day. Sometimes it lasted only a couple of weeks. Of the two methods to catch the fish, the 'float drave' was essentially a form of drift-netting where the nets hung down in the water from the boats which actively fished all night. At the 'ground drave', probably the commoner method, the nets were left anchored overnight, and the boats returned for them next day.

With hundreds of boats operating in a small area, the potential for conflict, confusion and even sharp practice was considerable, so the fishermen agreed to a form of regulation under the Scottish Admiralty Court,[22] whose formal interest in the matter was to see that official dues ('assize herring') were properly paid. The Admiral, usually a nobleman accorded the privileges and profits of the office, appointed in the fishing districts a deputy, who was usually a baillie of one of the burghs. The Admiral Depute then summoned the skippers, sometimes through a proclamation at the kirk doors on a Sunday, to a meeting to elect by majority vote, a jury of 15 persons. The Admiral Depute then read out last year's rules for the drave and the jury voted, again by majority, whether to keep them or to vary them.

Since the juries usually voted to keep all or most of the existing rules without comment, it is unusual to find a complete set in the records. In 1627, however, a new court was established in Fife to regulate the fishing west of Levenmouth, and a set was then described.[23] The first rule was designed to prevent profaning the Sabbath, and forbade laying floats or nets on a Saturday afternoon, selling herring on a Sunday, or setting nets again on a Sunday before 4.00 p.m. The second rule forbade attempting to catch herring by daylight until such time as the herrings were known to swim in daylight, at which point the Admiral Court would issue a licence to boats wishing to do this. It was believed that the herring could easily be scared away by daylight fishing. Other rules forbade boats from sailing to the drave before daylight (presumably those at the float drave would be out there already), or going to the drave with insufficient gear, or taking fish from other men's nets, or taking away for themselves loose floats or nets belonging to others. No one in time of ground drave was to stay by the nets at night, or to cut his neighbours' nets or gear, or ride too close to his 'witter' (which in this context probably means anchor rather than the more normal meaning of hook) or floats. No one was ever to allow his nets to float in the sea by daylight; every master was to give his oath that he and his company would keep the rules (the master accepting responsibility); every boat must have its floats marked with its own mark; and if any were found with different marks it was assumed they had been stolen. Finally, because 'honest men . . . are contenuallie opprest be ane number of maliciounes wicked insolent persones' who attend the drave more to steal from others than to profit by their 'awin industrie and labour', the Admiral would provide a boat with 'and sufficient number of honest men' to 'mell withe anie persone doing wrong' on the fishing grounds, and if necessary to bring them ashore and imprison them. This was a forerunner of the state fisheries cruisers of the nineteenth century. For this service every boat agreed to pay a fee.

This institution proved an effective way to regulate the fishing at least until the dissolution of the Scottish Admiralty Court in 1830. The surviving records of the Admiral Depute of East Fife for the late seventeenth century show essentially the same mechanisms as those of 1627, with an elected jury of 15 and an Admiral's boat in attendance, and similar rules, though laying nets on Sunday afternoon was no longer permitted. The court also acted to prevent fishermen selling any of the catch privately at sea: it had to be brought into Crail and offered in the market there.[24] Again in East Fife, when the herring arrived in 1739, the skippers were warned by 'tuck of drum' to attend the Admiral Depute Court in Anstruther Wester to 'hear the old laws and to make such new ones as shall best contribute to the advantage of the present fishing'. Fifteen of them elected a 'chancellor' as chairman, and determined that this year should be only a ground drave, that the boats shoot their nets at four in the afternoon at a signal from the Admiral's boat (this was usually a shot from a gun) and haul them in at four in the morning on another signal, no one to stay out overnight.[25] In East Lothian, David Loch reported in 1778 how he attended the Admiral Depute's Court in Dunbar with a jury of 22 skippers who similarly chose a 'chancellor' to chair the meeting, putting to a majority vote the rules 'to regulate an orderly shooting and hauling of their nets, the time when to shoot, and whether a float or a ground drave . . . then ends the court with a dram to drink the health of men and the death of fishes'.[26] James Miller in 1859 reported that at Dunbar even after the demise of the Admiral

Court, the fishermen still appointed 'one of their number, whom they style admiral, to arrange the order of sailing etc, and two chancellors, to whom all disputes are referred'.[27] Thus the fishermen before the late nineteenth century operated with a strong sense, not only of community, but also of the necessity for common action, mutually agreed by majority vote, to conserve and regulate the fishing.

3. Mixed Fortunes after 1730

After around 1730, the fortunes of the fishings in the Firth of Forth were much more uneven than they had been for most of the previous two centuries, apparently because the herring often failed to materialise for the Lammas Drave. This was also observed to have had a knock-on effect by greatly diminishing the numbers of cod, haddock and other white fish that fed on herring spawn and young. There were, however, good years and even good runs of years from time to time, as in the 1770s when David Loch visited Dunbar.[28] He found 52 boats operating here in September, and 43 on the Fife side. Dunbar had rediscovered its old skills in curing red herrings, once taught, he said, by Dunbar men to Yarmouth and Lowestoft, then forgotten in a period when the herring went away, but now recovered from the descendants of the very people who had once been taught in England. He found prosperous fish merchants, particularly Charles and Robert Fall, who ran the smokeries, and also had taken a contract to buy cod throughout the district from Eyemouth northwards, and along the east coast of Fife. A 'full-sized cod' he described as 18 inches long from the fin at the shoulder to the upper joint of the tail. These were fished between March and May, 'at which time there are also caught plenty of skate and halibut'. He also visited the Fife shore and found equal prosperity there (Fig. 2.5).

Loch was particularly impressed (as well he might have been) by the women of Fisherrow who accompanied their fishermen husbands to Dunbar for the drave, and carried the fresh herring back towards Edinburgh on their backs. This was a journey of 21 miles, carrying a basket with 150–200 herrings, made without stopping 'except to take some small beer if they feel thirsty or faint'. On arrival at Fisherrow they handed over the load to another relay of 'nimble-footed women who generally got to Edinburgh sooner than any carrier or cadger with a horse', to command premium prices because the fish were not only fresher but also less bruised

Fig. 2.5 – The plenty of pre-industrial fishing: see the variety and size of the fish caught at Pittenweem, early nineteenth century.

than anything carried on horseback or by boat.²⁹ From Eyemouth in 1822, however, the herring were sent the 50 miles to Edinburgh by cart, 'preferred to water-carriage, on account of its greater certainty, as the wind and tide would not always serve'.³⁰ That was just too far for the fishwives.

The Revd Dr Alexander Carlyle wrote a famous account of the strenuous women of Fisherrow and Musselburgh in the *Statistical Account* of 1792, in which he described their distinctive culture.³¹ It was nurtured by the fact that not just the fishwives but most of the working-class female population of the community were carriers, if not of fish, then of vegetables, salt and even sand for washing floors (they had the worst job), and laundry. Their husbands, apart from the fishermen, being weavers, shoemakers, tailors or sieve-makers, all sedentary occupations, were able to look after the children while the women were on the road. This gender reversal, according to Carlyle, led the women to adopt masculine habits, playing at golf and football. If one of their number got married whom they did not consider up to the task, they would observe 'Hout! How can she keep a man who can

Fig. 2.6 – One of the last of the Musselburgh Fisherrow fishwives, around 1950 – lineal descendant of the women who carried the herring from Dunbar in the eighteenth century. She is in traditional dress and carries a traditional doll dressed the same, in the annual 'box walk' gala procession.

hardly maintain hersel'.' The fishwives, though, married into their own 'cast or tribe, as great part of their business, to which they must have been bred, is to gather bait for their husbands and bait their lines'. But on four days a week they carried the fish in creels for the five miles into Edinburgh, sometimes bearing the fish in relays with three women to one basket, when they could reach Edinburgh in three-quarters of an hour. He says nothing of the long-distance carriers from Dunbar, of an earlier generation. Modern carts were replacing the other carriers in Musselburgh, but the fishwives went on as before, in the interests of providing quality (Fig. 2.6).

Reports from 1784 indicate a Lammas Drave still pursued successfully, for example from Eyemouth where some 40 boats followed both the ground drave and the float drave (the latter 'only known here within the last fourteen years'). By the 1790s, however, both white fish and herring were reported as scarce in and around the Firth of Forth. The authors of the *Statistical Account* generally gave a very gloomy account of the fisheries, and in a great deal of instructive detail. Probably at no time before the twentieth century did the picture seem so black. Whatever had happened must have had a natural cause, probably a climatic one affecting the inflows of warm and cold water into the North Sea, but it is unclear exactly what that was. It demonstrates the interlocking character of the ecosystem, where the temporary collapse of the herring stock deprived haddock, cod and flatfish of food from the spawn and fry, and led to their collapse as well.

All round, the story was the same. At Fisherrow, Carlyle said there had been no haddock fishery for seven years. At neighbouring Prestonpans, which also incorporated Cockenzie, haddocks were also reported as having disappeared. Two years before there had been great numbers of herrings caught in August near the town, and cod, skate, flounders, whiting, mackerel, lobsters and crabs were also available in season. Nevertheless, a general scarcity was reflected in prices, 'more than tripled within these twenty years'.

In the burghs of the outer Firth the picture was unrelieved doom and gloom. At Dunbar only 12 boats and 40 fishermen now operated, fishing for white fish and lobsters, since the herring fishing had become 'very precarious and uncertain'. At Crail, said the minister, where the Lammas Drave had been as regular as the farmer's crop, 'a sad change has now taken place: and we listen as to a fairy tale to the accounts given by old people of what they remember themselves or have heard related by their fathers'. The herring, once the mainstay of the economy of the town, now seldom came, and if they did, they failed to stay long enough to spawn. The haddock which fed upon their eggs and fry were also very scarce: eight or ten years before they could be bought for 25d the long hundred (132 fish), but now they cost 4d or 6d each. Fishing for great cod and skate was still good, but difficult for open boats as they had to be caught far out to sea. Lobsters were easier prey: what might seem to us the astonishing total of 20,000–25,000 was shipped yearly to London, but this, he said, was only half what it had been ten years before.

Confirmation of the disaster came from the coast north of Crail, though with nuanced differences. At Kingsbarns, it was said that some herring still appeared in early spring and autumn, but now 'such as are caught are sold at exorbitant prices' by merchants who sold them cured to the Mediterranean, and the poor felt the lack of herring as a cheap winter food. Few haddock had appeared for three years, but there were small cod, skate and ling to be got, and lobsters for the London market. At St Andrews, the departure of the Lammas herring from the coasts of Kingsbarns parish was emphasised: they had been 'in our memory . . . caught in immense quantities . . . but very seldom within the last twenty years, has that fishing there been worth mentioning'. Haddock were formerly supplied from St Andrews to Cupar and north Fife within a distance of ten miles. They had also largely gone, and the local fishermen in consequence become poor and lethargic. Yet the minister could give a list of other fish still in 'great plenty' in St Andrews Bay and adjacent coasts – flounders, soles, skate, halibut, turbot, cod and ling.

Inside the Firth, it was similar. The elderly minister of Kilrenny recalled how 'vast quantities of large cod, haddocks, herring, halibut, turbot, and mackerel' had been caught from Cellardyke, but the fisheries were now 'miserably decayed'. As a young man (he was born in 1710) he had counted 'no less than fifty large fishing boats, that required six men each, belonging to the town of Cellardykes, all employed in the herring fishery in the summer season'. The customs officer of Anstruther told him that in those days he had 'kept an account' of 500 boats gathered from all quarters at the drave and he himself had seen 'ten or twelve large boats come into the harbour in one day, swimming to the brim with large cod', and towing a string of up to 50 more behind them; you could buy the largest cod for 4d, and their offal was used as manure. It was an elegy for a lost past, no doubt losing little in the telling. His colleague at Anstruther Wester told of a continuing sale of fishes from the town to Cupar and Edinburgh, and even as far as Stirling and Glasgow. He listed cod, ling, skate, haddock, herring and flounders, and £1,000 sterling worth of lobsters that were sent yearly to London. Still, there had been substantial decline here too: the teinds of fish had been worth at one time £10–15 a year, but for the last 20 years they had never been worth more than 13s and sometimes sold for as little as 5s.

At Pittenweem, it was said only that there had been decline of the fisheries: there had been a plan recently to net turbot and send them to London but it had not come to much, though a considerable quantity of lobsters were exported there. According to a comment of 1803, turbot were newly fashionable: a 'general officer noted for his wealth and love of good cheer' had 'taught the people of Fife that they were eatable, and astonished the fish cadgers by offering a shilling a piece for the largest of them'. Until then they had been, in some places, considered unsaleable, and thrown on the beach for the poor to pick up.[32]

At St Monans, said the minister in the 1790s, there had once been a thriving fishing of cod, ling, haddock, turbot, and skate, but for the last four or five years the haddock in particular, once so abundant, had left the coast, 'not one being taken in a whole year'. The shoals of herring, once coming in both autumn and spring, 'are now become very precarious and of no consequence'. At Largo, where ten years ago haddock 'of a very delicate kind' had abounded, now there was not one in the bay; 'all that remain are a few small cod, podlies [saithe], and flounders.'

Finally, beyond East Fife, the minister of Wemyss parish said that although in East Wemyss and West Wemyss the fishing had fallen from six boats to one ('and the crew consist of old men'), at Buckhaven the number of fishermen had not declined. There were fewer fish than before, but higher prices compensated for smaller volumes. Buckhaven was geographically well placed for selling across the Firth in Edinburgh, and though formerly its fishermen had not been able to compete against the plentiful and larger fish caught and sold from the harbours of east Fife, this factor no longer applied. It also had its own squad of fishwives who went into the countryside to sell their wares. The list of fish from Buckhaven is familiar: haddock, cod, turbot, skate, whiting, sole, flounders, mackerel and herring. The problems were familiar too. The minister was told 'by a worthy fisherman' that 40 years ago as many as 25,000 haddock had been sold in a single day at 6d to 10d a hundred, but now that price was sometimes given for a single fish. There was also the usual story of boats that had formerly gone to the Lammas Drave of herring, at Dunbar, but now abandoned it as not worthwhile.

There are several things to note in this mass of detail. First, the crisis mainly hinged on the simultaneous disappearance of herring and haddock, the two main food fishes of the area. Other species had fared less badly, though prices had gone up. Secondly, the herring did appear occasionally, though not in the same numbers: their decline was seen as long term, but especially as having accelerated in the last 20 years. The troubles of the haddock were variously described as dating from between three and seven years ago. Thirdly, there was still enough fish to support internal trade to the Edinburgh market and elsewhere, but only in lobsters was there still a substantial trade outwith Scotland. The species most often mentioned in the *Statistical Account* were much the same as those that Robert Fergusson had mentioned in his poem, apart from the addition of turbot. However, it was the export trade in herring to Europe that had been the mainstay of prosperity in the burghs before about 1720; according to Lindsay in 1733 only about a tenth of the catch from the Lammas Drave had been sold within Scotland. The causes of the collapse of herring and haddock stocks in the outer Firth and beyond remain mysterious, but must have been natural rather than induced by man, as the catches then were so inconsiderable in relation to the stock. What they exactly were can scarcely be guessed at.

Meanwhile, however, higher up the Firth, the *Statistical Account* related something very different starting to happen. It was described by the minister of Queensferry in 1796. After saying that the fish normally to be found thereabouts were 'cod, haddocks, whitings, skate, flounders, herrings, crabs, lobsters and oysters' he goes on to relate how, beginning in the end of 1792, an enormous quantity of herring had been discovered to be coming up the

Firth to spawn in winter, between the middle of October and the end of February, in an area that stretched from Burntisland to above Bo'ness. 'It is a pretty general opinion', he said, 'that the herring shoals have formerly frequented this part of the Frith', and much to be regretted that they had not been exploited before. He said that the previous year he had observed from Queensferry, 80 to 100 boats 'almost every day busily and successfully employed'. The herrings supplied the Scottish market within a radius of 30 or 40 miles, and now 'vast quantities were cured and sent to the foreign markets'. The minister of Inverkeithing reported the same phenomenon, speaking of a single shoal that stretched in 1793 from Inchcolm off Aberdour to Inchgarvie off Queensferry. There were also hopes expressed at Buckhaven that the appearance of these newly discovered fish would also revive their fortunes.

These were later to be known as 'the winter herrin', and in the late nineteenth and early twentieth century they were a mainstay of fishing in the Firth. It had indeed been long known that some herring came to the area at this time of the year to spawn, but they had been regarded as inconsiderable compared to the fish of the Lammas Drave. Genetically, we now know them to have been a distinct stock.

The winter fishing off Burntisland engendered enormous activity for about a decade, and gave rise to a substantial, if short-lived, curing industry in the town. According to a government report of 1798, the fish arrived about 20 October, 'escorted to this Frith by a certain species of whale', and concentrated between Pettycur and Queensferry. The total fleet that assembled to catch them came from as far afield as Caithness in the north and Yarmouth, Kent, Bristol, Liverpool and Ireland in the south. It was said possibly to exceed 1,200 vessels, about a third of which were larger, decked herring busses from the west coast that came through the Forth–Clyde Canal.[33] According to the fishing historian Malcolm Gray, 'it was possibly the biggest herring fishing effort that had ever been made, regularly over a number of years, on any small sector of the Scottish coastline'.[34] But by 1805 the fish had vanished as suddenly as they arrived, and the fleets and industry dissipated. It was assumed to have been an 'Act of God', similar in kind to the appearances and disappearances of the earlier decades. Possibly, however, the extraordinary intensity of the fishing effort in this case depleted what may have been, after all, a comparatively small and discrete population.

Nevertheless, herring eventually returned again, this time in the shape of a revived Lammas Drave. From 1816 to 1822, and again from 1836, it flourished once more in East Fife, and at Dunbar, by 1819, some 280 boats employing nearly 2,000 men were again using the port. We can conclude with another poem, this by Anstruther-born William Tennant, celebrating 1816 with an account of the 'appearances' that presaged the arrival of the fish.[35] He described the seasoned fishermen scanning the sea for:

> Some well-known symptom of the herring race,
> Nor looks he long in vain; the enormous whale
> Spouts briny fountains from his nostril wide,
> Wheel-like, the awkward porpoise by the shoal
> Of countless millions rolls: the cunning seal
> Follows the multitude; the yellow solan,
> Stooping full frequent from his path aloft,
> On his defenceless prey, gives presage sure
> Of the long-wished for, happy herring drave.

Whale, porpoise, seal, gannet and millions of herring: they were indeed the signs of a return to a Firth of plenty. On this, the later nineteenth century was to build extraordinary prosperity, and then to sow the seeds for ultimate disaster.

CHAPTER 3
OYSTER WARS

1. Introduction

Autumn brought oysters into the taverns of eighteenth-century Edinburgh, and the oyster cellars were locations where inhibitions were forgotten. Robert Chambers, writing his *Traditions of Edinburgh* in 1835 when manners were less 'coarse' (his term) described the scene with a touch of wistful envy:

> A party of the most fashionable people in town, collected by appointment, would adjourn in carriages to one of those abysses of darkness and comfort . . . where they proceeded to regale themselves with raw oysters and porter, arranged in huge dishes, upon a coarse table, in a miserable room, lighted by tallow candles. The rudeness of this feast, and the vulgarity of the circumstances under which it took place, seems to have given a zest to its enjoyment.

It was one of the few places where men and women, even respectable women, were allowed to mingle, 'forgetting for a time the rules of good breeding and the customary formalities of society'. Both sexes indulged 'without restraint, in sallies the merriest and the wittiest' and jokes and remarks in doubtful taste 'were here sanctified by the oddity of the scene'. Sometimes the proceedings would end by the ladies persuading the 'oyster wenches', presumably the waitresses, to dance in the ballroom: 'though they were known to be of the worst character . . . the said wenches were always excellent dancers'.

One of the best known of these taverns was Luckie Middlemass's in the Cowgate, frequented by the young Robert Fergusson:

September's merry month is near
That brings in Neptune's caller cheer
New oysters fresh
The halesomest and nicest gear
O' fish or flesh'...
Whan big as burns the gutters rin
Gin ye hae catcht a droukit skin
To Lucky Middlemist's loup in
And sit fu' smug
Owre oysters an' a dram o' gin
Or haddock lug.

Whan auld Saint Giles at aucht o'clock
Gars merchant lowns their shopies lock
There we adjourn wi' hearty fock
To birle our boodles
An' get wharewi' to crack our joke
An' clear our noddles.

Whan Phoebus did his windocks steek
How aften, at that ingle cheek
Did I my frosty fingers beek,
And prie gude fare!
I trow there was na hame to seek
Whan steghin there.

The source of all this merriment and consolation grew on the bottom of the Firth of Forth, where the oyster scalps covered 50 square miles between Hound Point in West Lothian and Gosford in East Lothian, stretching over to Fife between Burntisland and Aberdour, but the thickest and best were situated off Granton and Leith south of Inchmickery and Inchkeith, and again off Prestonpans (Fig. 3.1). There is no good evidence that oysters existed even in the Middle Ages west of Queensferry, despite the extraordinary wealth of shellfish represented by the prehistoric middens of Polmont and other sites in the upper Forth estuary, though empty shells of indeterminate age could still be dredged up there. One witness, John Anderson, the Edinburgh fishmonger, claimed before the Royal Commission on Sea Fisheries of 1866 that Lord Auckland had at

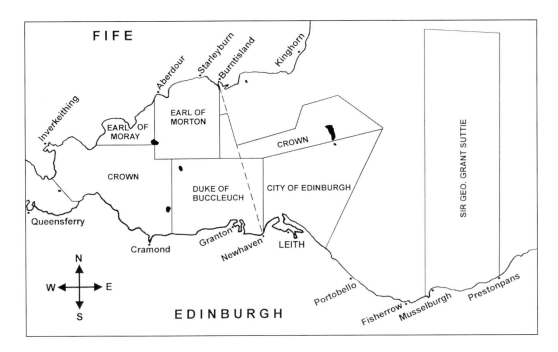

Fig. 3.1 – Firth of Forth oyster beds

one time possessed an oyster scalp there, but that it had been destroyed by peat floated down the River Forth from the agricultural reclamation of Blairdrummond Moss and adjacent mires, and then settling on the oyster beds and smothering them.[1] It has not been possible to find corroboration of this, and Anderson was not altogether a reliable witness. Oysters, to thrive and reproduce, prefer a water salinity of about 3.0–3.5 per cent, though there is some flexibility. Salinity averaged only 3.04 per cent when measured by the Fisheries Board between 1890 and 1894 off the Oxcar rocks in the midst of the scalps, so it could be that environmental conditions were unsuitable west of Queensferry during the historic period.[2]

The oyster scalps, as part of the seabed, originally belonged to the Crown, but by 1800 they had mostly become divided up between private individuals and a corporate owner, the City of Edinburgh, as illustrated in Figure 3.1. None of the claims were immemorial, as some of the owners liked to pretend, or beyond challenge, as we shall see. Most of them gave rise to violent disagreement over trespass and entitlement, partly because it was much more difficult to discern boundaries on the surface of the sea than on land.

They were mainly exploited by fishermen along the Lothian shore between Cramond and Prestonpans, and on the Fife shore between Kinghorn and Inverkeithing, but especially by those of Newhaven-on-Forth as a suburb of Leith (Plate 7), and the so-called 'east country' villages of Fisherrow, Prestonpans, and Cockenzie. The fishermen and their wives were often regarded as romantic figures of robust labour, especially by the early photographers (Figs 3.2 and 3.3). Because the trade was seasonal, beginning in September and ending in April (oysters are inedible in summer), the fishers were in fact often part-time. William Baxter from Cockenzie, who appeared as witness before the High Court of Admiralty in 1791, had been bred a slater, taking to the winter oyster fishing because he had a numerous family to support, and was at other times a sawyer. John Falconer from Cramond, another witness, had for 24 years worked at kelp burning in summer and oyster fishing in winter.[3] In Prestonpans parish in 1791, according to the minister in the *Statistical Account*, there were ten oyster boats needing a crew of 50, but only 23 of the fishermen were full-time: 'all the others work occasionally on land or sea, as they find most for their advantage'. The boats were divided into six shares, each man taking a share and the sixth being applied to its upkeep.

Oysters grew at depths of between two and 14 fathoms below high-water mark, but mostly at under six fathoms and many under four fathoms (36 to 24 feet). The very best were called 'pandores', supposedly because they grew close inshore near the doors of the salt pans. The oysters were reached by means of small dredges weighing less than 30 pounds each (Fig. 3.4), dragged at an angle of 35 degrees from a rowing boat with a crew of five. It was difficult to keep the dredge on the scalps except by precisely synchronised rowing, which is why the fishers sang their oyster songs to keep themselves in time. No doubt the fact that many of the fishermen were part-timers made it the more necessary. 'Long before dawn in the blackest season of the year', wrote the minister of Prestonpans in the

Fig. 3.2 – Hill and Adamson's classic portraits of the Newhaven community from the earliest days of photography reflect the romantic esteem in which it was held by the Edinburgh middle class. Here Sandy Linton, as confident as Brunel, is impressively posed against his fishing boat, his boys beneath, 1845.

Fig. 3.3 – Newhaven fishwives with their baskets, in traditional dress, c. 1847. Their strength, beauty and powers of doorstep barter were widely reported.

New Statistical Account in 1839, 'their dredging song may be heard afar off and except when the weather is very turbulent their music appears to be an accompaniment to labours which are by no means unsuccessful.'

Some liked to fancy that oysters would not allow themselves to be caught unless they were charmed into the dredge.[4] Thus Sir Walter Scott:

> The herring loves the merry moonlight
> The mackerel loves the wind
> But the oyster loves the dredging song
> For they come of a gentle kind.

The fishermen's songs were in fact improvised and completely nonsensical, 'a kind of musical ejaculation as they rowed':

> I rade to London yesterday
> On a cruiket hay-cock
> Hay-cock, quo' the seale to the eel
> Cock nae my tail weel?

Fig. 3.4 – An oyster dredge from the nineteenth century, recovered at Port Seton, 1960s.

Tailweel, or if hare
Hunt the dog frae the deer
Hunt the dog frae the deil-drum
Kend ye nae John Young?
Willie Tod, Willie Tay,
Cleikit in the month of May
Month of May and Averile
Good skill o'reasons
Tentlins and fentlins
Yeery, ory, alie!
Weel rowed five men
Asweel your ten.

Once landed and sorted, the oysters destined for sale in Edinburgh were carried up to town by the 'fisher lassies'. Those of Newhaven were especially admired. No less experienced a rake than George IV is said to have remarked that the women of Newhaven were the most beautiful he had ever seen.[5] But they were also extremely strong, and given to a sales banter with the housewives and their servants that was also said to be second to none. The oysters, their fishers and their sellers were together a treasured part of the city's picturesque cultural tapestry.

2. Squabbles in the Seventeenth and Eighteenth Centuries

Oyster shells are frequently found in excavations of Scottish medieval sites, and they occur as items in the kitchen accounts of medieval kings. No doubt most royal purchases would have come from the Firth of Forth, but there were few explicit references to the scalps until the second half of the seventeenth century, when the price suddenly rose.

An offer had been made in 1663 to supply Edinburgh with oysters at 3s Scots (3d sterling) per hundred (conventionally reckoned at six score and twelve), which, since it was rejected by the town council, presumably was not much below existing market price.[6] But by 1682, the council, concerned at apparent profiteering in the face of scarcity, fixed the retail price for vintners and ale-sellers at 8s per hundred: by 1687–88, it was 10s; by 1689, 12s; by 1694 and again by 1702, 16s. In 1706, it peaked at 20s for the best and largest, 'because of the present exorbitant price'. Plainly, the attempts at price control were not working, given a sevenfold increase over four decades at a time when there was no general inflation. Yet the scarcity did not last. Prices in the late eighteenth century were much lower – around 1770, the hundred sold for only 5–7d sterling (5–7s in old money Scots), still with little alteration in the value of money over that period.[7]

What is the explanation? One main concern, which was to be raised again and again in the next two centuries, was export. The two main destinations were England, primarily Newcastle at this time, and the Netherlands, with small quantities going to the Baltic. Almost all went out fresh. For instance, Conie Cornelie's boat, the *Tree of Life*, bound from Leith for Zierikzee in February 1672 carried 144,000 fresh oysters, but John Weaver's boat bound to the Baltic in March carried 30,000 'shell oysters' and six gallons of pickled oysters.[8]

In 1663, the town council of Edinburgh had expressed alarm at the 'new trade of the Zealanders sending their barks to Leith for oysters', fearing that their scalps would go the way of those of Holy Island in Northumberland, 'that was haillie harried and nothing left' after the Dutch had been allowed in. It was not expressly said, but it was implied, that they had been after immature oysters, presumably to lay down in their own exhausted beds: they had broken the 'very scapes', and ships' anchors might not be able to hold in the road of Leith 'if thes scapps be destroyed'. Henceforth it would be an offence to export oysters or to supply such ships with oysters from Leith.[9] There followed an order of Privy Council forbidding the export of oysters from Scotland, especially from the Firth of Forth. However, James Auchterlony, who had already contracted to supply 500 tons weight of oysters 'as occasion of shipping shall offer', was allowed to fulfil his contract. This was a considerable number, probably several million oysters depending on how many barrels were in a ton. Each barrel in eighteenth-century Prestonpans was reckoned to contain 1,200 'sizable oysters', and the boats that carried them to average 320 barrels each. If these were boats of about 30 tons, then each ton would be about ten barrels, so 500 tons would represent about 6 million oysters. This sounds an enormous number, but, as we shall see, it was well within the capacity of the Forth to supply, especially as it might be spread over more than one year.[10]

The response to the prohibition on export in 1663 came from a 'supplication . . . in the name of above three hundred families of oyster dreggers on the coasts of Lothian and Fife', explaining that they had earned their living for many years past by herring fishing at Dunbar and other places, but now that the herring had failed, they were 'necessitat to keep themselves from starving to dreg and sell oysters . . . the meanest of all trades.' If they were allowed to continue to sell to exporters they undertook to provide enough oysters to supply Edinburgh at 3s per hundred, and to charge the foreigners more. Privy Council consulted the magistrates on the offer, but on their advice rejected the petition. But six years later the prohibition was overturned at the request of the customs officials, because exports brought money into the country and were 'the maintenance of many poor fishers'. The export ban was reimposed for another three years in 1687, but after that it was apparently allowed to lapse again.[11]

So there was, after 1660, a new pressure of demand from foreigners, yet it seems most unlikely that it was on a scale to account, by itself, for the price inflation that occurred. Something else was happening. The years 1645 to 1715 span a period of unusual cold, the Maunder Minimum of climate historians, when sunspot activity was unusually quiet. The period was not equally inhospitable everywhere – the 1660s were actually a warm decade in Scotland and the surrounding seas. But weather here was at its bleakest between 1685 and 1710, and especially during 'King William's Ill Years', the famine years in the 1690s, when exceptionally severe winters and cold summers accompanied an incursion of sea ice from the Arctic into the North Sea, precisely at the time when the price of oysters peaked.[12]

Spawning occurs in oysters when the sea temperature rises sharply in spring and early summer. It used to be supposed that a minimum of about 15 ºC was required, but recruitment has been found at lower levels. When Fulton investigated growing conditions in the Firth of Forth between 1890 and 1894, he found that 15 ºC was only intermittently achieved, on a few days in some years, even within the area of the scalps and on the surface of the water, so it is reasonable to assume that, in a period of prolonged cold summers, oysters would have been unable to reproduce easily. Similarly, heavy mortality can occur after severe frosts in winter. On the other hand, oysters are relatively long lived and normally breed between the ages of three and ten, reaching the peak of their fertility at the age of five years. They are hermaphrodites, alternating between male and female phases in the same individual, those in the male phase ejaculating sperm into the water simultaneously, to seek out those in the female phase. They are tremendously fecund when conditions are good. The females release about a million larvae each: all but a tiny proportion of these are predated or otherwise die, but the capacity of an oyster colony to recover from a short-term climatic crisis is considerable.[13] This indeed seems to have been the case in the Firth of Forth. When Daniel Defoe came to write his *Tour* in 1726 he remarked of the Prestonpans area that 'they take great quantities of oysters upon this shore also, with which they not only supply the city of Edinburgh, they carry abundance of them in large open boats call'd cobles, as far as Newcastle upon Tyne'. Any sense of crisis was then entirely lacking.

For the remainder of the eighteenth century, the story of oysters in the Forth had the same familiar elements, of pressure of external demand and probable intermittent environmental stress, but now it gathered fresh elements, with new emphasis on property rights and trespass. It is probably no coincidence that the next significant round of tensions came in 1741, after cold and inclement weather since 1739 had threatened a return to the climatic conditions of the 1690s. Edinburgh town council that year again complained of 'foreigners and other strangers' dredging the town's scalps, particularly for seedling oysters, without authority, being guilty of 'theftuous practices' so that the 'very breed of oysters may be quite extirpate and carried off to the great and irreparable loss, both of this country and the community'.[14]

The council therefore in 1741 imposed a close season from 10 April to 5 September – oysters were inedible in summer, but this would not signify if they were being taken to lay down for breeding elsewhere. They ordered the fishermen of Newhaven not themselves to take, sell or export any seedling oysters and charged them with policing the banks, reporting to the Water Baillie of Leith any trespassing fishermen 'from Fife or otherwise', and any 'foreign sloops or vessels' caught fishing on the town's scalps, and then helping to apprehend them if necessary. The fishermen of Newhaven were to enrol individually in the books of the Admiralty of Leith, to give obedience to the regulations under the penalty of a fine and losing the 'privilege of fishing upon the said scalps in all time coming'.[15]

It is not clear if such enrolment actually took place, but the fishermen of Newhaven certainly assumed thereafter exclusive right to fish on the town's banks. In lawsuits and negotiations they were represented by the officers of their friendly society, which later styled itself the 'Society of Free Fishermen of Newhaven'. This had certainly been in existence since 1693 when the first box-masters are recorded, and probably since 1631, when the fishermen apparently signed a bond with the parish church to look after their own poor. Possibly it was even older. The fishermen liked to claim that their privilege of exclusive fishing went back to the foundation of Newhaven by James IV in 1510, but of this there is no good evidence.[16]

Another problem which arose in 1741 related to the scalps off Granton, at that time possessed by Lady Greenwich, wife of the Duke of Argyll, who had inherited them. They were to be passed in due course to the Duke of Buccleuch, and were among the most productive in the Firth of Forth. Edinburgh had complained unavailingly against the grant in 1695 as infringing their own liberties after they had been originally conveyed in 1692 by the Crown to Viscount Tarbert.[17] In 1741, the tacksman of the scalps, Robert Bowie, raised an action at the High Court of Admiralty concerning trespass and theft, ruinous to the complainer and the scalps, by a skipper of Kinghorn, David Sibbald, who had employed six boats of Burntisland, four of Kinghorn and three of Cramond to dredge for oysters to export to Holland. Sibbald indignantly denied trespass.[18]

It was common on these occasions for one side to accuse the other of taking immature or brood oysters for laying down elsewhere, to the long-term damage of the stock and the encouragement of foreign rivals. It was equally common for the other side to deny it. On this occasion there have survived detailed handling instructions from the buyers in Veere, the old

Scottish staple in Zealand. David Sibbald was told to be sure that the catch was cleaned and kept in the sea prior to shipping to the Netherlands, no nastiness of any sort was to be allowed near the oysters, the hold where they were kept was to be free of bad smells and bilge water, and cooking was to be done on deck as oysters could not abide smoke. Tellingly, the letter concluded 'we would rather have them small than large . . . pick out the big ones and sell them'. Clearly they were bought for restocking the Dutch beds.[19]

This pressure of external demand, combined with the harvesting of small oysters, was eventually heavy enough to cause obvious damage. In 1751 the Newhaven fishermen were stripped of their privileges by the town council because they persisted in selling undersized oysters to foreigners. The rights were conveyed for a time to a Leith merchant, on condition that he sold to Edinburgh at a fixed price, but the fishermen were back in their old position a few years later.[20] The scalps further east, and those on the south coast of Fife, were outwith Edinburgh's control, and mainly worked by fishermen from Cockenzie, Prestonpans and Fisherrow. The minister of Prestonpans, writing in the *Statistical Account* in 1793, told of serious over-exploitation. He recalled a time in the 1770s when the price had been 6d sterling per hundred as opposed to 15d when he wrote, and when boats returned with 6,000 oysters or more for a day's work, as opposed to 400–500 latterly. Formerly there had been 40 boats employed, including 16 from Prestonpans where there were now only ten. His colleague at Tranent corroborated his evidence of the price rise and said the 'common average' catch that had lain between 4,000 and 7,000, was now 700–800: the boats at Cockenzie had also dropped from 16 to ten.

The main demand was not now Dutch but English. There had always been a demand from Newcastle, and to some extent from Hull, whence mature oysters were taken in open boats that returned with Newcastle bottles, until the Scots began to make their own glass in sufficient quantities from the mid eighteenth century. This export continued though the boats returned empty, but now there arose a new and different demand from the London area. In 1773, a Leith merchant contracted to ship oysters on commission to the south, purchasing for ten different companies. He was said to have paid £2,500 a year to the fishers for ten years, and although the boats were issued with a gauge to ensure that no oyster was packed less than 1½ inches in diameter, this was widely disregarded.

This trade took place between January and May. Once in the south, the oysters were 'dropt in bays at the mouth of the Thames and Medway and other grounds [probably in Essex] to fatten until the fall, when they were dredged and sent to market'. Presumably in the interim they released sperm and larvae to replenish the stock in the English rivers. By 1786 the scalps exploited by the 'east country fishers' were, in the words of the minister 'greatly exhausted by this trade'.

Up to 30 cargoes were shipped in a year, and each vessel carried an average of 320 barrels containing some 1,200 oysters. As domestic consumption, including oysters sent by carts to Glasgow, was said to be just as big, there may have been 23 million taken in some years from these scalps. And if we add the catch of the Newhaven fishermen from the Edinburgh scalps and Lady Greenwich's scalps, the catch from the Firth of Forth could have exceeded 30 million. T.W. Fulton, using a slightly different way of calculation, similarly concluded that 'it is probably much under the mark to estimate the annual yield of the beds at that time at about 30,000,000 oysters'.

This was the context in which the events of 1788–92 unfolded.[21] It all began with a pitched battle in November, 1788, between, on the one side, the fishermen of the 'east country', led by Kelty Phinn of Cockenzie, 'with twenty small dregging boats under his management', plus four others and a sloop to take the catch away, and, on the other, the fishermen of Newhaven who had come out of harbour en masse suspecting that their rivals were fishing on the Edinburgh scalps. It is unclear how many boats Newhaven had, but 42 of their fishermen were charged with affray, as opposed to 55 on the other side. They fought with fists, oars and clubs, forcing the east country men into Burntisland harbour, where the fight continued.

The lawsuit that followed in the High Court of Admiralty lasted until March 1792. Appearing for the Newhaven fishermen and the City of Edinburgh was the leading lawyer in Scotland, Robert Dundas of Arniston, Lord Advocate and MP for Midlothian. Such a choice indicates that Edinburgh considered this to be something of a landmark case. Appearing for Kelty Phinn and his colleagues was 'Alexander Wright, advocate'. This is almost certainly a misprint for Alexander Wight, an experienced advocate who served in 1783 as Solicitor General for Scotland, and who apparently had a taste for defending the underdog – in 1792 he also acted as defence council for two booksellers charged with seditious libel. One wonders who paid his fees. The east country men were charged with trespass and theft, which they denied, countercharging assault. Furthermore, and unexpectedly, their lawyer denied firstly, that Edinburgh had ever been given a charter from the Crown that conveyed a right to the oyster scalps, and secondly, that even if it had, such a charter would be *ipso facto* invalid since 'fishing of oysters and other fish in the open sea is open and free to all the lieges and is not capable of a grant from the Crown. The fishing is held in sovereignty by the Crown as a public right, for the common benefit'.

This sounds like a dangerously radical argument, especially as it was maintained in 1791 at the height of the French Revolution. Dundas was taken aback, but without dealing with matters of fundamental law, he led a wealth of evidence to show that in previous years east country fishermen had acknowledged the rights of others by paying fees to fish, and in the case of the City of Edinburgh's scalps, by admitting trespass when caught, and paying fines. In the event Kelty Phinn and his associates were found guilty but dismissed without paying costs or receiving further penalties.

Yet there was more to Alexander Wight's argument for the defence than at first meets the eye. Edinburgh in support of its case could only produce a charter of Charles I of 1631, mentioning 'fishings', but not oyster scalps specifically, as part of the listing of various privileges or 'tendenda', what Wight called 'mere words of style'. Just such a case had come before the Barons of Exchequer in 1777, when the family of Ramsay claimed rights over oysters, on the same grounds that the deeds of their estate of Preston, which they had just purchased, had an early general reference to offshore 'fishings'. But in this case the judges had rejected the claim, on the grounds that it was oppressive to the poor fishermen of Prestonpans and Musselburgh, and liable to produce a monopoly not in the public interest.

There had been a recent and growing interest by private landowners in claiming novel rights over oyster scalps in the Firth. It began with the charter relating to Granton in 1692 to Viscount Tarbert, which seems to have been the first explicitly to mention oyster scalps ('solumodo aestraes'). Next was the Earl of Morton, who in 1735 startled the town council of Burntisland with a claim before the Admiralty Court to the oyster scalps lying off his estate adjacent to the town. Burntisland argued in reply that the seas were open to all, or at least to all local fishermen, to which the Earl replied that this was not the case with minerals or shellfish, which pertained to the landowner of nearby shores, and cited the case of mussel beds off Torryburn, planted by Lord Colville, from which his heirs drew a considerable income.[22] A later earl produced an instrument of sasine of 1786 investing him in the scalps. Then in 1773, the heirs of William Grant, the Solicitor General, secured from the Crown a grant of oyster scalps extending off their estate of Prestongrange, which according to Wight, proved the source of much litigation and 'and had been attended with no beneficial consequences whatever to the public'. There were at least four other landowners between Cramond and Cockenzie who in 1792 asserted some kind of claim to oyster fishing rights, and the Earl of Moray appears to have done so subsequently at Donibristle in Fife.

In fact the main outcome of the lawsuit in 1792 was to confirm rights of this kind, and to define the boundaries of the main concessions more carefully. It was a victory for private and corporate interest over public access to a natural resource. On the other hand, if Kelty Phinn had won, and the east country men been allowed to predate the City of Edinburgh scalps as ferociously as they had done scalps elsewhere, the Firth of Forth oysters would probably have been extinguished even sooner than in fact they were.

After this incident, there was peace on the scalps for 20 years, but in 1812 further fights broke out as the Newhaven men tried to arrest trespassers. Such incidents occurred sporadically until 1856, and the historian of the Society portrayed them as good-humoured horseplay. The minute book sometimes tells a different story: in October 1812 when six boats from the east country were arrested and fined £1 each, 'several of the Society were very much hurt'.[23]

3. Mutually Assured Destruction

At the start of the nineteenth century, the yield of the scalps was low, the price of oysters high and they were scarce. According to the recollection of James Wilson, box-master of the Newhaven fishermen, up to about 1828, 200 was a good day's catch, and prices ranged from 3s 9d to 5s a hundred. Reckoning a halving in the real value of money since 1770, the latter sum was equivalent to 30d, or about five times the price at the earlier date.[24]

Three factors were probably at play. Overfishing of the scalps, driven by the temptation of exports, continued to some degree, as is clear from the efforts of the town council of Edinburgh to stop it. There were two affrays in 1812 when a total of 18 boats of Fisherrow, Prestonpans and Cockenzie were seized, as well as other isolated incidents. The Newhaven men themselves continued illicit exports against the orders of the council, and from 1815 the council charged them a small rent in order to have a handle over their behaviour, and sometimes threatened to rent the scalps to others. It was all too easy for an English or Dutch sloop to moor in quiet waters near Inchmickery, close to the boundaries of several different claims as well as to the public grounds in the centre of the Firth, and load from local boats that could have been taking oysters from anywhere. In addition there was the burden on the fishermen from the Royal Navy, through the press gang and other forms of recruiting in the Napoleonic Wars. Finally, there was a renewed problem of climate: between 1782 and 1825 there was a series of bad winters and dreary summers, which though not continuous, were repeated frequently, especially between 1792 and 1815.[25]

After about 1825, in warmer conditions, there was a remarkable recovery in the oyster fishery. The scalps, despite having been badly damaged, were still capable of quick recovery. Heavy falls of spat (spawn) were noted in 1828, and the abundance of oysters during the 1830s led to the price falling dramatically, occasionally to as low as 3d a hundred (this was presumably a wholesale price, for bulk purchase), but usually from 6d to 12d. 'The natural replenishing of the beds and the consequent superabundance of oysters', says Fulton 'led to a repetition of the reckless fishing' that had characterised the later eighteenth century.

The council at first permitted exports, and 60 million were shipped out in three years between 1834 and 1836, some to the Netherlands, but most to Essex and Kent to replenish the beds for the London market. It was in 1836 that Charles Dickens represented oysters as the poor man's food in London, Sam Weller in the *Pickwick Papers* observing that 'poverty and oysters always seem to go together'. They were still sold in the expensive clubs and restaurants, but also now by costermongers in the streets of the capital. Probably at its peak in 1864, the sale at Billingsgate alone then totalled 496 million and the *Times* reckoned that London consumed 700 million that year.[26] The Scottish contribution to this huge consumption was therefore very marginal, but it brought prosperity to Newhaven, where 50–60 boats were busy throughout the season, with others working the outer scalps from Prestonpans, Cockenzie and Fisherrow. Consumption in Edinburgh itself was reckoned in 1839 to be 5 million a year:[27] it thus appears that consumption per head in Victorian London ran up to about 250 oysters per head per year, but in Edinburgh only to about 30, and London was roughly 15 times the size of Edinburgh. Exports were therefore crucial to the business as it had developed; as the minute book of the Free Fishermen recorded in 1834: 'oysters on our scalps being in exceeding plenty and little market at home for them, agreed to petition the town council of Edinburgh for liberty to export them to England, which was granted'.[28]

The council was pulled in two directions. On the one hand, they wanted to restrict exports to ensure a cheap and plentiful supply to the city, but on the other, they wanted to profit from the trade. In 1836, they laid down conditions restricting sales: none were to be sold except to Edinburgh without permission of the council; none were to be fished of smaller size than 1½ inches diameter (gauges were to be issued to the fishermen); and no boat was to catch more than 600 oysters per day for each man employed. Hotel keepers protested that they still could not get enough as long as any sort of export was allowed. The rent of the scalps to the fishermen, hitherto £25 a year, was increased to £74 after public roup in 1838.

Then, in 1839, the council again changed tack. The Treasurer's committee considered an offer from an English syndicate to rent the city's scalps for £300 a year, while guaranteeing a supply to the town of 60,000 large oysters a week 'at a price never exceeding 9d a hundred', providing that the syndicate was allowed to export such small oysters as they required. Recognising that hitherto they had been unable to prevent exports anyway, they found the offer irresistible. At the public roup that followed, a 'man of straw, who could not afterwards be found', ran the rent up to £600 a year, and the lease was finally taken for a term of ten years (later reduced to five) by George Clark, of Cricksea by Burnham-on-Crouch, Essex, in the name of the syndicate.

The Free Fishermen of Newhaven were outraged. Having failed to interdict the roup, they continued to dredge the scalps anyway, sending up to 50 boats to the fishing and refusing to work for the syndicate. Having paid twice as much for the scalps as he intended, and finding the town unable or unwilling to stop the Newhaven men, Clark brought between 60 and 70 large smacks up from England, each employing six dredges (as opposed to the single dredge carried by the Scottish boats). Thus the capacity that was going into dredging was suddenly multiplied about ninefold, and both sides had every motivation to get as many oysters of every size out of the scalps as quickly as possible. The town council raised an action in the Court of Session to restrain the Newhaven men from continuing to fish: the fishermen took out a counter-action against the council, on the grounds that they had 'the immemorial right of fishing on the City's scalps'. The Court declared the Newhaven men must not be prevented from fishing while the case was being heard, but judgement was not given until 1845.

Under these circumstances, the syndicate gave up after less than two full seasons. Following litigation, the council agreed in February 1841 to allow Clark to break the lease, which reverted to the Newhaven men. The Court of Session eventually found in favour of the fishermen, agreeing that, though the council had a right to the scalps traditionally considered theirs, the Society of Free Fishermen of Newhaven were entitled 'from time immemorial' by use and custom 'to enjoy the exclusive right or privilege' of dredging them, subject to such rules as the council might impose 'for insuring a proper supply of oysters at moderate prices to the inhabitants' of Edinburgh. The council had indeed formulated a fresh series of rules in 1841, 1842 and 1845, including limiting exports and the catch per boat, keeping records of the catch and allowing access to the books, using a gauge to impose a minimum size of the oysters marketed, and appointing supervisors – but the supervisors were never appointed and the books never kept. Apparently demoralised by the findings of the Court of Session, the council took little further interest in their scalps for the next 20 years, apart from other brief and ineffectual interventions in 1855 and 1858, and the fishermen continued in their old ways unhindered until 1865. Following renewed public outcry in 1863 and 1864 about the 'immense numbers' of young oysters under gauge size, kept awaiting export to England and the Netherlands, in the bottom of Granton harbour, or (after the outcry) more furtively in Starleyburn harbour on the Fife side, next to Burntisland, the council commissioned a report from Dr James M'Bain, an Edinburgh naturalist. He described a parlous situation.

There are conflicting accounts of the period between 1839 and 1865. The historian of the Society maintained that the foolishness of the council was entirely to blame: 'Mr Clark's ruthless dredging began the process that destroyed an industry which the fishermen had carried on faithfully and successfully from Time Immemorial'.[29] He admitted, though, that this disaster was compounded by the subsequent failure of the council to appoint supervisors or insist on record keeping. T.W. Fulton, the Scientific Supervisor of the Scottish Fisheries Board, maintained that the Clark incident, though damaging, was less serious than the behaviour of the fishermen afterwards, compounded by complete lack of effective action by the council: 'at the very time, therefore, when restriction would have been most efficacious, it gave way to unrestrained liberty'.[30] He had no illusions of the fishermen having being good stewards, though he seems to have seen the officers of the Society themselves as respectable and well-meaning men who were not able to control their members.

Dr M'Bain himself was unsparing in his language and conclusions:

> As it is illegal to remove the young brood oysters from the banks, the fishermen appeared to be so far ashamed of the proceeding that they shipped them furtively from Starleyburn Harbour . . . On Saturday, the 8th of April, Treasurer Callander and myself counted twenty large barrels on Newhaven Pier filled with young oysters and old oyster shells just brought on shore to be sent off as brood to England and Holland . . . this calculation amounts to 207,000 . . . the greater proportion of this enormous number in one day's dredging were young brood oysters torn from their proper breeding ground. It is quite obvious that unless efficient measures are taken to put a stop to this ruinous and shortsighted system of dredging, the oyster beds will soon cease to be productive, or be entirely destroyed.[31]

The council concluded that the city's scalps 'are at present nearly worthless, vast quantities of the seedling brood oysters having been dredged and sold for exportation'. They appointed an officer (for the first time) to oversee the landing and sale at Newhaven, where the Society itself also appointed two of its members to auction the oysters at the pier. The council proposed (but did not implement) a joint approach and a single management for all the scalps in the Forth, since the fishermen when challenged by one proprietor simply said they had been fishing on the grounds of another, or on the Crown grounds in the centre of the Firth.

But the scalps in 1865 were not yet 'nearly worthless' as the council said. Though badly damaged, it was still extremely profitable for the fishermen to continue to exploit them at the same rate or more intensively. In 1866–67, according to the box-master of the Society, 'almost everybody was at the oysters': the take was limited by the Society itself to 500 a day per man.[32] Up to 101 boats were at work, each now having three crew members and a single dredge (Fig. 3.5). With growing shortages now appearing in London, the price per hundred rose from 2s 6d in December to 5s in March, far above what it had ever been before, and quantities of brood oysters continued to be sold to the English and Dutch. Over the next two years, prices were

Fig. 3.5 – Boats dredging for oysters off Cockenzie: those in the foreground have begun to haul in the catch. From J.G. Bertram, Harvest of the Sea, 1873.

much the same, but exports to Holland became unprofitable. The number of boats fell. The council concluded in 1868 that their scalps were now so spoilt that it was 'advisable and necessary that the fishing should entirely cease to be prosecuted for some time', but instead yet another agreement was drawn up with the Society, to last for ten years, binding the fishermen to the usual conditions and increasing the size of the gauge to 2½ inches.

At this point the focus changes to the Duke of Buccleuch, who had inherited the scalps off Granton once in possession of Lady Greenwich. For some time these had been rented, like the town's scalps, to the fishermen of Newhaven, but the Duke also became dissatisfied with their ways of exploitation. He decided to let them to John Anderson, fishmonger of Edinburgh, who already had an extensive business, including salmon traps on the River Forth above Stirling, and who held leases from the Crown of scalps west of Inchmickery, and from the Earl of Morton of his scalps west of Burntisland. At one time, in 1847, Anderson had had a lease of the

Grant Suttie scalps north of Prestonpans, worked traditionally by the east country fishermen, but had given it up as unprofitable. Those scalps had been the first to be exhausted, but the Duke of Buccleuch's were reckoned to be the least damaged of those remaining. A condition was made that Anderson should employ Newhaven men at a fair wage, and supply them with the mussels that were dredged with the oysters, which were needed for line-fishing bait.

Anderson began by offering the fishermen derisory prices for any oysters they caught, thus abrogating the undertaking to employ them for fair wages. The Newhaven men responded, just as they had in 1839 in the dispute over the town's scalps, by continuing to fish as before and Anderson similarly responded by treating them as poachers and employing English smacks. These, however, for the first time in the Firth of Forth, used steam-powered dredges. The Society petitioned the Duke: after claiming that they had been 'oyster dredgers and cultivators all their lives and fully understand the mode of cultivating and improving the scalps and brood', unlike Anderson who was quite ignorant of the matter, they went on:

> If his system of dredging by heavy iron dredges worked by steam power be persisted in, within a short time the whole scalps will be utterly ruined ... the dredges worked by steam weigh 60 lbs, whereas the dredges of the Society are only from 25 to 30 lbs. ... the cultivation of oyster scalps pretty closely resembles the cultivation of a field, newly sown, by a harrow. The harrow properly steers and covers in the seed which a heavy plough would only uproot and bury ... the Memorialists' dredges are like a harrow ... those of your Grace's present tenant are like a heavy plough or grubber.

They concluded by saying that if the Duke had intended to ruin his scalps, he could not have found a better way than by leasing them to Anderson to work with heavy steam dredges: 'The Memorialists, who are all practical men, and many of them well advanced in life, are of one mind on this point'.[33] The Duke was not impressed, and considering the condition of the town's scalps, of which the Society had had sole custody from 'time immemorial', except for a brief interlude of less than two years, one can see why.

Fights subsequently broke out in the Firth, when the Newhaven men, defending one of their own trespassing boats, routed 'a steam tug strongly manned and Anderson himself on board', throwing stones and injuring the fishmonger's agent; later that day they attacked the English boats and ordered the crews ashore. The English came out four days later defended by a gunboat. The Newhaven men were accused of behaving like pirates and flying the black flag as they sailed the Firth in triumph, though they indignantly denied that the black flag was anything other than what they normally flew when they were fishing. When two policemen attempted to make arrests in Newhaven, they were attacked by 'the whole of the fishermen in the village with their wives and children'; the sergeant ran off towards Leith, but the constable was pelted with stones and starfish, struck on the head, cut on the nose and had his hat knocked over his eyes.[34]

All the miscreants escaped without legal retribution. It is interesting to speculate why. The citizens of Edinburgh had long held the Newhaven fishermen and their wives in high regard, as part of the cultural heritage of the city, and in 1867 the Society had gone to some pains to appear as loyal and respectable citizens. For the first time they held a procession to celebrate the appointment of new officers in the society, marching up to Edinburgh to George Square, Princes Street and Heriot Row with a brass band and bagpipes playing 'constant music all the way'. It was 'a grand display of banners and gilded boats' with six men rowing, one steering and another hauling lines, and another twelve men 'with long white batons walking and keeping perfect order'. The minute book of the Society reported that 'it was said by the inhabitants of Edinburgh and Leith that it was the best sight they had seen for years'.[35]

In 1868, the new Sea Fisheries Act empowered the Board of Trade, on request, to make an order for the establishment or improvement of oyster and mussel fisheries on public (i.e. Crown) ground, which order would confer the exclusive privilege of working such grounds providing it was done to the satisfaction of the Board. The Newhaven men applied for such an order for the public grounds lying north of the Town's scalps, which Anderson opposed, lodging a similar application of his own. His lawyer, at the public inquiry that followed, argued that the grounds would be better in private hands, though he exaggerated the number of the members of the Society by about tenfold:

> . . . the fishings would be better conducted by one man than a thousand. It would be in the interest of Mr Anderson not

merely to make an interest off it this year – it was his interest to cultivate it. He had no interest in indiscriminate spurging [*sic*] of the ground. His interest was all the other way. On the other hand it was quite plain that the interests of the Society of Free Fishermen of Newhaven, numbering 2000 individuals, could not be the same. The interests of each fisherman would be to take what he could get. The part he protected he did not preserve for himself – he preserved it for his neighbour. He was only one in a thousand.[36]

The Board, though, was not completely persuaded, and in 1870 awarded by order all the oyster fishings (including by agreement all those in private or corporate hands) east of a readily identifiable line from Newhaven to Burntisland on the opposite shore, to the Society, and that to the west to Anderson. Clearly it was hoped to stop the Newhaven men from trespassing on the other scalps that Anderson leased, while preventing the supply of oysters from being monopolised by a single interest. In the event, neither party satisfied the Board with their standards of stewardship, so, in 1877, following another inspection by the Board of Trade, the order was cancelled. The Newhaven men had simply carried on as they had always done, Edinburgh once again had failed to supervise them effectively, and Anderson (who complained of their continuing trespass) also sold great quantities of oysters to all the former markets and neglected to clean or maintain the scalps under his concession.[37]

Within a short time after this, all the scalps were effectively worked out. Catches landed at Newhaven from the town's scalps fell from 8,600,000 worth £10,782 in 1867–88, to 367,000 worth £963 ten years later and 5,640 worth about £20 in 1887. The other beds were no better off. Prices reached over 15s a hundred in 1890, but high prices were no stimulus to effort when there was nothing left to catch.[38] Oysters continued to be caught as by-catch with mussels or clams until around 1920, one was found alive by fishermen near Inchmickery in 1947, and in 2009 Dr Elizabeth Ashton of the University of Stirling discovered a few living on the south shore of the Firth, proving that the species was not actually extinct, as had been assumed. But the fishery had finally died in the last quarter of the nineteenth century. What had possibly been the most extensive oyster beds in Great Britain, which had provided food and delight since the Neolithic, were no more.

4. Conclusion

So where does the blame lie? Clearly it lies with overfishing and therefore with human greed, but that is only a superficial answer. It had been perfectly obvious for decades beforehand that unless restraint was shown the outcome would be ruin, so why was there no restraint, either self-imposed by the fishermen or externally imposed by authority? What roles did technology play in making exploitation easier on the supply side, and what roles did the market and infrastructure play in making the temptation greater on the demand side? Were there also external factors like pollution and climate at play, and if so, what was their role?

The oyster scalps of the Firth of Forth were analogous to a stretch of common land on shore owned by various proprietors, to which peasant farmers have varying degrees of access. Much discussion has taken place about how terrestrial commons have worked in the past and at present. The well-known theory of Garrett Hardin argues that commons inevitably lead to the tragedy of overuse and land degradation, because it is always in the interests of the individual to go on exploiting them to the utmost, as by doing so he gains the benefit of all the marginal gain but bears only a small proportion of the cost.[39] In addition, he is spurred by the fear that unless he acts to use the common as much as possible now, someone else will do so first and he will lose out. To this theory, others (notably Elinor Ostrom) have replied that history and modern practice are full of instances of the sustainable use of commons, because peasant societies are actually capable of organising themselves in ways that optimise mutual restraint for the common good, or because the owners or overlords of the commons can in fact establish rules that are recognised as fair, and obeyed either for fear of effective punishment or because of community recognition that they are for a common good.[40]

So the first question we might ask is why the oyster scalps apparently correspond closer to Hardin's model than to Ostrom's? It is not so surprising where the scalps were on Crown territory, that is, public ground where access was completely unrestricted. Individual fishermen from all the towns and villages around came and took their pick without interference. In those grounds in private ownership where specific fishermen or groups paid a rent to dredge, the threat of withdrawing permission in the face of overuse might have been an effective sanction. The problem, however, was to prevent trespass, which was extremely difficult at sea, where boundaries between concessions were invisible lines drawn between landmarks, and it was never going to pay even the Duke of Buccleuch to maintain gamekeeping boats at sea. The best chance of getting responsible use might appear to be on the City of Edinburgh's scalps, since these were rented to the Society of Free Fishermen of Newhaven, a community body which had sole right to the scalps and might have been expected to police them in a common interest, as Ostrom found so often happens on land.

There is much about the Society that would seem to fit it for this role. Their leaders were literate and articulate, conscious of their reputation in the city and anxious to reinforce it by respectable display. Although the activities of their members included fishing for herring and sprats, and acting as a pilotage service for merchant shipping using Leith, the oyster trade was often, as in the 1840s and 1860s, their main concern. They were aware of the need for predator control, by cleaning off 'all crabs, star-fish and others detrimental to the cultivation and preservation of the scalps'. They eventually themselves limited the members of the Society to take only a fixed number of oysters per day. In 1837, they passed a resolution to prevent any member of the Society exporting oysters to England or face prosecution. In 1857, they set up a mechanism to auction the oysters at the pier of Newhaven and to sell through a separate but allied 'union'. In 1865 they asked for help from the council in imposing the rules.[41] All these actions suggest an awareness of their conservation responsibilities.

The trouble was, however, that they were basically a friendly society, whose prime purpose was to levy dues to maintain the widows and orphans of the fishermen, and they frequently had difficulty in maintaining numbers and payments. Many of the fishermen either did not belong to the Society or allowed their fees to lapse. For example, in 1828 there were 220 members of the Society: but two years later 56 were found to be in arrears. The latter were charged in the sheriff court, but afterwards 'these disorderly members and debtors came to fight and maltreat the good members of the committee and did even strike and knock down the good members'.[42] In 1867 there was such a disturbance 'from a few individuals' at a meeting called at the pier to reward the two keepers of the oyster sales, that the troublemakers had to be expelled before business could be concluded, and next year the equivalent meeting had to be totally abandoned because of uproar, and reconvened in private.[43] At times the members were so bad at paying that 'the society lay asleep', as in the years 1848–49 and 1852–53, and membership fluctuated between 115 in 1851–52 and 348 in 1866–67.[44] The Society put on a bold face to the public, but in private admitted, for example in 1865, that 'conditions have been frequently broken and that great damage has in consequence been done to the oyster beds'.[45] It was weaker and less effectual than it liked to appear to the world.

If the Society could not impose order despite trying at least to some degree to do so, it is hardly surprising that the town council of Edinburgh similarly failed. It has been much criticised for not trying harder, and for making foolish decisions as in 1839, but presumably the councillors recognised the inordinate expense and probable futility of trying to police from outside what the fishermen themselves were failing to do from inside. Still, the council's parsimony in providing real or consistent supervision from above clearly sabotaged any chance that the Society might have had in trying to impose discipline from below.

Plainly, critical to the tragedy occurring when it did was a profound change in market conditions. For centuries, even millennia, the Firth of Forth had provided as many oysters as the inhabitants around could possibly need. There had been scares about shortages in the late seventeenth and eighteenth centuries, significantly associated with the rise of foreign demand for the first time, yet if the scalps were still capable of providing 20–30 million oysters a year in the 1830s, they were far from being fatally damaged at that point.

What happened? In terms of supply, the influence of new technology was probably small. True, the Newhaven fishermen used five men to pull a dredge in the eighteenth century but only three (sometimes one or two of these were boys) in the nineteenth, and it is not clear how they made the improvement – but the dredge was the same size and the overall improvement in the extraction rate could not have been dramatic. Much more potentially important was the steam dredge, and, as we have seen, this could also damage the bottom in dangerous new ways. But there is no actual evidence of this technology being used, except possibly in December 1864 when a 'Steam Fishing Company' based in Cockenzie advertised that it could provide dredged oysters (only for it to announce bankruptcy a month later)[46] and more certainly between 1867 and 1870, when Anderson gave up his lease of the Duke's scalps. No doubt it contributed to the reduction of those scalps to the condition of the town's scalps, but the latter had been exhausted without recourse to steam dredges.

A more significant effect was from the demand side. Though it is true that the Firth of Forth became drawn into the supply of immature oysters to England and the Netherlands initially without fatal effect, it was the intensification of demand from England after 1830 which probably did the greatest damage. The appetite of the Londoners was insatiable once the oysters had arrived at Billingsgate, so that even though Scotland provided a very small percentage of the total eaten there, it represented a very large proportion of the number raised in Scotland. The beds of Essex and Kent themselves survived the pressure only through reinforcing their breeding stock from outside, especially from France, and then they were badly damaged by the accidental introduction at the same time of parasites and alien pests.[47] The situation seems to confirm Fraser Darling's aphorism: 'Man does not seem to extirpate a feature of his environment as long as that natural resource is concerned only with man's everyday life: but as soon as he looks upon it as having some value for export, there is real danger'.[48] It is not that the fishermen became greedier or wealthy, but the opportunity offered to a poor people to make a better living from the marine commons than before, was immediately taken.

The structure and ineffective regulation of the commons then meant that the enhanced exploitation was likely to end in disaster, for the reasons that Hardin outlined. But it is worth noting that the remedy proposed by Hardin – to privatise the commons – did not work either. When the Board of Trade divided the scalps between Anderson to the west and the Society to the east, they were effectively privatising the western half. Its condition four years later was as bad as that of the eastern half: maybe privatisation came too late, but maybe Anderson, as the Society maintained, just had no clue what he was doing.

One intervention that might have worked was a much more radical form of state intervention, either to police all the scalps with armed fishery protection vessels, or, more positively, as Alexander Wight had proposed in 1792, to treat them all as Crown property, and then to apply conservation and modern cultivation methods to them. Such a thing was unthinkable in laissez-faire Britain, at least after the Royal Commission on Sea Fisheries in 1866 (see Chapter 4 below). But it worked in France, where the state owned the oyster grounds. Following near catastrophe in 1858, when in one department, 18 out of 23 beds were reported as completely destroyed, and in another, only three out of 13 gave even a few days' fishing a year, the state proactively took over, and pioneered oyster cultivation to save the industry. It was entirely successful. At Arcachon, where 20 million oysters had been raised in 1860, by 1907 production had been raised to 300 million. That was only part of the achievement, for there was then successful production all down the west coast. It also worked in the Netherlands, though to a lesser extent, where state supervision took over in 1870, following the usual story of overuse and collapse.[49]

Was pollution a factor in the catastrophe in the Firth of Forth, as has from time to time been alleged? Edinburgh began to pour its sewage into the sea virtually untreated from the middle of the nineteenth century, and by the middle of the twentieth there were very visible and horrible effects for at least a short distance offshore. Yet a fatal effect on oysters seems unlikely. Pollution would possibly have made the shellfish dangerous for human consumption, and might well have rendered their recovery from low levels more difficult, but would not of itself have been a prime cause of their near extinction. The mussel beds in the Firth of Forth have always flourished despite often having been subject to very serious pollution and for the most part being closer inshore than the oysters. There is clear evidence from the explorations of naturalists in the 1880s, when the oyster catch had already been reduced to less than 1 per cent of what it had been two decades before, that there was then still a much richer variety of molluscs and weeds along the Edinburgh foreshore than a century later.[50] The worst impacts of pollution probably came in the twentieth century, when it is possible that any recovery of the residual population would have been hindered. Fulton considered whether the dumping of ashes from steam boats, first complained of in 1825, or the dredging and dumping of mud from cleaning the harbours at Leith, Granton and elsewhere, could have had an effect, but concluded that it was in such a limited area that it could hardly be regarded as serious.[51] The same is probably true of dumping from Bridgeness and Bo'ness east of the bridges.

Climate may be a different matter. As we have seen, exceptional cold associated with the southern extension of sea ice in the late seventeenth century appears to have brought the scalps to a low ebb, even without undue human exploitation. Cold weather could have been a contributory factor again in the difficulties around 1740, and between 1780 and 1815. Early in the 1880s there was another incursion of very cold Arctic water and sea ice into the North Sea, which may have helped to administer the *coup de grâce*.

As many as 1,356,000 oysters were still being landed at Newhaven in 1870–71. Good spatting years were reported in 1874–75 and 1875–76, although the catch dropped to 426,000 by 1878–79. Nevertheless, in the latter year more than 2,000 dredging trips were made, and the catch was still worth £1,255 at the quayside. It was still business, though not what it had been. By 1882–83, however, after two severe winters, only 55,140 oysters were landed, and systematic exploitation was given up as not worthwhile – any landed after that were by-catch from mussel fishing. The cold trend continued in 1883, following the dimming of the sun from the ash clouds thrown up by the gigantic volcanic explosion of Krakatoa. Oyster landings next year were only 6,220, and never even approached that figure again before records were finally abandoned in 1894, after landings amounted to a mere 1,200, worth £5. Fulton systematically examined the state of the scalps in 1892, landing 317 oysters from 233 casts of the dredge over most of the former beds, but he found no spat, though he did find about 45 oysters only two or three years old. The 1890s were generally warmer than the 1880s, but, once an oyster population has been so severely reduced, the chances of fertilisation between scattered individuals declines.[52]

So, just as so often across the world in our times, the responses of human need and greed, triggered by transport improvements and market growth, unrestrained by effective governance and complicated by climatic factors, led to an environmental disaster – to irreversible resource exhaustion. We will see in later chapters how far this might also be true in other areas of the Forth.

CHAPTER 4
HERRING BOOM AND HERRING BUST, 1820–1950

1. Introduction

From around 1650 to about 1820, as we saw in Chapter 2, the lives of fishermen in the Firth of Forth who hunted the herring revolved around two episodes in the calendar year, the Lammas Drave, occurring in the months of late July, August and early September, and the Winter Herrin', which began in December or January and ended in March. Both were involved in catching spawning fish in inshore waters. The Lammas Drave went back many centuries, with good records from the sixteenth century onwards; sometimes it failed for a few years, or even for a decade or more, but it always came back in the end. The Winter Herrin' has a less clear history, but had also been known for a very long time. Around the end of the eighteenth century it had suddenly become a source of great prosperity in the upper parts of the Firth; these fish then apparently vanished, or at least could not be relocated, for some time. The story in this chapter is of the final phase of the exploitation of these two populations, which was to reach a phase of unprecedented intensity in the nineteenth and earlier twentieth century, before ending abruptly and probably for ever with the last disappearance of the fish.

For many centuries, great annual cycles of herring migration had supported the fisheries of the Dutch, the Norwegians and the Danes, as well as of the English and the Scots. In the North Sea, herring divide into distinct populations which spawn at slightly different times. The main focus of fishing down the British coast has been on three autumn-breeding populations, though there are others. The Buchan herring spawns off Shetland and north-east Scotland, peaking in July, the Bank herring

Fig. 4.1– Fishing at Dunbar in the earlier nineteenth century, showing how close the Lammas Drave might be to the shore.

spawns on the Dogger Bank and off the coasts of Yorkshire and Norfolk, peaking in August, and the smaller population of Downs herring spawns in the southern part of the North Sea and in the English Channel later in the year. Dutch and Scottish fishermen in the nineteenth century moved down the coast from Shetland to Norfolk as the season proceeded, giving the impression of following the herring as they went on migration, but in fact exploiting different stocks that spawned successively.[1] It is not clear to which population the Lammas Drave belonged, as it spawned closer inshore than most, within and around the mouth of the Firth of Forth, up to 12 or 16 miles out, but sometimes so close inshore that the nets were in danger of tearing on the rocks and sand (Fig. 4.1). Quite possibly, it was a small sub-population distinct from both the Buchan herring and the Bank herring. The Winter Herrin' was certainly distinct, described in 1957 as the only spring spawning population in the eastern North Sea.[2] It was also inshore. It has also now gone, at least as a commercial fishery.

Herring shoals were notoriously fickle in their appearances, especially these inshore populations or sub-populations. Like many other species of fish, they were much more numerous in some years than in others. This was, in part, a biological defence mechanism. An uneven cyclical production of spawn was a protection against excessive predation: if they had produced the same number of fry each year, they would have built up a proportionate number of predators to consume most of the recruits to the herring population, but by producing much more spawn in some years than in others they overwhelmed the ability of predators to consume them. The abundance of herring in a given area, however, also seems to have coincided with an influx of cold water from the Arctic; at times when icy Greenland currents flooded into the North Sea, the herring became plentiful around Scotland, but in warmer years they might desert the coasts entirely. Neither the cyclical pattern nor the dependence on the climate was understood, and such variations were ascribed by fishermen to the inscrutable ways of Providence. These great fluctuations and periodic absences also make it harder to detect whether, and at what point, overfishing or natural causes might be the reason for a decline.

Our concern is with the environment of the Firth of Forth and the waters at its mouth, and not with that of the North Sea as a whole, but, after about 1822, the fishermen of the Firth again regularly went much further afield than for nearly 200 years, when they had sailed to the Lewes. In larger boats and with better nets, they were no longer catching only the fish in the nearest waters, as they had concentrated on doing after about 1640. Of course, they still landed their prey as close to where they had caught it as possible. Anstruther men fishing off Aberdeen or Wick early in the season, or (by the end of the nineteenth century) even off Yarmouth and Lowestoft in September, would make for those ports as soon as they could, because herring decays quickly in the absence of deep-freeze; they would telegraph their wives with news of the catch and bring the money (but not the fish) back home. However, because Aberdeen or Yarmouth men working off Anstruther would do the same, statistics of local landings broadly measured depredations on local stock. Sometimes, though, the fish had to be carried a rather longer distance than the fishermen would have liked, either because it had been caught far out or because bad weather made the nearest port inaccessible. So local statistics usually, but not invariably, reflect the catch of local stock.[3]

Because they travelled in this way, after 1830 the immediate fortunes of the fishermen of Anstruther, Dunbar or Newhaven and all the other ports that specialised in herring did not any longer depend only upon the fish in the Firth of Forth. The comments in the *Scottish Fishery Board Reports* in the later nineteenth century often explain how the fishermen of local ports had failed to find much nearby, but had had a good year off Aberdeen or East Anglia. It was the income from the fish of the entire eastern North Sea that they spent on their new homes and boats, so the numerous masons, joiners, rope-makers, net-makers, net-tanners, sail-makers, shipwrights and oilskin manufacturers in the fishing towns also benefited from much more than the Lammas Drave and the Winter Herrin'. By the early twentieth century, many women and girls from Fife also joined the annual migration south to gut the herring in East Anglia, and their earnings also boosted the prosperity of the towns. In 1912 it was calculated that, in the Leith Fishery District, which then included not only Newhaven and Granton but Fisherrow, Cockenzie and everywhere on the Firth apart from the main ports of Fife, there were 2,100 fishermen and boys directly supporting over 3,000 jobs in subsidiary industries on shore (Fig. 4.2).[4]

Fig 4.2 – Danny and Mary Livingstone, an elderly fishing couple from Fisherrow. She wears the traditional fishwife's skirt and shawl and his smartness reflects service as a Firth of Forth pilot (c.1900).

Yet many local people in these industries did still depend mainly on fish within the Firth and in the surrounding parts of the North Sea. It was always local stocks of herring and white fish that supported the local fish merchants, and provided livelihoods for the many women who gutted and packed the catch from the Lammas Drave and the Winter Herrin', or who hawked it round the towns. Similarly, it was the local catch of fish of all kinds that supported the bait-gatherers, the makers of fish boxes and baskets, the coopers who made the herring barrels and the curers who dispatched them to the Continent, and the railwaymen who sent fresh fish to Edinburgh and London. In the last analysis, the prosperity of most ports (apart from Granton late in the nineteenth century with its trawler fleet following cod and haddock far out to sea) still rested to a considerable degree on the local marine ecosystem.

When commercial fishing vanished from the Firth of Forth and, indeed, from the whole of the southern North Sea in the later twentieth century, these communities had to find something else to do, or face decline. For example, though some fishermen in late twentieth-century Pittenweem would sometimes commute by minivan to work in Aberdeen, there were not many who were prepared to do this. What kept the economy of the burgh going by then was partly tourism and the annual arts festival, partly employment in the University of St Andrews, partly hawking fish brought down from Aberdeen and Peterhead around the west of Scotland and Perthshire by van, and partly still fishing out of Pittenweem for nethrops, lobsters and crabs, the final unexhausted resources in the sea round about.

The two local populations of herring represented by the Lammas Drave and the Winter Herrin' still, in the mid nineteenth century, thronged the Firth and the waters immediately beyond (Fig. 4.3). The main spawning grounds were then off the East Neuk of Fife and the Isle of May, though there remained others of less significance off the East Lothian coast and higher up the Firth. The Anstruther Fishery District covered more than the East Neuk, running from beyond Fife Ness to Buckhaven, and the graph overleaf represents what the Fishery Officers there recorded of the herring catch in that area between 1854 and 1954, the statistics being relayed to the

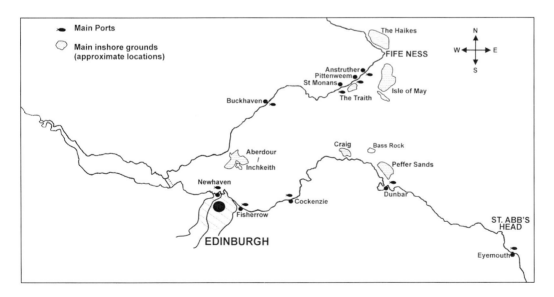

Fig. 4.3 – Herring Fishing, 1830–1950

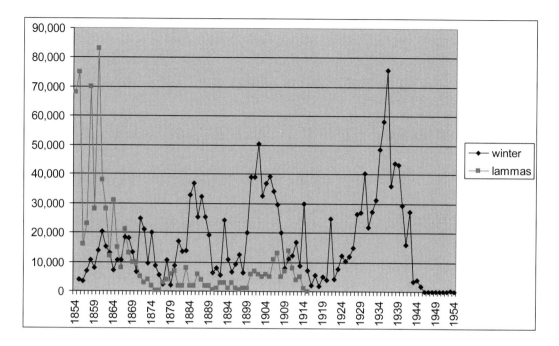

Fig. 4.4 – Crans of herring landed in Anstruther Fishery District, 1854–1954. Source: P. Smith, *The Lammas Drave and the Winter Herrin': A History of the Herring Fishing from East Fife* (Edinburgh, 1985), pp. 156–8. Six crans equals one tonne, approximately.

local historian, Peter Smith, in 1985 (Fig. 4.4). The totals are not always the same as those reported to the Scottish Fisheries Board for the same years, though they are usually close. They do not represent all the herrings caught in the Firth and its approaches, because the fishermen of Leith Fishery District also caught herring in the upper parts of the Forth, and sometimes in the lower parts too, which they landed at Newhaven or Granton, though until the 1930s this was a small proportion of the whole. Between 1876 and 1908 Leith District landings were little more than a quarter of Anstruther District ones. The fishermen of Eyemouth and Dunbar also registered large catches in Eyemouth Fishery District, but much of what they caught was off Northumberland, so their inclusion would distort the picture of what was happening to the Firth of Forth herring stocks. Even the Anstruther record may occasionally register herring brought from as far as 30 miles out in the North Sea. Nevertheless as a statistical guide to what was happening to the Firth stocks, this is good enough.

Before 1854, record keeping was less systematic, but the herring fishings are said to have failed in Anstruther between 1822 and 1836.[5] They suddenly revived. Between 1837 and 1839, about 25,000 barrels of herring a year were landed in Anstruther District; between 1856 and 1860 the number was about 57,000. A barrel is roughly the same as a cran, reckoned at 5.83 crans to a tonne, and although total catch may have grown faster than exports, most herring were exported at this date, and there can be no doubt of a great expansion beyond anything known in previous history, a phenomenon that occurred all over Scotland. The large peak of exports in the first decade of the eighteenth century, for example, equated to about 1,700 tonnes a year: 57,000 crans equates to over five times as much.

The Lammas Drave expanded dramatically from 1837, peaked in 1860, and crashed over the next decade, never recovering: the small-scale revival of summer landings that took place in the 1900s was probably from fish caught further out. The Winter Herrin', by contrast, increased from the mid nineteenth century in cycles of about 10–15 years, peak following slump, the peaks growing in scale until just before the Second World War. Had there been any boats at sea to catch them there would probably have been another peak in the years between 1914 and 1918. This fishing then also crashed, and well before 1954 the Winter Herrin' was also extinct as a commercial venture.

Catches from our two local populations formed only a small proportion of the total Scottish herring fishing effort. This followed a different trajectory, growing fairly steadily from about 200,000 barrels (or crans) at the end of the Napoleonic Wars to over 2 million before the First World War, before falling steeply to about 750,000 during the interwar years.[6] After the Second World War, overfishing became a general and recognised problem throughout the world, and herring fishing in the entire North Sea was suspended by international agreement in 1977, to allow stocks to recover from very low levels. That proved to be a rare success story for the Common Fisheries Policy, as offshore North Sea herring indeed made a comeback following the moratorium. The particular and melancholy interest of the Lammas Drave and the Winter Herrin' in the Firth of Forth, however, lies in the fact that neither of these two inshore populations ever recovered. Furthermore, they disappeared before sonar detection and Danish seine-net trawling, the modern villainous tools of overfishing, were known about. What is the story?

2. The Climax and the End of the Lammas Drave

Unexpectedly, Caithness holds the clue as to how the fishermen of the Firth of Forth were able to respond so fast and on such a scale to the opportunities offered by the return of the Lammas Drave to local waters after 1836. From the second decade of the nineteenth century, there had developed a sudden boom in the herring fishing of the Moray Firth, focused on Wick, and based on exploiting the Buchan population, which attracted large numbers of boats from the south of Scotland.[7] This prosperity enabled the harbours of the Firth of Forth gradually to build up a fleet of larger ships and a body of expert crews far in excess of anything seen in the previous century. The minister of Kilrenny parish explained in the *New Statistical Account* in 1843 how Cellardyke had a fleet of about 100 'large boats' of 13 to 18 tons, each manned by three or four regular fishermen and one or two 'halfdealsmen' (or two or three 'strong boys') – these latter owned no nets but assisted in hauling and rowing. Before 1837 or 1838 they had been accustomed to leaving for Wick or Peterhead in the summer, without a single boat left behind, but when it was discovered that the Lammas Drave was back, many remained at home and were as successful as those that went north. Naturally, after that, still more remained to fish locally, or returned quickly to add the Lammas Drave to their year's activities. Anstruther and Pittenweem together had about 100 boats around 1840, St Monans had 26 and Buckhaven had 154, of various sizes. Many of these had also been directed towards fishing in the north, and became available for the Lammas Drave when it reappeared. Crail was the only Fife town that opted out of this opportunity, and consequently never again reached its former prominence. At this time it had only 12 boats, concentrating on setting creels for lobsters for London, and for crabs for Dundee and Edinburgh, and on gathering whelks.

One of the reasons for Crail's decline was that the growing length of boats made its small harbour inconvenient. In 1826, a Cellardyke boat, 34 feet long, which had been tragically lost within sight of the Isle of May with all seven of her crew, was described as 'the leviathan of the fleet'. By 1849 something rather bigger than this, but still under 40 feet, was the standard for all east coast herring boats, now crewed by five men instead of seven, so the labour productivity and therefore the profitability of the catch to the crew had increased as well. Such boats were still under 20 tons burden and undecked, most remaining so until the 1880s. These boats also carried more nets, often 20 in place of the ten or 12 of the eighteenth century, and now made by machine following a technique first devised in 1828 by a manufacturer in Musselburgh.[8] The normal way of fishing now, at least in the Craig off North Berwick and in the Treath between Pittenweem and St Monans, was by drift-net, the 'float drave' of earlier times, though the anchored nets of the 'ground drave' were still used off Dunbar and in the Auld Haikes between Fife Ness and Kingsbarns, to capture fish when they were very close to the shore.

So the Lammas Drave, when it returned, was captured more efficiently than ever before. It could now also be carried to market by steamboat, or, after 1863 when the railway reached Anstruther, by train. The industry was encouraged by state regulation, especially by establishing a Crown brand for cured herring that was recognised as a mark of excellence by customers in northern Europe. 'From the mid-thirties rising prices emerge as the prime stimulant', observes Malcolm Gray, an indication that the foreign buyer as well as the fish merchant at home would pay well for whatever the fishermen could land.[9] So both supply and demand conditions help to explain how so many more fish could be landed and sold, even in the 1830s and still more in the 1850s, than in previous booms of the Lammas Drave.

A *Scotsman* report from Anstruther of 29 August 1839 conveys a flavour of the excitement:

> The herring fishery continues to prosper gloriously. All parties connected with it are now in the greatest activity from morning to night. Our usually quiet shore is now bustle and animation – man, wife and child busily employed . . . The present stock of our curers is fast filling up, but we trust curers at a distance will forward their stock without delay . . . Our general average is now about 130 crans, which is probably higher than any station in Scotland, when it is considered that our fishing commences about two weeks later than the stations in the north. On Monday three of our boats arrived from Helmsdale, having completed their fishing there, and are again busy with a second harvest.

The bonanza continued, with ups and downs, for 20 years, reaching its climax between 1854 and 1860. In 1855, which was to prove the second largest Lammas Drave on record, the prospect of a good season was suggested by the numbers of porpoises and whales off the Isle of May in late July, and the *Pittenweem Register* reported on 21 August that 'there never was a greater quantity of herrings brought into the harbours in this neighbourhood in any former period'. At the end of the season there were 'rumours of forty impending marriages' as the young men turned their profits into offers to their sweethearts. In 1860, the peak year, up to 400 boats were fishing in the

Treath off Pittenweem, and 'every boat seemed quite loaded as it entered the harbour at an early hour, and as soon as discharged immediately put to sea again for a second shot'. At Dunbar, too, great quantities were caught, and in 1861 the temptation to fish on Sundays proved irresistible there. Fish caught 'within gunshot of land' was transferred at sea to Irish sloops, as it was illegal to land it on shore on the Sabbath.[10]

When he was an old man, William Smith recalled the excitement and danger of the Lammas Drave in the 1850s:

> Whenever the Crail men saw the herring had set into the Hakes they sent word to Cellardyke, and through the town the cry was raised 'Herring in the Hakes'. Immediately all was bustle, and a rush was made to the fishing grounds at once ... I remember on one occasion on a Sunday when the news came to the town that there were 'Herring in the Hakes' but as the Cellardyke men would not fish on the Sabbath, none of them loosed a rope until after twelve o'clock on the Sabbath. My father went out early on Monday morning and shot a fleet of ten nets ... they hauled them in and were at Cellardyke harbour with forty-six crans by ten o'clock in the morning ... they returned and hauled the nets they had left, and were at Cellardyke by ten at night with seventy-five crans. That was good work for a Monday.

In calm weather the boats were filled 'sinking them to the number forward and up to the name aft in the water', and occasionally overloaded boats met a swell and foundered, drowning the crew, as the fishermen could not swim.[11]

After 1860, the catches at the Lammas Drave suddenly declined sharply, and by 1870 summer fishing for herring in the Firth of Forth was practically a thing of the past. At the time, this seemed regrettable but hardly surprising. It had done so before, for example in the late 1650s, in the 1720s, the 1770s and the 1820s. It was assumed that it would go away for a while, and then return as it had always done in the past. This time it failed to return to the traditional inshore grounds, ever.

Yet could an inshore fishery of this sort that had been going for centuries, albeit with interruptions when the herring disappeared of their own accord, really become extinct from overfishing when the peak catch in one year seems in modern terms relatively modest? It was no more than

14,000 tonnes in the Anstruther District in 1860, with possibly as many again from this population caught in the Leith and Eyemouth Districts combined. It might indeed have been possible, if the population was small enough, and genetically distinct from others. The fish, rich, oily and highly esteemed, were captured round the Firth of Forth on the point of releasing eggs and milt on the spawning grounds. Others, and we cannot tell how many, would have fallen prey to other fishermen during their migrations round the North Sea, and as their fishing effort was also increasing this would have added to the pressure on the stock. But to kill the fish just before reproduction was to inflict maximum damage on the population.

One incident showed how vulnerable they might have been. The principal grounds at the mouth of the Firth were the Peffer Sands near Dunbar, the Craig off North Berwick, and in particular the three famous grounds off the East Neuk of Fife: that between Fife Ness and the Isle of May; the Auld Haikes, which lay offshore between Fife Ness and Boarhills; and the Treath, also known as California (after the gold rush) or the Fluke Hole, off Pittenweem, stretching west almost as far as Elie Ness. The herring, when they were in the Firth, returned here year after year, attracting also hordes of haddock, codling and turbot, plaice and other flatfish to prey on their eggs and fry: hence 'Fluke Hole', fluke meaning flatfish. The herring were traditionally caught by their gills in drift-nets, which were played out from the boats as they fished by night, and the flatfish and white fish were caught by baited 'long-lines' (Fig. 4.5). At some point after 1854, fishermen began to experiment in catching flatfish and white fish, using beam trawls from open boats, dragging purse-shaped nets along the gravelly, shelly floor of the Treath, a technique that seems to have been started in Buckhaven and to have been quickly copied elsewhere in the East Neuk. David Boyter, a witness from Cellardyke, long afterwards, in 1884, recalled how he had been among the first to trawl the Treath:

> We got a great deal of spawn in the net, and I believe the whole spawn bed was destroyed. Even the St Monance men came over to our town of Cellardyke and they advised us to drop it, for the great evil that it was doing. Well, we dropped it and sold all our trawls and they began themselves. And what are they doing? They have hauled the very bottom out of the place. At Pittenweem there is never a herring goes into bay, summer or winter.[12]

The grudge against St Monans for this behaviour was to continue for a century.

According to other eyewitness accounts, the trawler-men had brought up quantities of herring spawn, 'taking and trampling it underfoot, and then taking a shovel and heaving it over the side'. One press report in 1861 referred to sales of the spawn to farmers to use on their fields as manure at a shilling a cartload.[13] The angry fishermen of Cellardyke and Pittenweem in 1861 persuaded the Scottish Fisheries Board to impose a legal moratorium on all trawling in the Treath, but subsequent diving investigations by Board scientists two years later could not find significant quantities of spawn, so they concluded that there was nothing to save. Indeed, George Allman, Professor of Natural History at the University of Edinburgh and their principal investigator, was sceptical of all the fishermen's stories and doubted that even if it had been disturbed that the spawn would die as a result, providing it was returned to the water. In 1865 the ban was lifted.[14] The summer herring never returned to breed, and the valuable turbot and other fish which fed on the eggs and fry disappeared as well. In the words of James Marr, in 1884, when he had had 30 years experience of fishing out of Pittenweem, the Treath 'was once termed California, but is now like the walls of Jerusalem; it is nothing but desolation'.[15]

In 1866, the Royal Commission on Sea Fisheries, which included two of the leading scientists of the day, Lyon Playfair and Thomas Huxley, reported.[16] It reached the conclusion, more based on the laissez-faire political ideology of Cobden and Bright than on science, that fishermen should be allowed to fish, in Huxley's phrase, 'where you like, when you like and as you like'. This was later enshrined in legislation that repealed all previous attempts to regulate or restrict sea fisheries, dismissing the evidence of the fishermen as ignorant and partial. A later Royal Commission in 1884–85, on which Huxley also served, remarked that 'fishermen are apt to rely on tradition rather than observation and are naturally opposed to any method of fishing which they conceive may interfere with their livelihood'.[17] Yet their evidence was in fact first hand, informed by personal experience, and

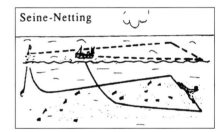
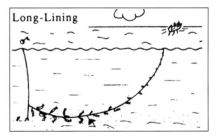
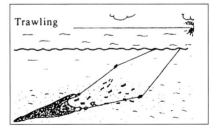

Fig. 4.5 – Methods of fishing – drift-netting, long-lining and trawling.

generally Fife fishermen of the time were quick to embrace the new when it was useful to them. Their observations certainly seemed better founded on observation than Thomas Huxley's own profoundly unscientific faith in non-intervention. Yet if trawling did ruin the Treath as the fishermen believed, the same could not have been true of the Auld Haikes and the other grounds where the Lammas Drave operated, as early trawls could not work on a rocky bottom without tearing the net. Here we must assume that the decline, if it was not natural, must have been due to the very heavy pressure of drift-netting on a limited stock.

Twice later on, a small peak in recorded summer landings encouraged a belief that the Lammas Drave might be about to return. In 1883 Anstruther District reported the best summer since 1865, but the herring had been caught 20 miles or more east of the Isle of May, and the old inshore grounds were not reoccupied. Similarly, between 1906 and 1910, there were some years when landings were above 10,000 crans, but again the fish came from beyond the Isle of May, some of the boats getting a tow into Anstruther from steam trawlers. Possibly the grounds were the White Spot, 33 miles to the east-north-east of the island, where summer herring had been known to spawn since 1890.[18]

Herrings, once caught, had to be gutted packed and dispatched before the profits came in (Fig. 4.6). Belle Patrick, a girl of about 12 at the time, recollected the scene at Anstruther when the last of the summer herrings were landed around 1910:

> The Drave was the most exciting fishing of all for the children, for it coincided with the long summer holidays from school – nine weeks in which to haunt the harbour and piers. Once again the open space along the shore side of the harbours was occupied by boxes and barrels belonging to the various fish-buyers, but there was added interest for us in the presence of the 'crews' of women in the curing yards.

She described how many of these were Gaelic speakers brought from the west by the buyers, and how they worked in crews of three, one packer to two gutters. They could gut 60 herrings a minute, using a very sharp knife:

> Because of the nature of the work the slightest cut could become infected and very painful, so the hands were protected

Fig. 4.6 – Women gutting and packing herring at Dunbar, c. 1905: not as improbably cheery as sometimes portrayed by visiting photographers.

by rag bandages and we children used to beg rags and hang around ready to render first-aid when it was needed. The third member of the crew collected the full baskets and packed them between layers of salt in barrels. If a packer was of short stature she nearly disappeared into the barrel as she packed the first few layers, and she would emerge bit by bit as the barrel filled up. When it came to within a few layers of the top, the children would swarm around and compete for the job of finishing it off . . . we used to watch fascinated by the quickly moving fingers which just seemed to flicker over the heap of herring to transform them into neat layers, each layer crosswise to the one underneath.[19]

It may serve as an elegy, for the Lammas Drave never came back again in any form.

3. The Winter Herrin'

After 1865, most of the fishing towns of the Firth of Forth grew apace, enriched and rebuilt with new prosperity. The opinion of the Royal

Commission about Scottish fishermen being slaves to past tradition was quickly belied by the storming growth of the industry. The Scottish herring cure as a whole increased from about 600,000 barrels in mid century to over 2 million on the eve of the First World War, driven by a dramatic growth in exports to eastern Europe which was taking three-quarters of the total after 1900, when the fishery had become the largest in the world. Much of this growth centred on ports like Lerwick, Wick, Peterhead, Fraserburgh and Aberdeen in the north, but innovation diffused everywhere. Boats became much larger, and by the 1880s most of these were decked (Fig. 4.7). By the 1890s some were over 60 feet in length, carrying two sails and a crew of eight.[20] These boats were huge and beautiful, like the *Reaper* berthed in Anstruther today, of a type known as a Fifie, though actually built on the Moray Firth in 1901 (Plate 9 and see also Plate 10). She is 70 feet long, 61 tons burden, carried a crew of eight and a steam capstan to manipulate her nets and 250 square meters of sail, and her mizzen mast is 47 feet high – the apotheosis of sailing boat design.[21] Only after 1900 was there a serious move towards steam drifters, and it had not gone far in the Firth of Forth before the First World War.

The advance in nets more than kept pace with boat design, notably the introduction of light cotton netting of smaller mesh from the 1860s, allowing a fivefold increase in the area of netting shot for the same weight carried. One consequence of this was that far more immature herrings, or matties, were caught in Scotland: only 40 barrels in 1859, but 200,000 by 1880. By the 1890s the biggest boats could shoot 70 nets and a net train could extend to over two miles, compared to a maximum of a quarter of a mile early in the century.[22] Yet, until almost the end, the general principle of drift-net fishing remained no more than a greatly enlarged version of the old float drave, still pursued with a dash of the hunter-gatherers' knowledge. Stewart Dick described the St Monans fleet in 1910, carrying 50 or 60 nets apiece, going out in the evening:

> Each net is about ten yards deep, and the line of nets may stretch about two miles or more in length. A 'pitch' having been selected – and the fishermen are wise in all the signs that show the presence of herring, among the most notable of which is a sort of phosphorescent gleam on the water – the nets are thrown out. The boat is pulled across the tide as the nets are 'shot'. They sink down to a considerable distance under the surface, twenty

Fig. 4.7 – A fishing fleet, still all sail and wood, waiting for the evening tide, c. 1880.

or thirty feet, and are kept in an upright position by sinkers on the lower edge and corks on the upper. At intervals a large float is attached by a line to the nets, and lying on the surface indicates their position, as they drift with the tide. At dawn the nets are lifted.[23]

None of the advances in herring fishing technology even in Fife and Lothian were distinctively motivated by opportunities in and around the Firth of Forth. The great bulk of the herring continued to be caught much further away during the summer peregrinations of the boats between Shetland in the north and East Anglia in the south. Nevertheless, Winter Herrin' gave an occupation to boat and crew between January and March which was unique in Scotland.

After the 1880s, this fishing took place increasingly at the mouth of the Firth of Forth, though with spawning beds of winter herring also accessible from Newhaven, Burntisland and Buckhaven. Higher up above Queensferry, they either declined or wore out: possibly they had been overfished, perhaps they suffered from pollution from sewage and industrial waste, or perhaps some of those highest up the estuary which declined earliest were smothered by the sinking mass of waterlogged peat that floated down from the reclamations above Stirling. The ground between Inchkeith and Aberdour was still attracting local boats in the first half of the twentieth century, but the main areas were near the Isle of May, and off Fife Ness. The Treath slowly and partially recovered from its problems, to become again a ground for winter spawning herring by 1890, following a new prohibition in 1885 on trawling throughout the Firth. James Smith of Cellardyke, yet another fisherman witness before a different government committee in 1888, rather grudgingly conceded that 'the poor man is making a little livelihood now'.[24] It never completely recovered its old reputation.

The number of fish captured in the Winter Herrin' in Anstruther District increased greatly before the First World War, though unevenly, in a series of cycles where each peak was higher than the one before. It was helped by the construction, paid for by the Scottish Fisheries Board and executed by the Stevenson family of engineers, of the Union Harbour in Anstruther, begun in 1866 for the Lammas Drave, just as that fishing was failing (Fig. 4.8). It was only completed in 1877, following many technical and financial setbacks, after the Lammas Drave had very clearly failed. It cost the Board over £80,000, but was a success in providing good winter refuge even for the biggest boats, unlike Dunbar harbour, where the Board spent £35,600 on the Victoria harbour from 1842 without at all solving the problem of safe and easy access in stormy weather.[25] Interestingly, St Monans paid for its own very successful new harbour, which by 1912 had the biggest fishing fleet in the Firth of Forth.

An even greater boon was the arrival of the railway at Anstruther in 1863. The winter herring was nothing like as oily as the summer herring, so it was more difficult to cure, but it could be sold fresh to Edinburgh and even London if carried away quickly. 'The railway brings within the morrow the sea harvests at Anstruther pier, on the dinner table in London or Manchester', wrote George Gourlay in 1879, 'the first whistle of the train across the Border gave such an impetus to demand as to double the price within the year'.[26] Activity was intense in a good winter. In just one week in early February, 1884, 800 wagonloads of fish were dispatched by rail from Anstruther, and no fewer than 3,400 telegrams sent from the local post office.

The following year, the entire winter catch of 40,000 crans, employing 220 boats, was sold fresh.[27] Kippers and bloaters were sometimes made, but in 1887 accounted for little more than a tenth of the total. 'Scottish taste leans much farther in the direction of fresh fish than of fish in a cured state', observed the Scottish Fisheries Board in 1900, though five years later they also said that a taste for smoked fish was growing rapidly. By then it was easier to market fresh herring by using ice in the railway wagons. It came by boat, and Belle Patrick as a girl used to hang around the ice ships to hear 'the foreign sailors talking in their own lingo' and to pick up splinters from the quay; 'we got as much pleasure out of the slithers of ice melting in our not-too-clean fingers as the youngsters of today get from their hygienic dollops of frozen fruit-flavoured water nibbled off a stick'.[28]

Fig. 4.8 – The Union Harbour, Anstruther, in 1887: big boats in a big modern harbour constructed with help from the Scottish Fisheries Board.

After the First World War the entire Scottish herring fishing industry fell into deep crisis, as export demand for cured barrelled fish in the markets in Russia and Germany, tumbled away to very little. One of the few relatively bright spots seemed now to be the Winter Herrin' of the Firth of Forth, with its tradition of providing for the home market. Even before 1914, it had developed from being a local enterprise to one that had attracted boats from all down the east coast of Scotland. Now they were joined by boats from as far as Lowestoft, and particularly from the west coast coming through the Forth–Clyde canal. The type of boat changed too, steel hulls on the drifters replacing wood. Steam engines using coal, and, increasingly, motor engines using oil, replaced the beautiful energy economy of sail. By 1930, there were 313 boats working the Winter Herrin' in the Firth of Forth, 43 steam drifters (Plate 8), 268 motor yawls and only

Fig. 4.9 – Boats of many kinds leaving Anstruther to fish for herring at dusk, 1934.

two sailing boats (Fig. 4.9). As well as supplying the home market, the fish merchants developed ways of 'klondyking', selling lightly salted herring on ice to Hamburg, where it proved more acceptable than the old product. For this reason, after 1926, no more gutters and packers worked in Anstruther.[29]

More important still, the boats from the west introduced a new and more destructive way of catching herring, by ring-netting, a technique sometimes also known as seine-netting, which had been developed in Argyll in the nineteenth century. It arrived with boats from the Clyde coming through the canal to work the upper part of the Firth, and was quickly adopted by fishermen in Newhaven and other harbours in Lothian, no doubt partly because the method of surrounding the shoal with a ring of nets and drawing them towards a pair of boats, was not very different from what had always been used in the area for sprats. Ring-netting caught far more fish, and a much higher proportion of immature ones in the total.[30]

In 1924 when they first encountered the new method, the fishermen of Cellardyke were so angry that they passed a resolution that no more herring caught in this way were to be landed in Anstruther harbour, arguing that ring-netting would lead to the complete destruction of the Winter Herrin'. In 1930, the fishermen of St Monans similarly ordered away two Campbeltown boats and declared a boycott on ring-netters, but the boats then made for Pittenweem where they received a warm welcome and landed their catch. It is hard to believe that this was not in revenge for what St Monans boats had done to the Treath 70 years before. By then, too, ring-net fishermen from Cockenzie, Fisherrow and Newhaven had begun freely to land their catch in Anstruther harbour. Before long, local towns, even Cellardyke, began to build 'nobbies', a type of west-coast boat well suited to ring-netting, and 1937 'was the first year that Pittenweem landings made a significant contribution, due to their fleet adopting the ring net'.[31] Soon Anstruther boats were taking the new technology to England. But herring continued to be caught in several different ways: the Scottish Fisheries Board reported in 1933 that drifters did best at the winter fishing in the Anstruther District until 10 March, when due to 'the greater density of the shoals', the ring-netters did better, while some small boats were still setting anchored nets close inshore. In 1934 the fleet consisted of 60 steam drifters, 96 motor drifters, 135 motor ring-netters and 26 motor boats equipped with both types of gear – 'considerable friction has resulted'.[32]

All this is a telling example of how some fishermen might resist an innovation for conservation reasons, but be unable to do so for long because others were taking advantage of it to out-compete them on the same fishing grounds. In these circumstances, higher authority might have stepped in on the side of conservation. But in 1930 the Scottish Fisheries Board, in their own words, 'was asked to intervene by restricting the method of seine-netting in the Firth but after careful enquiry reached the conclusion that restrictive action was not necessary'.[33]

The race to extinction quickened. The Scottish Fisheries Board reported 13,000 tons (75,800 crans) landed in Anstruther District, with another 5,100 tons (29,700 crans) landed in Leith District. Next year, landings were up in Leith but down in Anstruther and in 1938 the pattern was repeated with ring-netters from Newhaven accounting for well over half the total (8,700 tons compared to 7,700). Newhaven was now selling herring on Monday mornings, implying a Sunday catch which the more Sabbatarian Fife ports still would not countenance, led by St Monans. Fishing on Sunday deprived the herring of their Sabbath of uninterrupted reproduction, the one day in seven when they could spawn free of extirpation. The fish were brought into Leith from the main grounds between the Fife coast and the

Isle of May, with only occasional catches now from the inner reaches of the Firth.[34]

Britain slid into war in 1939, but much of the Firth herring fishing continued: it could take place relatively close inshore, and the contribution to the national food supply was very welcome. The catch still amounted to 8,700 tons in 1942, so the respite that most other fish stocks experienced in wartime was only partial. A take of this size might have still represented a very high proportion of the population, if the number available was shrinking in the down phase of a natural cycle. In peacetime, redoubled efforts produced fewer and fewer fish. The catch quickly became negligible, though as late as 1958 some shoals were located that could not be caught, on the Mar bank off the Isle of May, 'lying on the bottom heavy with spawn'. Effectively, commercial fishing for spawning herring, the classic Winter Herrin', was dead by 1946. However, a relatively unimportant winter herring fishing continued from Newhaven for a few years more, based on a different stock of small, two-year-old fish that had been bred elsewhere and only came inshore to winter in the Firth of Forth (Fig. 4.10). For example, in 1963 some 400 tons were caught off Burntisland, and sold mainly for canning and for pet food. For the rest of the decade they were a by-catch in sprat fishing, seldom exceeding 50 tons a year.[35]

Scientists from the Marine Laboratory at Aberdeen argued in 1971 that, as the decline had taken place in wartime when fishing effort was reduced, the failure of the Winter Herrin' was unlikely to have been due to overfishing, and suggested that it was 'almost certainly because of some change in the environment making it unsuitable as a spawning area'.[36] But in fact the catch in the early years of the war was still substantial, and a sudden unspecified alteration in the environmental quality of the Firth in the 1940s seems unlikely. A fish population that has once been reduced below a critical level may find it hard to reverse the decline. This may be due to a 'predator trap', where generalist predators that depend on a variety of other foods but which also will take herring when they get it, eat enough of the young herring to prevent a significant recovery of the stock. Or damage by overfishing may remove the most productive of the population, that part most capable of producing a sudden burst in numbers of eggs and fry that overwhelms the ability of predators to consume them. In any case, the heaviest episode of exploitation ever was followed by the end of commercial fishing.

What happened in the Firth of Forth prefigured the fate of the herring in the entire North Sea, with one critical difference. In the 1960s and early 1970s, the development of Danish seine-net trawling and its variants, 'the most destructive devices ever known',[37] aided by electronic sonar to detect the exact location of the shoals on the bottom, enabled the fleets of many nations to concentrate on the immense spawning shoals on the Dogger Bank and around the English Channel. Landings of North Sea herrings peaked in 1965 at over 1 million tonnes, four-fifths of which were juvenile fish, and decline after that was very rapid, as catch far exceeded recruitment.[38] By 1975, the famous Dogger Bank herring fishery was fished out by targeted seine-netting moving along the bottom, and the grounds are no longer used today. In 1977, the Common Fisheries Policy agreed to a four-year closure of the North Sea to herring fishing: it partly worked, and the fishery was reopened under a system of 'total allowable catch' with quotas for different nations. But the regulations were too generous for true sustainability, and in 1996, the quotas were halved.[39] The herring fleet still goes to sea from a few ports around the North Sea, but consists only of a handful of very sophisticated modern trawlers capable of catching a large tonnage in a short time. 'Two modern herring or mackerel purse seiners with thirty men aboard can outfish a thousand boats with five thousand men aboard of the kind used in the 1840s'.[40] The difference between the situation in the wider North Sea and in the Firth of Forth is that controls were imposed in time to have had an effect on the total stock in the former, but not in the latter. Some small, juvenile herring bred off the west coast of Scotland do still come into the Firth of Forth in winter, but of the countless hordes of mature fish, one population in summer and the other in winter, that once lay spawning in tiers above the gravels and stony bottoms of their favoured grounds, not a trace remains.

Fig. 4.10 – Pouring out the last of the herring: Fisherrow, 1955, just before the final collapse of the fishing in the Firth of Forth.

CHAPTER 5
LINES AND TRAWLS

1. The Sma' Lines and the Great Lines

In the nineteenth century, the Firth of Forth and its approaches swarmed with large fish of many kinds, cod, ling, haddock, whiting, plaice, turbot, halibut, skate and much besides. Herring never provided the fisherman's only means of subsistence, though it might be the most profitable part. John Mitchell in 1864 gave an account of the yearly round of the Newhaven men. From January to May some were busy with the winter herring, others with the oysters, some with sprats, and others went 'to the cod, ling, haddock, and skate fishings'. In May and June 'the majority of the enterprising fishermen' sailed north to the herring in their open boats, some as far as Stornoway, staying away for two months, 200 miles from home. On return in August, they continued to fish for herring off Dunbar and Eyemouth, even down to Northumberland. In September they returned to the oysters, 'a never-failing but scanty means of support ... at the cheerful but laborious dredge'. Others set creels for crabs, or netted skate east of Inchkeith, or sailed 'to the fishing-banks off the mouth of the Forth to fish haddocks, cod, ling, halibut and skate'. Many united the occupation of fisherman with 'the equally hazardous but more responsible profession of pilot', beating about in their open boats at the approaches to the Firth for a week or more, looking for ships to escort to Leith. This was a Newhaven speciality, rather as caddying golfers was a speciality, much less dangerous, for St Andrews fishermen in the mid twentieth century. Mitchell concluded in respect of the Newhaven men:

> Their mode of life is certainly of a very exciting nature: the fisherman spends his nights in the middle of winter on the ocean, in an open boat, happy to snatch a nap on the boards if

the spray or the waves will allow him, – and this not only for one night, but for many nights in succession. We have often seen the fishermen landing from their boats with their faces incrusted white with the salt of the ocean, and yet we hardly know of an instance of a fisherman or a fisherman's son preferring another profession, so much have example and habit wedded them to their dangerous and exciting avocation.[1]

Some communities were hardier than others. The *Pittenweem Register* in 1846 talked respectfully of the Buckhaven fishermen at the herring drave, as 'like the planets, they never tire or stop or rest, when one eye is shut in sleep the other is wide awake watching the fish. They require no lodging – their boat is their nest night and day'. The paper concluded, in a painful attempt at rhyme, that 'All work and no play is a Buckerman's joy, but all work and no play makes a Pittenweemer a bad boy'.[2] Small fishing communities which specialised in a very local resource, like St Andrews, with its plaice and flounders from the bay, Crail (at least from Victorian times) and Cove in Berwickshire, with their lobsters and crabs from a rocky shore, or Alloa and Kincardine, with their sprats and sparling at the head of the estuary, though they all went after other fish as well, had more constricted horizons. There was no point in their fishermen staying out on the waves for nights on end. Other places did not even have proper harbours, but used the beach. Ross, near Burnmouth on the Berwickshire coast, for example, fished from natural runnels on a rocky shore, around 1850 sending 'significant quantities of cod, ling, haddock and salmon to Edinburgh by sea, while lobsters and crabs were shipped alive in smacks to London'.[3] Such creeks were unusable to boats longer than about 15 foot.

Flatfish, like the halibut and skate talked of by Mitchell, and round fish like the cod, ling and haddock, also whiting – together termed demersal fish, living near the bottom – played a larger part in the lives of most fishermen of the outer reaches of the Firth than they did at Newhaven, because here there was no opportunity to go after sprats and oysters, or for piloting, though almost everywhere had a chance to set creels. In the outer Firth round fish were more plentiful than they were higher up, and there were also more varieties of flatfish, especially plaice, flounders, sole and sometimes turbot, wherever there were sandy or shingly bottoms, as in St Andrews Bay, or off Aberlady, Tynninghame and Largo, or in the Treath by Pittenweem. These fish all generally spawned out in the North Sea, releasing their eggs to float in the waters and not, like herring, dropping them in clotted masses on the seabed: but many came inshore to feed, especially on herring spawn and fry, or on other small fish, so they were often plentiful in the same places in which the herring chose to spawn. They were also abundant 20–70 miles out beyond the Isle of May, on such grounds as the Mar Bank and the Wee Bankie, where the larger fish fed on sandeels. Eventually, when big-time trawling and great lining came, boats from the Firth of Forth sailed out in search of cod into the Norwegian Sea and even into the Atlantic.

The way to catch most demersal fish, before the start of trawling in the later nineteenth century, was by 'lining', also known as 'long-lining', and this method also continued after the spread of trawling, because the quality of line-caught fish was much superior to that of trawled fish – they were not smashed up in the net. Lines were of two types, the 'sma' lines' and the 'great lines' (*alias* 'gretlins' or 'gartlins'). Sma' lines were used for the smaller fish, above all haddock and plaice in the Firth, and the great lines for larger fish that often swam further out, like adult cod and ling, as well as for haddock. The difference initially was as much in the size of hook as in length of line. Sma' lines, used inshore, accounted for most of the effort at least in the earlier nineteenth century, and were particularly used in early winter, great lines more frequently from April to June when it was safer to go further out. Both in the nineteenth century were given up by most fishermen at the times of the year when the more lucrative herring became available.

All fishing could be appallingly dangerous, but it was an episode at the sma' lines that brought about the infamous Eyemouth disaster. On 14 October 1881, 300 men in 40 boats put to sea to catch haddock from the Berwickshire port, on a fine and clear morning after a period of storms, despite the warnings of a falling barometer. Eyemouth was a community that had not done well that year, and there was a level of hardship in the town that was unusual among the fishermen of the south-east of Scotland; it was also a community accustomed to taking risks when others stayed in port, for the sake of the premium paid in the market for fresh fish in a time of bad weather. The fleet was suddenly caught in a virtual hurricane on the fishing grounds 12 miles offshore: 189 men were drowned as they tried to re-enter the difficult and inadequate harbour. It was the worst disaster known in Scottish fishing history, leaving 73 widows and 263 fatherless children.[4]

Such exceptional events were mourned as a national tragedy, but less remarked was the steady toll of drowned fishermen that happened every year. The Scottish Fisheries Board kept the grim reckoning: in 1898, for example, they reported 73 deaths among fishermen, of which 53 were from the east coast.

Little capital was needed to set up an open boat for the sma' lines, but great labour was required to bait them. In Pittenweem, in the early twentieth century a boat would set to sea at the sma' lines with five men, each man deploying two lines with 600 hooks spaced 3½ feet apart, 60 fathoms (120 yards) long: so the boat would deploy 6,000 hooks on about three and a half miles of line. Around Anstruther in 1857, the boats at the great lines were larger than formerly, partly decked, so able to catch fish up to 70 miles out and to fish in winter as well as in summer. They carried a crew of eight, each deploying 1,800 hooks, so 14,000 hooks per boat, on lines maybe totalling eight miles in length. In 1873 this was said to be three times longer than they had been in the days of open boats, 'our haddock fleet, now of sixty decked boats ... could, with our combined length of line, stretch from the Isle of May to the Naze of Norway'.[5] There are signs that increasing competition over time drove the fishermen to bait more hooks per trip. Thus in 1889 it was reported that the number of hooks per line at St Andrews had increased from 800 to 1,000 in winter and from 1,500 to 1,800 in summer, and at Coldingham from 920 to 1,200. It is unclear if these latter figures are strictly comparable with Pittenweem, but in any case practice varied from place to place.[6]

After 1890, some steam boats were fitted out for great lining, and this changed the scope and nature of the fishery, though at first the capitalist owners of the new boats could not keep the skippers and crew to work them when the summer herring fishing came round. Steam liners were never numerous compared with the steam trawlers, but they could sail great distances, and when steam gave way to diesel engines they could and did reach Rockall. They also had much longer lines than the nineteenth-century boats; John Muir, skipper of the last boat to go to the great lines from Fife in the 1970s, had a line 14 miles long, compared with the 12 miles that was common just before that date. Such enormous lengths were shot and hauled by a powered capstan, even on sailing boats.[7]

For the sma' lines, the burden of baiting fell entirely on the women (Figs 5.1 and 5.2). Belle Patrick described how in Pittenweem before the First World War they would bait the lines with mussels, a daily task between two tides, in the 'cellar' of their houses, a stone-floored room that opened at ground level onto the street:

Fig. 5.1 – Women baiting lines in St Andrews, c.1847: Hill and Adamson's picturesque arrangement of their photograph could be for the stage of an opera house, and belies the sheer hard work.

There, in summer or in winter, in fair weather or in foul, they sat 'redding' the tangled lines and baiting each one of hundreds of hooks with a mussel taken from a bucket standing beside each woman and shelled with a special knife. As the hooks were baited they were laid evenly along the side of an oval-shaped shallow basket, layer upon layer, until the line was finished. This was a skilled occupation, as the lines had to run out free, and one hook laid askew could foul a line as it was being 'shot'.[8]

The great lines were, at least at that date, more often baited by the men at sea as they sailed north, giving the women a respite over the weeks that this type of fishing lasted. Then the women turned their attention to mending old nets and mounting new ones. So laborious, essential and particular was the work of women in fishing communities that it was rare for the men to marry outside them.[9]

It is not easy to know how many local demersal fish were landed in the ports of the Firth of Forth, either by the sma' lines from inshore waters or by the great lines from the banks off the Isle of May, because by the time that data collection was properly established in the 1850s, trawling was beginning and the great-liner boats were also going further out. As the

Fig. 5.2 – Baiting lines in North Berwick, c.1920: the mussels are put on hooks on the line, which is kept in a wooden container called a scull. It was not uncommon for a fisherman to set to sea with 1,200 hooks on 240 yards of line, all baited by his wife.

decades passed, the trawlers, especially those from Granton, and latterly the steam and motor liners too, operated further and further out in the North Sea until they came to be fishing off Faeroe or Iceland, landing fish in Leith that had originated hundreds of miles away and been carried home on ice. But the quantity of line-caught fish originating in or near the Firth certainly must have greatly increased in the nineteenth century.

Even after the coming of the railway, some of the haddock landed was sold cured. The traditional 'Crail capon' was dried whole in a chimney (presumably over peat) and not split like an Arbroath smokie. William Tennant in his poem *Anster Fair* (1812) described how it had been eaten in the hand, as a fast food:

> Next, from the well-aired ancient town of Crail,
> Go out her craftsmen with tumultuous din,
> Her wind-bleach'd fishers, sturdy-limb'd and hale,
> Her in-kneed tailors, garrulous and thin:
> And some are flushed with horns of pithy ale,
> And some are fierce with drams of smuggled gin,
> While, to augment his drowth, each to his jaws
> A good Crail's capon holds, at which he rugs and gnaws.

It is not clear when this particular delicacy became a thing of the past, but it was probably when the local peat deposits were worked out, perhaps even in the eighteenth century. But curing houses that produced a more familiar type of smoked haddock were set up in many towns and operated until well into the twentieth century. In 1862, one Anstruther curer bought, in one day, 1,700 dozen white fish for the Glasgow market.[10] The new railways made the west of Scotland easy to reach, and the fact that any fish landed could be sent fresh or cured to all the British centres of population within 24 hours was a stimulus to bigger and bigger catches. Belle Patrick recalled the Saturday landings at Anstruther around 1914:

> The piers were so closely covered with cod, ling, skate, halibut and turbot that there was scarcely room to put a foot down. Most of these fish would be sold fresh, packed in ice and dispatched to the big markets, even as far as Billingsgate.[11]

There were similar scenes at Pittenweem, St Monans and Newhaven, though much of the fish itself eventually came from far beyond the Firth of Forth.

The ecological pinch-point in the Firth was obtaining bait, especially for the sma' lines. The main bait for them was mussels, though scallops and limpets might be accepted; the great lines used herring. In other parts of Scotland bait gathering was often yet another task for the women, but around the Firth of Forth it was more often done as a full-time job by men, with the shellfish sold on to the sma' liners. The quantities used were immense. 'It is almost impossible to exaggerate the importance of a good and cheap supply of mussels and bait for the Scottish fisheries' declared the committee appointed by the government to investigate the bait supply in 1889. At Eyemouth it was calculated that in the three years 1885–87, on average 38 tons of mussels had been needed each year to catch 42 tons of fish. The cost of bait was one-eighth of the gross value of the fish. Over 80 million mussels were used yearly in Eyemouth alone, and this was only one of half a dozen comparable sma' line fleets in the area. Mussels were not eaten by people to any great degree; they were too useful for the fishes.[12]

Mussel beds were extensive and valuable, particularly along the south side of the Firth between Bo'ness and Musselburgh (Fig. 5.3) and in the Eden estuary by St Andrews, also on places along the north shore of the Firth. Some had apparently been planted artificially by human effort: a lawsuit of

Fig. 5.3 – Loading sacks onto a cart from the mussel scalp at Musselburgh, c.1920.

1735 said that it was 'well known' that the mussel beds off Torryburn had been planted by Lord Colville, from which his heirs had drawn considerable wealth.[13] The scalps were under strain at an early date. *The New Statistical Account* for Bo'ness related how in 1803 a large fleet of Newhaven fishermen arrived and 'almost wholly removed an extensive scalp or bed of mussels'. They were pursued by the locals who 'recovered part of the plunder', but not enough to restore the bed. Similarly, a scalp at Tynninghame had been worked out before 1889 by men from North Berwick with shovels.[14] In 1894, the town of Cockenzie was the beneficiary of a legal order allowing it to take possession of the local scalp on behalf of the community, 'owing to the depredatory instincts of a local carter who was fast clearing all the mussels from the beds between Port Seton and Cockenzie'.[15] Attempts were also made at mussel conservation by seeding exhausted beds with mussels brought from deeper water, as in the Eden, or by constructing dredges so that the smallest fell through them, as on the Buccleuch scalps off Granton, but these efforts did not fill the gap between demand and supply. The Newhaven Fishermen's Society declared a close season from 1 May to 1 September, and won praise from the Fisheries Board.[16] On the Eden, despite conservation efforts, the mussels so deteriorated in size that four or five were needed on one hook where one would have done before.[17]

The pressure on the scalps rose inexorably as the quantities of fish landed increased, and by the 1880s there was a deepening sense of crisis. The east coast drew on the west for supplies, and the railways shifted vast quantities: in 1878–89, the Caledonian Railway moved 27,000 tons from Port Glasgow, but by the end of the next decade it was reported that the scalps there had been largely dredged out, as they had elsewhere down the west coast and in parts of Ireland.[18] Mussels were imported from England, Ireland and even from the Netherlands. Substitutes like scallops were tried particularly for cod bait (no one regarded scallops as human food) but could hardly fill the gap: 801 tons were landed in 1887 from 'the most prolific clam bait beds' that covered several miles of seabed off Cockenzie and Prestonpans. The Scottish Fisheries Board noted that nothing was being done to protect them, in this case blaming the Newhaven fishermen for overfishing.[19] It was competition from trawling that eventually saved all the shellfish, by undermining the profitability of the sma' lines, and from 1897, a decline in mussel landings in Scotland due to diminished demand began to be reported.[20]

The consequences for the other creatures that lived on and around the Firth of Forth of these depredations on the shellfish are unrecorded, but must have been considerable. Birds like the eider duck and the oystercatcher depend to no small degree on mussels, and when Victorian bird recording began they did not seem to have been very numerous compared with today. By the middle years of the twentieth century both were reported to be spreading rapidly as mussel stocks recovered. Certainly in the Firth of Clyde the spread of the eider was seen to be connected to the recovery of the mussel.[21] By then, supplies of haddock and cod landed in the Firth no longer depended on the sma' lines. They had been supplanted by the trawl, and many demersal fish were coming from distant waters anyway. That is our next story.

Because of the mussel problem, the sma' lines proved, in fact, to be another way of fishing that was unsustainable under the pressure of increased demand, much as the oyster fishery and the inshore herring fisheries had proved to be the same. Lining rested on an unhealthy ratio between the wild biological resource needed to catch the fish (the bait) and the quantity of fish that could be caught with it, analogous to (though not as extreme as) the way that modern fish farming rests on wild biological inputs of food that equal or exceed outputs. As demand rose under the pressure of national rather than local markets, the pressure on the shellfish became intolerable, and a whole ecosystem suffered. Fortunately, once the pressure was removed, and in contrast to the oysters and the herring, the mussel beds recovered of their own accord.

2. The Start of Trawling

The idea behind trawling was very simple (see Fig. 4.5). With a drift-net or an anchored ground net, the net was kept vertical and moved (if it did move) with the wind and tide; with a ring-net, also known as a seine or (in the 1860s) confusingly as a trawl net, a net was used to encircle a shoal, often with more than one boat. With a beam trawl – the trawl properly so called – a net was dragged along the sea floor, using a weight, and the fish scooped up into the 'cod end' at the back. The drifter or ring-net boat was ideal for catching herring and other pelagic fish that swam in the middle water, the trawler for catching the demersal fish that lived close to the bottom. It was concisely described by an English observer in 1883 as:

> A large bag-net dragged along the bottom of the sea by a boat or vessel, where the ground is free from rocks: and it captures a great variety of the best quality of fish found in our seas, namely turbot, soles, plaice, brill, dories, red mullet, skate, hake, cod, dabs, etc., haddock, whiting, and not a few large oysters, crabs, scallops, queens closely allied thereto, – in fact nothing comes amiss to it.[22]

Modern trawling began in the south of England around the middle of the nineteenth century, with the spread of a simple 'beam trawl' (the net was held open by a rigid wooden or metal beam) pulled behind a fishing smack. It originated at much the same time in several centres, initially Brixham in Devon, Barking on the Thames, and Great Yarmouth. From there it spread up the east coast to the Yorkshire and Northumberland ports, and English sailing trawlers began to appear in Scottish waters. The first port to adopt the sailing trawler in the Firth of Forth was probably Buckhaven, perhaps because the fishermen there already had some familiarity with manoeuvrable nets in the sprat fishery of the upper Firth, though the principle there was to encircle the fish rather than to drag the bottom. Before 1860, it had been tried in several places, Cellardyke, St Monans and

Fig. 5.4 – Thomas Huxley: fishermen did not like his manner.

St Andrews among them. Some of them gave it up almost at once because of the perceived damage that it was doing.

We have already seen the uproar in East Fife over damage to the herring spawn on the Treath off Pittenweem, but in England as well as in Scotland, trawling was widely opposed by those who feared destruction of fish stocks. Henry Fenwick, MP for Sunderland, who introduced the debate in the House of Commons calling for a new Royal Commission in 1863, said that there was 'an almost universal cry that our fisheries were falling off year by year' due to 'the neglect, perhaps the ignorance of Nature's own laws', in particular by the destruction of spawn by trawling.[23] The subsequent Royal Commission on Sea Fisheries included Thomas Huxley (Fig. 5.4) and Lyon Playfair, influential scientists, though not marine specialists. After hearing evidence from up and down the country, they recommended the opposite to what Fenwick had expected, calling for the abolition of all previous legislation that had imposed any restrictions whatever on sea-fishing, which indeed happened in the legislation of 1868.

The scientists held that neither trawling nor any other known method of exploiting the riches of the sea could have any impact upon its natural abundance. Fenwick had assumed, as fishermen also did, that all fish deposited their spawn on the sea floor, just at a time when the Norwegian scientist Georg Ossian Sars was discovering that cod eggs, at least, float in the open sea. Huxley made no assumption, but accepted the opinion of Professor Allman of Edinburgh that even where fish were known to spawn on the bottom, disturbing and even lifting their spawn would have no ill effects, and in any case he believed that the reproductive abilities of sea fish were so enormous that it was beyond the power of man to harm them. Trawling predominantly killed the predator fish of herring – cod, haddock and plaice. By removing these, the volume of herring in the sea should actually increase. For all the scientific prestige which any opinion of Huxley carried, these were not scientific opinions, since they were not based on systematic data-gathering but on general impressions. The Royal

Commission passed ideology off as science, reflecting the political laissez-faire of the age of Cobden and Bright more than knowledge of marine biology. Trawling was encouraged to continue unhindered.[24]

For the next dozen years trawling was carried on exclusively from sailing boats, which were limited in their capacity to trawl effectively against the tide and to operate in bad weather. Scottish boats were not much involved, preferring the tried and tested techniques of catching cod, haddock and flatfish with lines, despite the growing problem of procuring bait. There had, however, been an experiment with steam trawling in Scotland as early as 1862, when a Forth Steam Fishing Company had been formed. It adapted 'at considerable expense' for trawling, a 62-ton steamer, the *Ariel*, bought in Glasgow, screw-propelled by two engines of 16 horsepower, a donkey engine for the capstan to haul the nets, and a saltwater well in the hold to keep the catch alive. On its trial run out of Leith, according to *The Scotsman* of 3 July, it carried a number of guests, including Sam Bough, the celebrated landscape painter, but he left no record of the gentlemen's jaunt, or of the 'substantial dinner' enjoyed in the afternoon. The *Ariel* lowered the trawl first off Aberlady Bay and steamed to North Berwick, catching nearly a ton of fish, and later shot the trawl again off the Bass Rock and caught nearly as many again. Probably the first steam trawler in the world, though not recognised as such, it had all the features that made later models so successful – screw driven, able to work against the tide, a steam capstan, a well for the catch – yet within less than four years the company was bankrupt and the experiment apparently forgotten. It is hard to say why it failed.

The next sustained application of steam to trawling came 11 years after the dissolution of the Forth Steam Fishing Company, and was pioneered on the Tyne, first at North Shields by adapting a paddle tug to carry nets in 1877, and then by the construction there in 1881 of a purpose-built, screw-driven steam trawler.[25] Almost at once, Edinburgh business interests became interested again. On 2 December 1881, *The Scotsman* announced the formation of the General Steam Fishing Company based in Granton, with an initial capital of £100,000, headed by James Wilson, Esq. of Restalrig, 'manufacturer of Leith and Berwick', and Robert Hutchison, Esq. of Hillwood, builder and Dean of Guild of Edinburgh. Expressing surprise that steam had not been used more extensively in the industry, and attributing this to the lack of capital among ordinary fishermen, the directors declared their intention to acquire a 'Fleet of Steam Trawling Vessels of the most approved description', enabling one or two to return to port daily to land the catch of the rest of the fleet, which could remain 'far out to sea where more and better fish could be got'. This was the first such company in Scotland, and though Granton was soon to be outpaced by Aberdeen, it remained for many decades easily the second most important trawler port north of the Border. Within two years there were about 30 steam trawlers, English and Scottish, operating off the east coast of Scotland.

Instantly there was conflict between traditional fishermen-owned line-fishing sailing boats and drifters, and the new company-owned trawlers. The latter did not remain 'far out to sea', but frequently fished close inshore in direct competition with the sma' lines and herring boats, and whether close or far, they had scant regard for the gear and nets of other fishermen. Steam trawlers had great power, with chains on the ground-rope and steel cables, to 'drag and roll rocks along the bottom, crushing, pulverizing, and stripping the living matrix and liberating the mud and sediment underneath'.[26] Trawls ripped up the bait beds of scallops and mussels, disturbed the spawning grounds of pelagic fish, and the fishermen believed that trawling also killed other spawn and caught too many immature fish. So intense was the resentment that Parliament was prevailed upon in 1883 once again to set up a Royal Commission to investigate whether or not trawling was damaging to the British fishing industry.

The General Steam Fishing Company of Granton responded to the campaign by circulating to every Member of Parliament a document setting out their side of the case, in what *The Scotsman* of 9 June called a 'decidedly aggressive' way. First, they said that conventional fishing in small open or half-decked boats was unsafe, while steam trawlers were 'comparatively safe and comfortable'. Then they averred that trawling caught many fish 'of a kind not taken by hook and line at all, thus adding greatly to the food supply of the kingdom', giving employment to many and much business to the railways. They went on to state:

> that it destroys no spawn, and takes no immature fish: and that it does not seem to affect the quantity of fish in the sea, as there is no observable falling off in the fishing of any kind: further that it does not affect the prosperity of the hook and line fishermen, as they now catch as many fish as before: and lastly that it is carried out without appreciable injury to the lines of these fishermen.

Finally they observed that it was discreditable to Scotland that its fishing industry was carried on in its present primitive style. Even in the long history of Parliamentary lobbying, this is a remarkable set of misrepresentations.

The new Royal Commission was headed by the Earl of Dalhousie, said to be sympathetic to the case of the poor fishermen but suspicious of unnecessary regulation, and at Dalhousie's insistence it again included Thomas Huxley. Though, due to ill health, Huxley was not able to be present at the end of the hearings, he played a vigorous part in the meetings in the Firth of Forth ports. He was, however, increasingly aware of criticism by other scientists of the unsubstantiated nature of his own opinions on the inexhaustible nature of the sea, and asked for the Commission to be assisted by William M'Intosh, newly appointed to a chair at the University of St Andrews, and the first professional academic marine biologist in Britain (Fig. 5.5). The Commission also drew on the scientific expertise of the recently created Scottish Fisheries Board, and with them devised a series of experiments to test the effects of trawling, ultimately involving the purchase of the first fishery research vessel in British waters, the *Garland*. The Commissioners were aware that definitive results might take years of experiment and observation, but their initiative did much to set fishery science on a proper long-term basis.[27]

Fig. 5.5 – William Carmichael M'Intosh, Professor of Natural History at St Andrews and protégé of Huxley, the first full-time fisheries scientist in a British university.

Meanwhile, they set about gathering information from witnesses in the fishing ports as in 1863, but this time they interviewed fishermen themselves, as well as officials, merchants and various other experienced middle-class gentlemen. On arrival at St Andrews in September 1883, they assembled at the town hall a range of local men from the surrounding communities on the Tay and in Fife, all of whom testified to the damage caused by trawls. Feelings were running high. At one point the spectators in the hall broke into applause, and were asked by Dalhousie to refrain from either hissing or cheering the witnesses.

Perhaps the most interesting witness was Alexander Wallace Brown, line fisherman of St Andrews who had also worked on a sailing beam trawler, who began by saying that 'we were all trawling ourselves about ten years ago in sailing vessels...we cleaned the bay'. Like other fishermen, he believed that all fish eggs (not just herring) sank to the bottom, and that their spawning grounds were being disrupted by the action of the trawlers. He had decided to investigate fish spawn in more detail himself, and over a period of 18 years undertook amateur scientific investigations of his own by capturing spawn and keeping it in a small tin box bored with holes by a needle, in a tidal rock pool by the castle in St Andrews, examining it with a microscope that he had bought for 7s 6d. Thomas Huxley cross-questioned him closely, even aggressively. Huxley's style was later to be criticised by a fish merchant from St Monans, who said 'he has a way of speaking, and we do not hear every word he says, and then the fishermen might say a thing really they do not mean'. On this occasion he asked the witness if he knew the difference between fish eggs and whelk eggs, or if his eggs had been alive and fresh when obtained. Huxley went on: 'I want to know how you are able to tell a young haddock from a young herring when the two are just hatched?' Brown replied: 'Well, sir, the difference between a haddock and a herring is this. The haddock has got a head almost like a bull dog and a herring again has got a head almost like the sharp point of a pin. In the case of a haddock after it is hatched and you see the fin that lies down in here, it lies down by its side; with the herring again it sticks even out'. Huxley: 'It is quite clear that you have thought very carefully over the matter . . .'[28]

Huxley also eventually said: 'Your method is capital. I am not prepared to accept your conclusions but I can respect your method of operation'. He knew that Sars' earlier findings that cods' eggs were free-floating were also likely to be applicable to other demersal fish like haddock and plaice. Nevertheless, he asked that M'Intosh, in taking on a research role for the Commission, should use a trawler to fish the same piece of ground at regular intervals for the next six months in order to 'register the nature of the products of each haul'. Huxley explained that the Commission especially needed 'more precise and specific information . . . as to the action of the trawl-net in bringing up "spawn" and immature fish'. M'Intosh's work for the Commission was greatly to reinforce his reputation and led the following year to the establishment of the first marine laboratory in Britain at St Andrews.[29]

Not a good sailor, M'Intosh undertook voyages at fortnightly intervals between January and August 1884, on a Granton steam trawler in St Andrews Bay and off North Berwick in the Firth of Forth, but also (in an apparent departure from the original plan) as far afield as Caithness and Scarborough. Sometimes Lord Dalhousie was on board; he clearly believed in hands-on experience. He witnessed M'Intosh's suffering from sea sickness, telling the House of Lords that the professor had displayed 'very great heroism . . . not being accustomed to the seafaring life he had endured very great hardship'. They made 93 hauls in all. M'Intosh confirmed what Huxley had expected, that the eggs of cod, haddock and other pelagic fish did indeed float and were not affected by the trawl, and also that there was no proof of injury to herring spawn. He also formed the view that trawling did not result in as many immature fish being discarded as unfit for human consumption as many fishermen thought (they said about a quarter by weight), and though he agreed that trawling had caused a drop in catches especially of large fish in St Andrews Bay, he also believed it would recover in time if rested.

He gave evidence in London before the Commission in November 1884. It was reported, and closely examined in Fife by the local fishermen. They did not like what they read, and formed an opinion that the Commission would come to a conclusion unfavourable to them because of Professor M'Intosh. As he relates, he began to find them muttering threats and using 'opprobrious epithets' as he passed their houses on his way to the laboratory. Then on 6 March 1885, preparing to leave home for the students' Liberal dinner, he saw in the dusk 'a great crowd of people with an effigy in their centre' coming down the street. He quickly locked the front door and the garden gate, and retired inside. The crowd halted at his house, rattled the gate, 'and yelled and shouted like demons'. They then burned the effigy in the street, 'adding tar and other combustibles'. A friend and colleague, Professor Scott Lang, coming down the road, in his words, 'gathered it was a demonstration against Professor M'Intosh because of his views on trawling which as I understood was to the effect that it did not do the damage to the fisheries that was alleged, and, in particular, did not destroy the eggs of fishes'. The unfortunate man was then put up against the wall by the rioters, who demanded to know what he had to say: he replied, wisely, that M'Intosh was an honest man, that it was possible he might be wrong, but had only said to the Commission what he believed was right. They let him go at that, and after burning the effigy, 'shouting and gesticulating, they dispersed'. Some local police were present and one young officer made as if to draw his baton. He was told by his inspector not to do so, Scott Lang adding 'had he not done this, I believe there would have been a bloody riot'.[30]

Next day, M'Intosh's students, offended by the insult to their professor, marched to the St Andrews fisher-town and offered to fight. But the local fishermen, who the previous night had been reinforced by others from Cellardyke and other neighbouring communities now gone home, sensibly stayed indoors. M'Intosh put the trouble down to unspecified 'mischievous agitators in the background' and 'blamed not the susceptible fishermen'. But of this there is no evidence at all. A clue to the strong feelings of the 'susceptible fishermen' lies in a long letter that M'Intosh received from one of them, apparently Alexander Wallace Brown, demanding to know the basis of his statement to the Commission. 'Having read very carefully the evidence you gave', the writer put three questions: firstly, 'I do not doubt that you might get spawn floating, but did you prove it to be either haddock, cod or whiting spawn?': secondly, did M'Intosh bring the spawn that he caught to maturity and hatching point?; thirdly, if so, at what stage were the eggs obtained, newly spawned or at the point of hatching? 'If you would kindly state how many days you had the spawn before it hatched and what month you received it you would much oblige . . . I ask these questions for the interest of my self and my brother fishermen'. It reads much like a re-run of the questions to which Huxley had subjected Brown in September 1883.[31]

M'Intosh apparently did not reply to this letter, but some of the research workers at his marine laboratory later tried to mend fences by demonstrating to the fishermen how fish eggs indeed floated in a bucket. Eventually the professor became an acceptable character locally, and the local fishermen undertook work for him. In 1913, one even saved the laboratory from destruction by reporting a fire which had been started by indignant suffragettes, but that is another story.

The Commission concluded that 'the recollections of fishermen extending over a long period of years are not sufficiently precise to allow any conclusions to be drawn from them'. They complained of 'the vague character' of their evidence, and the way 'fishermen are apt to rely on tradition rather than observation'.[32] In the matter of the eggs of demersal fish this was justifiable enough, but they also encountered many fishermen with practical recent experience of the ways in which trawlers behaved. Some were very angry. George Heugh of Pittenweem had come before the Commission to complain about the trawlers carrying away his lines:

> They come from Leith every morning and they do not seem to care; they just seem to act as if the sea was their own. 'Just let me go this way, keep away from me.' That is just what the trawlers want 'Keep out of my way'.

Up to two years ago, he had been able to make 30 shillings a week from his sma' lines, but now only 14 shillings.[33]

Others had information of the environmental effects of trawling on the fish and the seabed. John Wilson of Newhaven met the Commissioners in the town council headquarters in Edinburgh, and described how he had worked on a trawler which had fished north-east of Inchkeith, getting 45 boxes of fish the first day and 108 for the week:

> When the rest of the trawlers heard about it, there were fourteen of us next week on the top of it: and in three weeks time we had that spot of ground raked and torn up and in such a condition that we had every fish of it. I believe it was as bare as the council chamber. I believe there was nothing escaped. I have seen it as high as three or four cartloads of ground in the trawl . . . horse bait, and stones and shells . . . and clams.[34]

William Hunman of Cockenzie invited, 'any fisherman or any member of this Commission' to come with him to the grounds off Prestonpans where 'the trawlers come regularly down and put away their beam trawl' to see where 'they have taken away the upper crust of the ground, and mark you, it is the upper crust of the ground that the scallops of clams live amongst'. John Murray from North Berwick spoke of the damage done to the bottom:

> Now the ground is cleaned of this sort of shellfish by trawling, and now we have no large fish because their food is all taken away. I quite disagree with some of the evidence that has come before your lordships stating that the turning over the ground makes fish more plentiful. I consider this statement is quite un-natural. The less ground is disturbed the more plentiful fish is.[35]

Witness after witness said the same thing: the ground was being torn up, fish were disappearing or diminishing in average size, and fishermen took more effort to achieve the same catch. The Royal Commission was not unaffected. Its final report has been described as a 'whitewash',[36] and it certainly resulted in no restrictions being put on the trawlers operating offshore. However, the Commissioners, and also M'Intosh, did accept that serious damage had occurred to the inshore fisheries, though it was expected to be only short term. Trawlers were forbidden to fish within three miles of the coast, and the Scottish Fisheries Board was given powers to close other waters. In 1886 the Firth of Forth itself was closed to trawling inside a line between Tantallon Castle, the Isle of May and Fife Ness, and also within St Andrews Bay. Recovery was confidently expected.

3. The Slide to Catastrophe

At first very little seemed any different. There were no immediate signs of recovery, but the Firth and its environs were still full of the wonders of nature. In 1889, the *Garland* came upon a shoal of immature whiting, three–five inches long, stretching 36 miles from the Oxcars rock, in the middle of the Firth west of Granton, to the other side of the Isle of May. The Board scientists estimated it to hold 230 million fish.[37] In 1886, Professor M'Intosh was sent a 'fine old male tunny' (blue-fin tuna), weighing six cwt, as a present from the Granton Steam Fishing Company, caught with their 'huge beam trawl then in use off the Firth of Forth'. He compared it, 'swimming alone in northern waters', to a solitary male rogue elephant: such a fish would have come north in pursuit of the herring, mackerel and other fish still plentiful east of the Isle of May:

> We ate part of the muscle and sent gifts of it to all members of the senate in the city. One of these complained of being ill after the diet, but he had eaten too much of it, for it was oily, and a heavier diet than salmon.[38]

Otherwise life went on much as before. There were still complaints of trawlers fishing close inshore, especially at night. There were requests for better protection by small fishermen. There was continued dislike of the trawler-men. James Smith of Cellardyke described them to a Parliamentary Committee in 1888 as: 'not honest men, but a parcel of roughs, men that neither fear God nor regard man', fishing on Sunday as well as on Saturday: 'Now, how can they expect the blessing of the great God to support them?' You won't get much dust after you sweep a room, he said, 'neither will you get any fish if you go after these trawlers have been working night after night'.[39] Trawling interests were still accused of misrepresenting the facts.

'It is absurd to say that trawlers have been of benefit to the white fishing grounds', wrote George Collen, fisherman of Eyemouth, in a letter to *The Scotsman* on 5 March 1890, 'it has been the experience of the line fishermen that their usual fishing grounds have been spoiled by trawlers'. He cited a fall in the catches by line fishermen from 60–100 stone a trip, down to 9–15 stone, following trawling.

Furthermore, some of the things which the line fishermen had been saying that had hitherto been regarded as exaggerations were now found to be an understatement. The Scottish Fisheries Board in 1893 said that, in June and July, trawlers operating within 40 miles of the Isle of May were on their own admission throwing overboard as offal half to two-thirds of the haddock caught, because they were considered too small to market. In 1903 the Board undertook more investigations and found that 30 per cent of trawled fish caught were discarded, most 'belong to forms which are quite edible and marketable and are rejected merely because of their small size'. In 1905 they found that 59 per cent of fish caught by the Scottish trawlers which they could study was treated as refuse and spoke of the 'enormous destructive power of the modern otter trawl' operating in shallow water and at certain seasons.[40] The problem of discards hardly originated with the Common Fisheries Policy.

Very soon, even after protection had been improved, the Scottish Fisheries Board scientists began to observe an unexpected lack of recovery in the numbers and size of demersal fish within the Firth, and to attribute it to excessive catches further offshore. For example, in 1892 they reported that haddock caught off Anstruther in the closed area fell from 18,400 cwt in 1888 to 11,700 in 1892.[41] The Board agreed with the fishermen that, however measured (per ton, per yard of line, per square yard of net), catch per unit of effort was falling all the time. Their principal scientist Thomas Fulton became an outspoken critic of Huxley's view that the seas were inexhaustible, notably in the Board's reports of 1891 and 1892:

> It has now been made clear by statistical and scientific investigations that the seas round our coasts are not the inexhaustible store houses of food material that they were thought to be less than a generation ago. The doctrine that the operations of man cannot disturb the balance of life in the sea, and diminish or exhaust the supply of valuable food fishes is now abandoned by fishery authorities almost everywhere.[42]

He believed that man could intervene with a programme of fish rearing, such as the Board was beginning to sponsor at Dunbar, planning the release of several hundreds of millions of plaice. This, however, proved a blind alley when restocking failed to produce results.

Fulton's opinions brought him into immediate and furious conflict with M'Intosh, who, after first publishing a number of critical papers, in 1899 brought out *The Resources of the Sea*, containing an exhaustive critique and influential refutation of the work of the Board. He argued that Huxley had been right all along, and indeed that it had been a mistake in 1886 to close the Firth of Forth and St Andrews Bay to trawlers. The appearance of decline, he maintained, was solely due to bad science, and such was his academic prestige at this point that these criticisms were taken very seriously. However, evidence had been building up elsewhere that fish stocks in the North Sea were declining. As early as 1893 a Parliamentary Select Committee on Sea Fisheries 'for the first time in the modern period . . . revealed clear and unequivocal evidence of the impoverishment of the fishing grounds over extensive areas'. Even the British trawling interest was worried then, and, in a move that set the tone for the next 100 years, blamed the foreigner, initially the Dutch, the Germans and the Danes.[43] Then, immediately after M'Intosh published *Resources*, Walter Garstang in England demonstrated that there had occurred a halving in catch per unit of trawling effort in British waters over the previous ten years.[44]

Gradually, the international scientific consensus turned against M'Intosh. He was completely unrepentant, publishing a second edition of *Resources* in 1920, claiming that its original conclusions were unassailed. The revised book should 'aid in dispelling the ever recurring fears as to the diminution of the marine food-fishes'.[45] By the time he died in 1931, however, he was a lonely and isolated figure, still going to work, but in an unheated laboratory that his own university no longer supported with funds.

Meanwhile, trawling had become even more destructive. In the early 1890s, the spread of otter boards, replacing the beam across the opening of the trawl net by two boards that acted as hydroplanes to keep it open, enabled trawlers to work rougher ground and greatly enlarged the types of ground over which they were effective.[46]

Pinched on all sides between a cheap supply of fish from the offshore trawlers, problems with obtaining bait, and a diminishing supply of fish within the Firth of Forth, the smaller or more vulnerable fishing communities were starting to go to the wall. In 1897 and again in 1901, the Board reported that Newhaven fishermen had given up line fishing, 'so scarce have haddocks become in the Firth'. In 1901, it was reported that at Buckhaven some fishermen were becoming coal miners for part of the year, and at Kinghorn that fishing would be abandoned by most of the fishermen if there had been anything else they could do: though at Largo there were still good fishing grounds for plaice. In 1904, the fishermen at North Berwick had become part-timers; fishing for codling at Limekilns, Inverkeithing, Aberdour, Kinghorn, Kirkcaldy, Dysart and Wemyss, and for codling and flounders at Burntisland, was now mainly for local consumption and of little importance, and the same was true at Buckhaven, where fishing for herring, cod and haddock was 'gradually falling off'. As late as 1914, there were still 35 boats registered in Buckhaven, but only eight were above six tons. In 1907, fishermen were also becoming miners in Prestonpans. Though there was some revival of line fishing from Newhaven in 1907, by 1911 it was said that this was now restricted to Cockenzie, at least on the south side of the Firth. Even at Dunbar, by 1913 the principal earnings were from crabs and lobsters: 'the other fisheries are not now of much importance'.[47] In short, by the eve of the First World War, outside east Fife and Cockenzie, there was hardly any fishing taking place from Firth of Forth ports apart from the activities of the Granton trawlers, the winter herring boats mainly from east Fife, numerous boats setting creels and perhaps doing some hobby fishing to earn a little on the side, and the sprat boats at the head of the estuary.

A glimpse of the Firth of Forth fishing fleet on the eve of the First World War is afforded by the *British Fishermen's Nautical Almanack,* published in St Andrews in 1912 (see Table 5.1 and Fig. 5.6), though it gives no details of boats in the inner estuary above Queensferry. No fewer than 914 boats are listed, totalling 23,587 tons, but 37 per cent of these were of under ten tons, little yawls. Half the total tonnage was based in the three east Fife ports of Anstruther, Pittenweem and St Monans, the last having most. These were predominantly drifters, for much of the year fishing herring well away from the Firth. The trawler ports of Granton and Leith also stand out, even without reckoning the fleet of whalers which was based in Leith and sailed in Arctic seas. The trawlers also operated increasingly far out. The four biggest boats were steam trawlers of around 180 tons based in Leith, but there were only 12 boats of more than 90 tons in the entire Firth. Three ports of renown in earlier centuries, Crail, Dunbar and Newhaven, now had scarcely any boats above ten tons, their fishermen reduced to pottering

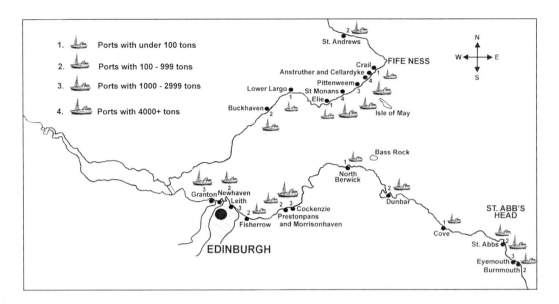

Fig. 5.6 – The fishing fleet in 1912

around inshore, mainly after crabs and lobsters, or seeking mussels for bait for the long liners. Others, like Buckhaven and Fisherrow were little better off, but Cockenzie remained a busy place, though mainly with rather small boats that specialised in lining.

After the First World War, the Granton trawlers, though banned effectively from fishing within the Firth itself and hampered by bad industrial relations and strikes both before and after the war, had a double boost. Firstly, four years when submarines kept the fishing fleets at home had allowed the depleted fish stocks in the North Sea to grow again. Secondly, the fish-and-chip trade really took off. Even as early as 1920, it was said that there were 25,000 outlets consuming 4,000 tons of fish a week 'in the country' – presumably the United Kingdom. Fish and chips restored the catch to popularity: earlier, in 1909, there had been complaints before that the catch was kept too long on board the trawlers, so that the public was being put off the fish by its tainted quality. It also provided a market for small haddock that would otherwise have been 'shovelled overboard'. Even in the depths of the interwar depression, in 1933, when fish hawkers found trade difficult in the poorer parts of the cities, fish fryers were less badly hit: 'the suburbs of the principal towns are now well catered for by motor fish shops and fish and chip kitchens'.[48]

Most of the trawlers from Granton at first still concentrated their efforts within easy auk-flight of the Isle of May. In 1920 there were 70 at work

	Numbers of Boats						Total	Boat Tonnage
	1–9 tons	10–29 tons	30–49 tons	50–69 tons	70–89 tons	90 + tons		
St Andrews	31	6	4	–	–	–	41	375
Crail	30	1	–	–	–	–	31	82
Anstruther and Cellardyke	12	22	47	14	15	3	113	4,605
Pittenweem	29	9	25	7	5	3	78	2,345
St Monans	7	12	72	26	5	–	122	4,892
Elie	2	3	–	–	–	–	5	67
Lower Largo	17	–	–	–	–	–	17	43+
Buckhaven	12	16	7	4	–	–	39	888
Granton	–	8	38	9	3	–	58	2,347
Newhaven	45	–	–	–	–	–	45	167
Leith	5	2	3	–	–	5	15	1,024*
Fisherrow	9	18	10	–	–	–	37	896
Prestonpans and Morristonhaven	5	–	3	–	–	–	8	152
Cockenzie	38	35	40	2	–	1	116	2,598
North Berwick	7	–	–	–	–	–	7	15
Dunbar	23	2	–	–	–	–	25	105
Cove	15	–	1	–	–	–	16	69
St Abbs	13	6	12	–	–	–	31	623
Eyemouth	8	19	38	–	–	–	65	1,911
Burnmouth	35	6	4	–	–	–	45	383
TOTALS	343	165	304	62	28	12	914	23,587

Table 5.1 – The fishing fleet in 1912. Source: *British Fishermen's Nautical Almanack* (St. Andrews, 1912). Thanks to Robert Prescott for bringing this to our attention.

* Leith figures exclude another 17 vessels, 1238 tons, which were clearly whalers working in arctic waters.

+ Also one boat in Largo registered as ½ ton.

from Leith District (as compared with 40 in 1919), mostly fishing 'well inshore near the entrance to the Firth of Forth or off St Abbs Head'. There would have been others, from English ports or Aberdeen. In 1925 it was said that two-thirds of the catch landed at Leith and Granton came from between the Bell Rock and the Farne Islands, and in 1930 that 60 per cent came from this area. Only from 1931 was it said that more than half came from more distant grounds in the north. This implies not so much greater enterprise as fewer fish close to home. The problem of overfishing in the North Sea began to be recognised, and in 1933 legislation set restrictions on the size of mesh and on the minimum size of haddock and most flatfish allowed to be landed.[49] It was the first such restriction since the abolition led by Huxley of all the earlier legislation in 1868.

During the interwar years, sail was rapidly becoming obsolete, and diesel engines were powering even small boats. An important innovation reported in 1921 was the use of the Danish seine. The early models involved a stationary boat that hauled a net, up to 80 yards long, by means of two warps, which might each be a mile long, stirring the ground to shepherd the fish into the net. It was very effective for demersal fish, cheap to operate and gave smaller diesel boats from the east Fife ports a chance to compete with the trawlers.[50] At first it was not allowed within the Firth, but from 1936 the Board gave permission for seine-netting boats up to 50 feet in length to work in an area on the south shore. After the Second World War use of this method grew quickly. Landings of white fish taken by seine in Anstruther District grew from 51 cwt in 1938 to 27,909 cwt by 1949. Much of this would have been taken outside the Firth of Forth, and it was inconsiderable compared with the 571,000 cwt of fish landed at Leith by the trawlers in that year.[51] It is hard to say how much of the latter was caught even within 100 miles of the Isle of May.

The years after 1945 were dominated by a growing dislocation of fishermen and politicians from the bitter reality of fish biology. Although the evidence of declining catch per unit effort was widely known, and although there had been growing signs that the North Sea was in trouble since at least the 1930s, nothing effective was done to protect it. Instead, governments embarked on a programme of subsidising larger and more destructive trawlers and purse-seiners, which could reach further out to waters with relatively undisturbed populations: now they could use echo-sounders to find the shoals, more efficient nylon nets to catch them and hydraulic stern-loading gear to draw them aboard. In 1966 government grants were available for new fishing boats, covering 35 per cent of cost for boats of over 80 feet, 40 per cent for those below.[52] In the 1930s it had been discovered that Icelandic waters yielded ten times the volume of fish for the same effort as the North Sea was now capable of doing. The Granton trawlers joined the trek to the distant seas, landing much of their catch in the west of Scotland, and what they landed at Leith often came from the Arctic. In 1975, Iceland, in self-defence, and after years of dispute with Britain and Germany, and with the support of the United Nations, declared a 200-mile exclusion zone to foreign trawlers.[53] Predictably, by 1978, the Granton fleet was bankrupt, its remaining trawlers withdrawn from service, but they had had a profitable innings (Fig. 5.7).[54]

The European Common Fisheries Policy to which Britain now subscribed, had acted effectively to declare a moratorium on herring catches from 1977 to 1980, but that was almost its only sensible decision in the gathering crisis. As Callum Roberts has put it:

Fig. 5.7 – Money to be made and fame to be had: the Golden Haddock Awards presented by fish merchant Joe Croan in Edinburgh, 1973.

Instead of responding to mounting evidence of overfishing by easing pressure to allow recovery, Europe's politicians sought to prop up an ailing industry with overgenerous quotas. More fish stocks declined. Like gamblers desperately seeking a change of luck, they spent from their savings with inevitable consequences – exhaustion both of luck and of fish.[55]

In respect of North Sea cod stocks, the International Council for the Exploration of the Sea (ICES), which advises governments on fisheries conservation, noted a fall in the spawning stock biomass from 157,000 tons in 1963 to 38,000 in 2001, a 76 per cent decrease. Since 2003 it has recommended that no cod at all should be caught in the North Sea, to allow time for stocks to recover. European governments acting through the Common Fisheries Policy, however, have disregarded this advice, and fixed quotas that were much too high: 82 per cent of the cod landed between 2005 and 2010 were juveniles aged under four years. Unless new fish can be allowed to reach breeding age, there will be a collapse as absolute as that which has overtaken the cod fishery on the Grand Banks of Newfoundland.[56]

As for the Scottish fishermen, their invariable reaction to the negotiating stance of British politicians has been to complain of having been sold down the river, but only because the quotas awarded to them were not as big as they wanted. Should they sometimes succeed in getting a figure closer to what they wanted, the media portrays this as a negotiating triumph for Scottish ministers. Yet under the Common Fisheries Policy, quotas were routinely set 25–30 per cent higher than levels advised as safe even by national scientists, because the political need to prop up landings in the short term has been considered more pressing than the ecological danger to the entire stock.[57]

By the new millennium, less than a fifth of all the fish stocks in the North Sea were considered to be in a healthy condition. In 2010, it was calculated that landings of demersal fish per unit of fishing power, a measure of the commercial productivity of fisheries, had fallen by 94 per cent in England and Wales since 1889, and because, where Scottish data can be added, it shows an almost identical trend to English data, one may assume the same here. The collapse is entirely associated with trawling and refers to the areas fished in the North Sea and the North Atlantic: it 'implies an extraordinary decline in the availability of bottom-living fish and a profound reorganization of seabed ecosystems since the nineteenth century industrialisation of fishing'.[58]

As for the fish stocks within the Firth of Forth, or within 50 miles of its entrance, the story was the same. Increased exploitation was as severe as elsewhere. Only British vessels were allowed within the Firth, but both British and foreign vessels were at liberty to operate beyond. Danish vessels during the 1990s trawled on the fishing banks beyond the Isle of May and caught enormous quantities of sandeels, an important food for other fish as well as for seabirds and seals. In 1999, this was finally stopped by order from the European Union. Local motor-driven seiners and pair-trawlers could always work just beyond the Firth, and if the boats were less than 50 feet long, their operations were declared legal even within the area previously forbidden to trawling in 1885. Landings of demersal fish at Pittenweem, the main port of landing in east Fife from 1950, rose from an average of 15,700 cwt in 1950–52, to a maximum of 94,400 cwt in 1980–82, falling back to 8,700 in 1998. Now, in 2011, there are no landings of demersal fish at all. However, landings of nethrops (alias Norwegian lobster or prawns) which were formerly regarded as bait, or at best eaten as a snack by crews at sea but certainly not saleable, began to rise. A market for them as scampi was found first in Spain and then at home. Landings of shellfish at Pittenweem rose from nothing in 1957, to 26,000 cwt in 1999, and now trawling for nethrops is the principal fishing activity in Fife.⁵⁹ Otherwise there is only creeling for lobsters and crabs, a little dredging for scallops which is unquestionably harmful to the sea floor, and some fishing for razor-shells in Largo Bay and St Andrews Bay, which often involves illegally electrifying the seabed and sending divers down to pick up the razor-shells among the debris of all the other creatures killed there. Fishing for nethrops in the Firth of Forth is subject to a quota and is officially regarded as sustainable, despite some misgivings about the by-catch of immature cod and other fish.⁶⁰

Since the late 1970s, the trawls of the Scottish Environment Protection Agency have monitored fish diversity within the Firth of Forth. The number of different species recorded has tended to increase (it is now around 36), which is attributed by SEPA to the improving quality of the water in the Firth as pollution has been brought under control.⁶¹ However, welcome though this is, the health and naturalness of an ecosystem can hardly be measured by a simple count of its biodiversity. There is an important difference between the number of fish species recorded and the numbers and size of large fish swimming in the Firth, particularly as in the recent past there has been a famous plenty of the latter. Unfortunately, there is every indication that the deterioration in the quality of fish life within the Firth of Forth did not come to an end in 1970 or in 2000. For a very long time, sport angling has been a pastime for a great many in the Firth, and a source for income for those with small boats willing to take the sportsmen out. This is how the situation appeared to one of the most experienced anglers in 2011:

> I have fished in the Firth of Forth for nearly thirty years: over this period I have seen a big decline in the quality and quantity of the fishing here. Up until the mid-90's cod were plentiful, with specimens of over ten lb being caught on a regular basis in the outer Forth and good numbers of fish of four to five lb being caught as far up as Burntisland. A friend had a theory that about this time large factory ships were sitting out in the North Sea hoovering the mass shoals of sandeels heading for the inshore waters, thus taking away a valuable source of the food chain for the fish. The mackerel shoals also have been on the decline and I can only assume that the lack of sandeels has had a major impact on this species as well. Back in the 80's huge shoals would make their way right up the Forth as far

as the road and rail bridges at Queensferry. Over the last five years there has been a big decline with the mackerel. Fishing for flatfish is still very popular off many of the beaches, with good numbers of flounders up to a pound or a pound and a half in weight still being caught. Again, between the early 80's and 90's fish over two 2lbs were more plentiful. Nowadays, a flounder over the 2lbs mark would be regarded as a specimen. The other popular flatfish in the area i.e. the plaice and the dab, are very rarely caught now in the Forth.

The thing I believe has affected the flatfish, is the prawn and scallop boats working in the outer and sometimes the inner Forth, as they plough up the seabed and kill a lot of immature fish and marine life in the process.

In general the fishing in the Firth of Forth has declined drastically. In the early 80's numerous boats from Pittenweem, Anstruther and Crail would be taking anglers out at the weekend morning and afternoon from May until September. I don't know if there are any doing this now. Also it was not unusual for competitions fished from Dysart to Buckhaven to attract up to 400 anglers: sadly this number is now below 100, but whether this is down to the decrease of fish or to the age of computer games and technology who knows?[62]

So this is what we have done to the Firth, which two centuries before was so well known for its abundance – for its turbot, its plaice and flounders, its skate, its halibut, its cod and haddock, as well as for its saithe and its herrings and oysters, and for the sprats, sparling and salmon that we shall consider next. We can look at the blank face of the sea, and see what our eighteenth-century ancestors saw, but can we look beneath it and not be appalled by the changes we have brought to pass, and be ashamed?

CHAPTER 6
TRAPS AND NETS IN THE ESTUARY

1. Introduction

In 1838, a young Edinburgh doctor, Richard Parnell, won a prize from the Wernerian Natural History Society of his native city, for a remarkable pioneering 'essay', of over 400 pages, with illustrations, on the fishes of the Firth of Forth. *The Scotsman* correctly declared it to be 'one of the most notable contributions which the Zoology of Scotland has received within the last fifty years'. He enumerated 125 species, one or two of which (like the whitebait) might not be recognised by science today. To do his research, he frequented the fish-market of Edinburgh to get specimens, and he must have gone out on trips with the Newhaven fishermen, as many students

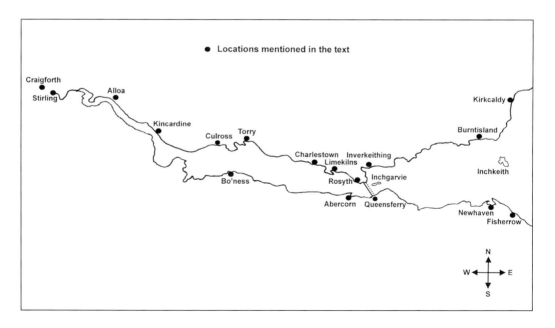

Fig. 6.1 – Fishing in the upper Firth

had done before him, including Charles Darwin. He certainly had good connections with Alloa at the head of the Firth.

He seems to have known the muds and gravels of the upper part above Queensferry, described today as the estuary, particularly well.[1] Most of the species on his list are still there, though none of the commercially valuable ones are so common or widespread now. A few have disappeared completely from the modern Firth, like the twaite shad and the sea lamprey, or have always been occasional (a swordfish was stranded in Alloa in 1826, and another became entangled in lobster pots there in 2009). Many of them were tiny creatures, like sticklebacks and blennies. One was enormous: the conger eel could grow ten feet long and weigh 130 lb. A number were well worth eating, and his essay contained information on their culinary and commercial importance. Not least in the estuary above Queensferry, there was in the early nineteenth century a great diversity, ranging from the obviously oceanic like herring, sprat and small cod frequenting banks and shores, to those which visited the rivers to breed, like common eels and river lampreys, and above all, from the commercial point of view, salmon and sea trout.

Normally, different forms of fishing predominated in the estuary compared to lower down the Firth, partly because the big demersal and pelagic fish were not numerous enough to attract the concentration of effort and capital characteristic of the outer Firth. The only exception to this was for a few years around 1800, when the winter herring swarmed off Queensferry and attracted boats from far and near (see Chapter 2). Normally, the local fishermen did not venture far, and they did not play a dominant role in the small towns in which they lived, such as Alloa, South Queensferry, Limekilns or Bo'ness. At one time in the seventeenth century the town of Stirling had planned a special Fisher Row, which came to very little.[2] The main implements of fishermen here were nets and traps operated from the shore, or small boats using relatively modest nets, and some using the sma' lines. Generally speaking, they had less power to influence the abundance of the fishes than the fishermen lower down, though when fishermen from Newhaven joined them they brought more capacity from outside, especially to the sprat fishing. What seriously devastated the ecology of the estuary from the nineteenth century onwards was not overfishing but pollution, a topic that we shall discuss more fully in the next chapter.

By the nineteenth century the main purpose of the nets and traps operated from the shore was to catch salmon, as will be discussed below

(Section 3). But some, at least, of the earlier traps had a more general purpose. A characteristic form was described in the *Statistical Account* for Alloa in 1791:

> They erect what are called *yares*, a sort of scaffold projecting into the water, upon which they build little huts to protect them from the weather: from these scaffolds they let down, at certain times of the tide, their nets, and are often very successful in taking the smaller fish, such as herrings, *garvies* or sprats, *sparlings* or smelts, small whitings, haddocks, sea trouts and eels. In this manner salmon are sometimes caught as well as Congo eels, sturgeon, soals, turbots, cod, gurnet or piper, and skate. Sometimes at the end of September, there comes a vast shoal of fish called *gandanooks* or Egyptian herring. They have a faint resemblance of the mackerel, but with a long sharp bill like a snipe.

Today we call gandanooks, saury or skipper fish. The commonest form of yare (or yair) at this time consisted of a leading fence made of wattle, built between the high- and low-water marks, directing the fish towards the platform from which nets could be lowered. There were also other and older traps, sometimes also called yares, which involved an enclosure built of stone, or of wattle palisades, and perhaps also with leading lines of wattle, where the fish could be caught on a falling tide as they tried to move out. The footings of a number survive as archaeological remains, of which the most impressive is a stone-built D-shaped structure, visible above the encroaching mud in Torry Bay (Fig. 6.2).[3] Its scale would have demanded substantial resources to construct it, suggesting that it could date from medieval monastic times, or from the ambition of a wealthy landowner like Sir George Bruce of Culross, who late in the sixteenth century dug the underwater mine nearby.

The estuary has not everywhere stayed the same. The trap at Torry Bay is covered today with several feet of thick mud, but when it was in operation this part of the north shore of the Firth of Forth was famous for the beauty of its 'clean white sands', still remembered with bitter regret at the end of the nineteenth century.[4] The present mud is presumably the consequence of the drainage of the mosses in the Carse of Stirling, begun at Blairdrummond by Lord Kames in 1766 but carried on by his descendants and neighbours

Fig. 6.2 – Remains of a fish trap near Culross, once among 'clean white sands' before peat from agricultural reclamation was floated down from Stirling and changed them into mud. Note the scale of the footings of the stone trap.

with still greater energy until the 1860s. Vast quantities of peat, possibly millions of cubic yards, were thrown into drainage channels and allowed to drift down the estuary. H.M. Cadell in 1913 recalled seeing lumps of peat still littering the shore at Bo'ness, and an Admiralty chart reproduced in 1883 marked 'drifted peat' in Torry Bay and near Longannet.[5] The Earl of Dundonnell salvaged some when it came ashore, to use as agricultural fertiliser on his own land, but most of it turned to mud or sank to the bottom of the river. The damage to fixed traps of all kinds was apparently considerable and was certainly resented at the time. The *New Statistical Account* of 1845 for Tulliallan (effectively Kincardine) attributed the decline of fishing there to the floated peat, and when William Cobbett came to Torry in 1832 he found the fishers' houses along the shore poor and derelict, though he did not say why.[6] Yet it cannot have inflicted lasting damage to the salmon run up the river, or much more would have been heard about it later on, and it would not have been possible to have had the thriving netting business that was still being carried on round Alloa and Kincardine at the end of the century.

Just how much damage the peat also did to fish spawning beds within the estuary and how extensive the deposits were are largely unknown. The most alarming account of this was given in evidence to the Royal

Commission on Sea Fisheries in 1866, by John Anderson, fishmonger of Edinburgh, who had extensive interests in the salmon fishing and other activities. He described the peat coming down the River Forth in large lumps, clogging the nets, ebbing and flowing with the tide and gradually dissolving and sinking: 'some banks are eight or nine feet deep in solid sludge, and that covers the banks where the garvies and herrings used to spawn'. He said it came down as far as Queensferry. Anderson, however, was not the most convincing witness: garvies (sprats) never spawned in the Firth of Forth as he said they did, he described damage to oyster beds belonging to Lord Auckland that cannot be traced and he declared that there had been an Act of 1856 to stop it, which also cannot be traced, accusing 'higher powers' of obstructing his efforts to get it enforced. But at least his testimony indicates the wrath that some felt about the pollution from the carse, and while his statements may be exaggerated there is no reason to think that his account is completely wrong.[7] The Scottish Fishery Board's survey of fishing grounds within the Firth of Forth in 1890 noted that two or three grounds once used by spawning herrings had disappeared, and that a good gravel bank off Charlestown used by codling had become barren: they thought the cause, at least of the latter, was 'either the dredgings from Bo'ness or the muddy silt of the Firth'. When Thomas Fulton in 1895 used the *Garland* to examine the bed of the estuary off Charlestown and Queensferry, in his investigations into the decline of the Edinburgh oysters on behalf of the Scottish Fisheries Board, he found the bottom 'duffy', consisting of lumps of hardened mud with old oyster shells (but of unknown date), above Queensferry. He concluded that this was consistent with Anderson's account.[8]

In the middle and later nineteenth century, after the older yares and stone traps on the coast had mostly fallen out of use, the main prey of fishermen in the estuary came to be sprats, and salmon and sea trout, with sparling as a minor target species. It is with their exploitation that this chapter is particularly concerned.

2. Sprat Fishing

Sprats are shoaling, pelagic fish that breed in the Atlantic and North Sea in summer, but younger fish move in enormous numbers into sheltered firths and estuaries around the Scottish coast in winter. They are closely related to the herring, although very much smaller, typically three–six inches long, and much oilier. In the nineteenth century they entered the Firth of Forth in immense but very variable numbers, mainly between November and February. They were known locally as 'garvies', and the island of Inchgarvie, today beneath the Forth railway bridge, marks the spot around which they were usually most plentiful. An investigation in 1861 described them as being numerous west of Queensferry, occasionally common as high as Kincardine and Alloa, and in some years going as far east as Inverkeithing or even towards Aberdour and Burntisland, 'but this is rare'.[9] In some years they went right up to Stirling, in others they were caught off Inchkeith, and they could be so numerous in Granton and Leith as to be a nuisance in the harbours. After the Second World War, when echo-sounding and sonar made the location of shoals moving up the Firth easier, they were also caught in large quantities off Inchkeith and even Kirkcaldy. They still occur in substantial numbers in the Firth of Forth, swarming in some years, almost absent in others.

Fishing for sprats had always been part of the life of the estuary, but around 1827 or a little later, a new way of netting them began, probably devised at Newhaven, involving several boats co-operating to encircle a shoal with a ring-net (also described as a 'seine') and hauling them in together. When this fishery really got underway around 1836, there were a number of things which made it violently disliked by the established herring fishers lower down the Firth, and posed problems of regulation. First, it was not scientifically clear whether the sprat was a different species from the herring – perhaps these were young herrings, and the nets might be illegal, as they were below the mesh-size which was permitted for herring fishing. Even if they were indeed a different species, if herring shoaled with sprats, these nets would still be illegal, as they would take herring as by-catch. Then, if the sprat fishers caught great quantities of immature herring, the viability of the whole herring fishery even as far as east Fife and Dunbar might be put at risk if young fish that wintered high up in the Firth were being killed before they could reproduce. Lastly, the way the sprat fishers operated at sea got in the way of the traditional herring fishers, and they were accused of deliberately or carelessly damaging their drift-nets. Captain Samuel Macdonald, commander of the fisheries cutter *Princess Royal*, did not like the sprat fishermen at all: 'there is only a poor class of men engaged in seining in the Firth of Forth', he said in 1866, contrasting them with the herring fishers of east Fife, 'they are not genuine fishermen in fact'.[10]

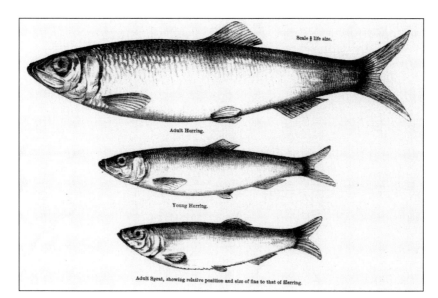

Fig. 6.3 – Herrings and sprats: the differences between a sprat and a young herring was long disputed – were they the same species or not?

The question of the identity of the sprat took a surprisingly long time to resolve. In 1836, the Fisheries Board employed the Edinburgh anatomist Robert Knox to investigate, undeterred by his notorious reputation after the trial of Burke and Hare in 1829, when he had been found to have purchased murdered corpses for his dissection table. Knox discovered small but consistent differences between sprats and young herrings. Perhaps the public had no more faith in him examining fish than they had in him cutting up people, because in 1845 the Board called a second expert, James Wilson, 'the well known naturalist', to demonstrate to them again the differences between sprat and herring. Nevertheless, scepticism continued, and as late as 1865 James Bartram suspected that they might be the same species and that to fish for sprats was 'killing of the goose for the sake of the golden eggs'. The question was raised again before the Buckland Commission in 1877 and only definitively settled in 1883, when the Scottish Fisheries Board published a paper on the differences, which were mainly in the relative positioning of the dorsal fin and in serrations along the belly (Fig. 6.3).[11]

Even if they were different, how many herring swam with the sprats? Opinions on this also varied, with the sprat fishermen maintaining that they seldom mixed, and that the difference between a shoal of herring and a shoal of sprats could be detected simply by putting an oar in the water and judging the resistance, an observation which serves to demonstrate how thick the shoals were. They did not wish a mixed catch, which fetched less in the market. Herring fishers maintained, on the other hand (for example to the Buckland Commission) that most of the 'sprat' catch was sometimes herring.[12]

Scientists initially argued that herrings made up only a trifling percentage of sprat shoals, though exceptionally bad weather or spring tides might mix them up. Between 1883 and 1885, Professor James Cossar Ewart of Edinburgh University and a collaborator purported to show that normally herring made up less than 1 per cent of sprat shoals. It was established empirically later, however, that this figure could rise much higher. By 1905, it was put as varying between 1 and 20 per cent, and in 1965 it was found that a quarter of the entire 'sprat' catch that year was young herring.[13] Possibly by then, echo-sounder and sonar, when used to identify sprats in the water, were considerably less discriminating than the oar and other traditional methods had been in 1860. Sprat trawling in the Firth of Forth was finally banned by the European Union in the 1980s, because of this perceived danger to young herring, and that ban remains in force despite a campaign by Fife fishermen between 2005 and 2009 to have it lifted.[14] The debate then was much the same as it has always been. The scientists said there was too great a risk of a substantial herring by-catch, and the fishermen said there was not.

So sprat fishing was highly controversial from the start. The Fisheries Board first put a stop to it with the help of a gunboat in 1837, when both herrings and sprats were numerous in the Firth at the same time, and groups of rival fishermen came into violent conflict.[15] The biggest row, however, reached its climax in 1859 and 1860, and the evidence at subsequent enquiries throws much light on the character and organisation of the fishery at this time.[16] An Act of 1851 had confirmed the minimum mesh for a herring net at one inch, and gave the Board new powers to confiscate offending nets – a sprat net had a half-inch mesh. A new law became necessary because the Court of Session in 1842 had ruled at that time that it was an unreasonable interpretation of a law of 1809 to confiscate sprat nets when so few herring were caught in them.[17]

In 1859, the Board moved to confiscate the offending nets, and unwittingly found itself fighting some of the poorest members of the fishing community. The cost of setting up a sprat boat was only £20, with £6–10 for the nets. The fishery supported 40 boats from Newhaven, 16 from South Queensferry, five from Burntisland, three each from North Queensferry, Limekilns and Bo'ness, and two from Inverkeithing, and each boat had a crew of four. About 160 out of the 350–400 fishermen of Newhaven depended entirely on the sprats, either the poorest or those too young to have accumulated capital:

> The stoppage of the sprat fishing has caused much suffering. Their small savings are exhausted. The credits at the shops are small and are now being closed. Many of the men whose occupations require a nourishing diet, have ceased to live on beef, broth and potatoes, and take stirabout instead: while money for fuel has become so scanty that a subscription has been established to aid them.[18]

At South Queensferry, about 40 fishermen and some 'common labourers', an eighth of the total population, went to the sprats: they had longer credits than the Newhaven men, but these were also nearly exhausted.

Faced with seizures and at their wits' end, the fishermen began to fish illegally at night, hiding their nets 'even in coal pits' by day. Public opinion in Edinburgh, not for the first or last time in respect to the Newhaven community, rallied to their side. In 1859, the men failed in a legal challenge to the Act, and paraded the streets opposite the Board's headquarters, demanding to send a deputation inside. This was refused, 'in consequence of their numbers and threatening appearance'. Disturbances threatened at South Queensferry. At Bo'ness, an attempt to seize the nets was foiled by fishermen armed with stones and aided by their women. The commander of the *Lizard*, called in to help, said he could not seize any nets without arming his boat, and would not be responsible for lives lost. The Board's guardships were pelted with stones and threatened with ramming: 'serious riots became imminent'. Then the Lord Advocate intervened, stopping the seizures, and sponsoring a new Act in 1861 which resulted in the establishment of a zone west of Queensferry where the sprat fishermen were allowed to use their small-mesh nets. Even this did not end the trouble, as two years later the sprats themselves did not migrate above Queensferry, but stopped and became plentiful immediately to the east, so the Board tacitly allowed them to be fished as far west as Inchkeith. Sprat fishermen entered Granton harbour in pursuit of the shoals, and the harbour authorities wanted to call in the navy to evict them, 'but the gunboats did not consider it expedient to interfere'. The Board was blamed on all sides for a situation not of its making.

After 1868, despite the general bonfire of legislation that followed the Royal Commission Report,[19] it remained illegal to fish with small nets except in the Firth above Queensferry. Resentment, however, did not stop with this decision. In 1870, Buckhaven herring fishers again petitioned against the sprat fishing, and again the Newhaven Society of Free Fishermen with their colleagues in South Queensferry leapt to its defence. In 1877 a letter to *The Scotsman* on 18 April disparaged the interests of the sprat fishermen compared to the herring industry as 'comparing a mole hill with Mount Blanc', and an unsuccessful campaign was run again that year by herring curers and by Buckhaven fishermen before an unsympathetic Royal Commission on the Herring Fishing.[20]

The last big inquiry was carried out by the Scottish Fisheries Board in 1908 in response to a demand by fishermen in Fife and East Lothian to halt sprat fishing altogether, even above Queensferry, on the usual grounds that it destroyed immense quantities of immature herring wintering at the head of the Firth that would later breed off the May. This was firmly rejected by the Board, who argued that it had been repeatedly shown that very few herring occurred in sprat shoals, and that the business was essential to Newhaven, where the fishing fleet had shrunk while that of the east Fife ports had grown. They also rejected allegations that sprat fishermen deliberately and illicitly went after herring. Finally, they tartly observed that the Anstruther men actually took the mature herrings on the point of spawning, which 'as has been pointed out by high scientific authority, is a far greater interference with the reproductive power of nature than the capture of halfling herrings'. In fact the wintering immature herring in the Firth of Forth came from the west coast and had nothing to do with those that bred in and around the Firth, but no one knew that at the time.[21]

One of the characteristics of sprats is to be incredibly abundant inshore in some years, and almost absent in others. Thus in 1866, 'the shoals were so numerous and dense that little search or skill was required to catch them; the boats had only to go upon the fishing grounds, and cast and haul their nets, when they were at once filled to the gunwale'.[22] After a run of good years, however, the fishing might fail entirely and the sprat fishers then had to seek other employment, if they could find it. This occurred in 1882, 1897, 1905, 1910, 1922, 1930, 1935, 1960 and so on. Such fluctuations were reflected in prices. In 1836 prices had been 2s a cran, but next year rose to 15s; even in one year (1895) it swung from 5s to 21s. In years of glut, merchants would not buy any more and sprats were sold to farmers to be spread on the land as manure. Professor M'Intosh recalled an incident when 17 railway wagonloads were sent from the Forth to farms in the Carse of Gowrie: like many Victorians, he found such waste shocking and thought 'the sprat should be preserved for human consumption'.[23]

As with other fish, the railway widened the market. A small number of Firth of Forth sprats ended up as whitebait in London restaurants, but primarily they were the food of the poor, described in 1838 as 'a cheap and agreeable food' in Edinburgh, and in 1867 as 'very extensively used by the working class and much prized'.[24] A letter to *The Scotsman* of 29 December 1860 from South Queensferry, described how 10,000–15,000 tons had been landed the previous year and distributed by cart to Edinburgh and Glasgow, but 'chiefly [by] the railways, which remove the great bulk of the fish to London, Birmingham and Manchester and other inland towns, where the fresh fish are much esteemed and fetch a high price'. Yet by the 1890s this was no longer the case: tastes had apparently changed, and both demand and prices in Britain were low.

This altered again after the start of the new century. Canning sprats, 'after the manner of sardines', had been referred to as early as 1867. But it only became significant later, when Scandinavians became involved. In 1905, a Norwegian fish curer employed Newhaven men to catch sprats on a fixed-price basis. Next year, Swedes entered the market, and in 1908 three Gothenburg merchants bought nearly the whole catch and exported 4,725 barrels from Leith to Sweden, half cured. The process was 'partially secret', but the intrigued officials of the Scottish Fisheries Board reported that the sprats were well washed and freed of any herrings that might have got accidentally mixed in; then they were placed on a large table where they were 'roused or mixed with a brown powder', thoroughly covered and barrelled. The powder was said to contain 13 ingredients, of which 'it is easy to see that salt, sugar, pepper, cinnamon and bay leaves form a part'. They received 'the finishing touches' in Sweden, where they were canned and marketed as anchovies (a quite different species) and 'some perhaps come back to this country as sardines'. By 1911 most of the catch at Newhaven was exported to Sweden and Norway in this way, with some going to Hamburg.[25]

For a few years after the First World War, at least until 1923, buyers from Norway and Sweden were operating in Newhaven again, though the small sprats caught at Kincardine and Alloa were only good enough for manure. However, the British then seem to have learned the tricks of the trade and in 1938, nine-tenths of the sprats were being sold to canning factories in

Fig. 6.4 – Newhaven sprat boats around 1950: a fair living was sometimes made from sprats, between the 1830s and the 1950s.

Dundee, Leeds, Colchester and Yarmouth, and supply was inadequate to demand. Then, following the Second World War, the public lost interest again, and sprats were little used for human consumption. They were sold for oil and fish meal, or were processed into pet food. The Newhaven fishermen always were the main catchers of sprats in the Firth of Forth (Fig. 6.4). Sometimes they imposed a quota on themselves in an attempt to keep up prices, but in the 1960s and 1970s the catch was marginally profitable and often came close to being junked. There was still a very minor outlet as whitebait, for those prepared to pay for fine dining.[26]

The sprat today cannot be trawled legally in the Firth of Forth, though it can be lawfully caught in the small boom-nets operated by a handful of fishermen at the head of the estuary. This prohibition is because of the need to protect herring shoals, but numbers of sprats in the North

Sea have been increasing again since the 1980s, probably because larger fish that predated them have been overfished. They sometimes still come into the Firth in incredible numbers, as in 2001, when an immense shoal, estimated at 15,000 tons of fish, rushed into the Royal Naval Dockyard at Rosyth when the gates were opened to allow the entry of HMS *Liverpool* for a refit. Here the fish were trapped, died from lack of oxygen, and rotted, making life so difficult for the navy that they had to call in nanobiological technology from California to get rid of the smell. The Firth of Forth has its wonders yet.[27]

3. Salmon and Sparling

The sprat was a little fish, usually regarded as of little note. By contrast, the estuary once had a great reputation for the queen of fish, the salmon. In 1521 John Major, listing the merits of his native land, declared the Firth of Forth to be 'abounding in salmon'. In 1656 the Englishman Richard Franck, visiting Stirling, declared that 'the abundance of salmon hereabouts in these parts is hardly to be credited'. He told the usual tall story (nowhere verified) about a law that prevented servants being fed salmon more than three times a week, and said that the price of sixpence per fish did not seem high to a southerner. In 1661, the French visitor, Jorevin de Rocheford, spoke of Dunbar as 'famous for its great fishery of herrings and salmon', presumably netted at sea on their way to the river.[28] These were all words of praise but not hard statistics, and in particular they do not allow us to compare one Scottish river with another. At the end of the nineteenth century, when for the first time we can make comparisons, the reputation of the Forth does not seem outstanding at all. The rateable value of the Forth salmon fishing was only £3,700, and it was far eclipsed on the east coast by the Tay (£19,100), the Tweed (£15,200) and the Dee (£17,200). Even the Spey and the North Esk were worth much more.[29]

It is likely, however, that by then, through pollution and water abstraction, the Forth catchment had already begun to lose some of its pristine appeal to salmon and sea trout. Nevertheless, today it still has important populations which, every year, in a miracle of migration and instinct, swim down from the Arctic seas and enter the River Forth and its tributaries to breed in the freshwater gravels. Systematic and reliable records of the take were not kept before 1952, but the average total annual catch in the late 1950s was over

5,000 salmon, and in the early 1980s over 4,000 sea trout. In the first five years of the twenty-first century, this had fallen to only 2,000 salmon and 1,000 sea trout, though there have been better years since. However, before 1990, more were trapped commercially by various kinds of net, either in the Firth or in the river, than by anglers. Now, fly fishermen on the fresh waters are almost the only human predators left. Catch numbers are obviously related to the size of the salmon population but for many reasons may not be a very close reflection.[30]

We have already seen how, traditionally, there were traps, 'yares', in the Firth that caught salmon, but were not specifically designed to target only this species. However, there were also other types of traps, eloquent of human ingenuity, which were indeed aimed primarily at salmon and sea trout. These included Craigforth cruive, blocking the River Forth two miles above Stirling and close to the tidal limit. It consisted of a stone dam built along a natural rock shelf, with one gap in it, filled by a wooden box to catch the fish. This was supposed to be opened for some of the fish to pass up the stream, at least on Sundays. This famous trap had existed since the Middle Ages, and was often a bone of contention between the owner and other river proprietors. In 1898 it was described by the Scottish Fisheries Board as eight or nine feet high when the tide was out, with the box six feet by 13 feet. The Board considered it of doubtful legality. It was bought out by the angling proprietors upstream in 1904, and then deliberately allowed to go to ruin.[31]

Then there were nets on the 16-mile tidal stretch of the Forth between Craigforth and Alloa. Many of them involved catching salmon by their gills in a drift-net, with the help of a small coble. Sir Robert Sibbald in 1707 said that 'many of the gentry get salmonds in their powes' and the *Scots Magazine* in 1746 spoke of 'pock, stop and herrie-water nets which they should find people making use of above the pow of Alloa'.[32] A device, popular in the nineteenth century, was the 'hang net', strung across the river from side to side at slack water and left for three-quarters of an hour, but attended by a boat that would gaff any fish caught in them. Other forms of net did not entangle but encircled the fish. Prime among these was the net-and-coble, where a 'sweep net' was slung between a boat in mid-stream and men who walked along the shore.[33]

Finally, there were two developments of the old yare, designed more for the estuary, and for the coast beyond. These diverted fish swimming near the shore by lines of net (not wattle or stone as in the old yares), leading

Fig. 6.5 – Salmon stake nets, 1983. The long net in the foreground is a 'leader' that diverts the fish into the 'fish court' at the seaward end, where they are held.

into traps a little further out. First, there were stake nets (or 'fly nets') introduced around the beginning of the nineteenth century to be used on soft and sandy shores such as Largo Bay (Figs 6.5 and 6.6). The *New Statistical Account* mentioned them on the south shore at Bo'ness, Abercorn, Queensferry and Inveresk (Musselburgh). They consisted of rows of poles up to 800 yards long, erected between high- and low-water marks, fastened together with ropes from which curtains of nets were suspended; these were set at an angle to the shore so as to form leaders towards other enclosures of netting, or 'courts', with entrances designed to admit the salmon but so labyrinthine and protected by net bottoms and lids so that that they could not find their way out again. The nets were lifted on Sundays, and during the salmon close season, which until 1824 ran from mid August until the end of November, afterwards extended to mid February. Allied to them, but more suitable for rocky or deeper shores like those off Crail and Kincraig Point, were 'bag nets', suspended from buoys on the surface of the sea, and held in place by anchors, but also attached to the shore by leaders that directed fish towards them. They were fished from flat-bottomed cobles but were legally required to be within 1,300 metres of the shore.[34]

Of course, other fish than salmon and sea trout, both common and rare, often found their way into these traps. Parnell mentioned in 1838 that

Fig. 6.6 – Salmon stake nets drying on the beach at Kinghorn c.1930.

lumpfish were commonly caught in the stake nets, and though not saleable in the market, they were boiled by the fishermen for their own food. Today lumpfish are more esteemed when manufactured into scampi. Parnell also said that about once every three years the fishermen caught a sturgeon in the nets at Musselburgh or Queensferry, and that it had also occurred recently at Alloa, which suggests that sturgeon may still have attempted to breed in Scottish rivers at this time.[35]

All these forms of trap, except net-and-coble, were described legally as 'fixed engines' (in the old sense of fixed devices). Following a notable lawsuit in 1900, all such were declared to be illegal in the river above Alloa, which left just the net-and-coble for use in the tidal river, and the stake net and bag net on the coast. The Scottish Fisheries Board in 1906 made a count of net-and-coble 'stations', and found 30 between Craigforth and Alloa, using 37 'shots' and 63 'hauling grounds'. In 1899 they had made a similar count of the coastal nets, finding 175 bag nets and 26 stake nets in the Firth, and concluding that this was 'the most productive and important factor in the purely commercial aspects of salmon fishing'.[36] The most valuable of these were in Fife beyond the estuary, as in Largo Bay. Today, a combination of competition from salmon farms and a policy by the anglers to buy up the net operators has left the rod fishers without a rival. They have the salmon and sea trout to themselves.

Poaching salmon was also a pastime that irritated proprietors, though perhaps it did not destroy as many fish as the lairds and their keepers feared. Some of it was done in the streams by cunning men just using their bare hands. On the Allan Water, for instance, W.L. Calderwood, inspector for the Scottish Fisheries Board in 1904, said that 'much snatching and stroke hauling of fish is reported'. On the coast it mainly took the form of illegal netting. In 1907, 21 men were arrested 'for drift or hang net fishing in the estuary, five of them for breach of interdict'. The poachers probably sold their catch to hotels and fishmongers, and sometimes had a success that eluded the legitimate fisher. In either 1901 or 1902, a great black salmon, heavily infected with sea lice and the ugliest the catcher had ever seen, was landed at the tidal junction of the Forth and the Devon. It weighed 103 lb, and Calderwood privately interviewed the 'quiet, self-reliant man' who had landed it with two friends:

> No visible record of the fish was retained, since the possession of the fish was fraught with a certain amount of danger to the captors. I have, however no reason to doubt the guarded statement of my informant, who is well known to Messrs. Anderson, and personally believe that in reporting the matter I am recording the greatest known weight of any British salmon.[37]

There is no doubt, however, that the deadliest enemy of the salmon and of the salmon fisher was neither the trap nor the poacher, but the water polluter, abstracter and obstructer. That story, however, belongs to Chapter 7. By way of a preliminary, however, and to show what pollution could do to another fish population in the Forth estuary, let us briefly consider the history of the sparling.

The sparling (also known as spirling, sperling and smelt) is a little salmonoid fish, usually six–ten inches long, that breeds at the mouth of the River Forth and remains within the estuary all year.[38] They are also found on the Tay, and on the Cree in Dumfriesshire, but since the nineteenth century have vanished from at least 12 other Scottish tidal rivers where they were once common. At high spring tides between February and April they come upstream to spawn a little below the tidal limit, on rocks, gravel and vegetation. Good quality water is essential to their successful breeding. Sparling were referred to by the poet William Dunbar early in the sixteenth century – he begged the king to come home to Edinburgh from Stirling, 'quhair fisch to sell is nane but spirling'. In 1707, Robert Sibbald said that they were found 'in great quantities near Stirling'. They were also well known to Richard Parnell in early Victorian times. He described small ones as 'taken in great numbers' near Alloa from November to January, when they were replaced by larger ones (presumably more mature) until March:

> In the month of March these fish ascend the Forth in large shoals to deposit their spawn in the fresh water: this they shed in immense quantities about two miles below Stirling Bridge, when at that time every stone, plank, and post, appear to be covered with their greenish-coloured ova.[39]

This run of breeding fish supported a local net fishery. Despite Dunbar's disparaging remark, sparling have a delicate flavour and a fine cucumber smell when very fresh, but to make the best of them in the days before refrigeration, they needed to be enjoyed within a few hours of being caught. Parnell called them 'much esteemed as a luxury for the table', adding that 'numbers are sent to the Edinburgh market where they receive a ready sale'. They were so abundant and cheap near the catching place that they were also a food for the poor round Stirling and Alloa. The fact that they were still plentiful in 1838 shows that any damage done by floating peat down from the mosses upstream had not extended to altering the chemistry of the water.

Sparling were still flourishing at the end of the nineteenth century, when the Scottish Fisheries Board in 1895 spoke of the fishing as 'a most remunerative branch of the industry for those involved in it'. This was a period, however, when water quality in the Forth was beginning to deteriorate, and only four years later the fishery was described as 'now of comparatively little importance'. For Alloa in 1900 it was nevertheless still the most important part of their fishing, pursued with pouch and drag nets, though apparently also followed as a cover for poaching: sparling fishers here were said to catch 'numbers of salmon, and to keep them when not seen by the watchers'. Bo'ness and Kincardine were the only other local ports participating, and they also went after codling and sprats. Sparling catches for the whole of Scotland on the eve of the First World War seldom reached 500 cwt, most of that landed in the Forth, but prices in 1914 were eight times as high as for sprats, so even a small catch could be disproportionately worthwhile for the fishermen.[40]

In the interwar years, the description of sparling fishing as 'insignificant' was repeated again in 1920 and 1923, and volumes caught in the whole of Scotland in the 1930s, and specifically in the Forth, were half what they had been before the war. The rivers continued to get more polluted. There was a slight increase in catch after the Second World War, followed by a decline to nothing in the mid 1950s, another small resurgence in the 1960s and then apparent extinction in the 1970s. Its collapse was attributed by scientists to overfishing, silting in the spawning grounds and particularly to gross pollution leading to a shortage of dissolved oxygen in the water.[41] The very worst episodes of oxygen depletion on the River Forth occurred in the post-war period, as we shall see in the next chapter.

The river began dramatically to improve during the 1980s as modern pollution control came into play, but there was no immediate sign of the reappearance of the sparling, which was assumed to have gone forever. In view of improved water quality, scientists began to consider a trial reintroduction from the Tay, but before they could act, in 1989, a single specimen was recorded again in the Forth, and then more. By 1994, thousands were appearing at the screens of Longannet power station, and now it is a common species once more in the estuary. A few are captured again in the boom-nets of the Kincardine sprat fishers, but most are sold for pike-bait, as gourmets seem to have forgotten how tasty they are.[42] But the return of a species thought locally extinct is a great cause for celebration. As we shall see in the next chapter, it is possible in environmental history for things to go from bad to worse, and then to go to good again.

CHAPTER 7
POLLUTION

1. Inventing Pollution

Pollution, either of the rivers or of the sea, was not a serious issue in the catchment of the Firth of Forth before the nineteenth century. Of course, Edinburgh famously stank. Even before 1700, some 40,000 people lived in an area about a mile long and half a mile wide: this made it, along with Norwich and Bristol, the second most populous city of Britain. Though small compared with London, it must have been more closely built than any other. Some of its tenements went up six or seven storeys on the side facing

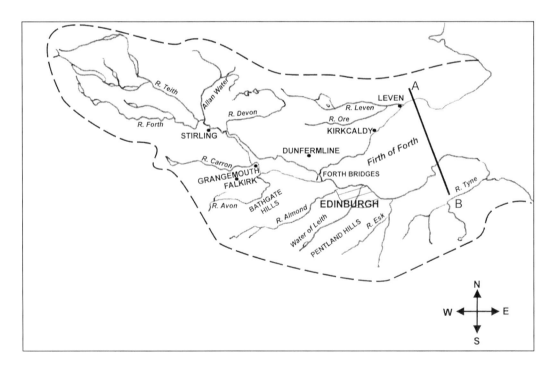

Fig. 7.1 – The Forth Catchment Area. Line A–B is the approximate limit of fresh water influence on seaweed distribution.

the High Street, and ten storeys or more at the back where the ground sloped away, and the inhabitants either deposited their refuse in middens at the foot of the tenement stair or threw the sewage out of their windows, to the ancient cry of 'gardyloo'. Here it flowed down the runnels into the Nor' Loch on the site of what was to become Princes' Street Gardens, or down the open sewers to the foot of the Canongate from the High Street, and thence into burns flowing towards the sea. Dr Johnson, walking slowly arm-in-arm with James Boswell up that street on an August night in 1773, 'grumbled in my ear, *I smell you in the dark.*'

In fact, ever since the end of the seventeenth century, the city had had a well organised system of refuse collection which disposed of the human ordure to the farmers of the surrounding area: it did not prevent the smell, but it stopped the sewage accumulating. The farmers in a ring of parishes around Edinburgh used it, by the late eighteenth century, as a source of nitrogen to fertilise continuous crops of wheat in their fields, which the millers and baxters of the city turned into bread, which the inhabitants ate and recycled again. This is why the gardens of Morningside and Newington, then deep in the country, often contain fragments of eighteenth-century clay pipes, testimony to the addictions and relaxations of the citizens, and which got caught up with the rest of the detritus. The manure passed toll-free on the roads, and when in 1787 the turnpike operators tried to charge for it, the farmers organised a boycott and left Edinburgh to cope as best it could. The price rose, according to the minister of Duddingston in the *Statistical Account*, from as low as 2d a load in the 1730s to as high as 1s 6d ('in some circumstances') in the 1790s, at which point the farmers formed cartels to bargain with the city to keep the price down. And so it was throughout the towns and villages of the Firth of Forth: middens were a valuable property sold to the highest bidder, and generally the waters ran clean and sweet.

This began to change under two related forces. One was the rise, along the banks of the rivers, of industrial plants which carried out some kind of polluting activity. The worst of these to start with were distilleries, breweries, paper-mills, tan works and textile mills. To these were added, in the later nineteenth and early twentieth centuries, coal washing plants, paraffin factories, especially in West Lothian, heavy engineering, especially around Falkirk, and from 1924, the petrochemical refining complex at Grangemouth. We think of Edinburgh itself mainly as a service centre – for banking, insurance, retail, law, religion and education – but there was a whole range of manufacturing industries that fed the needs of these businesses (like printing and book-binding to service the banks, schools, university, churches and the general public, or the manufacture of fine biscuits to go with Melrose's afternoon tea). By 1900, Edinburgh was the second manufacturing city in Scotland, and most of the trades flushed their untreated waste down its drains.[1]

But to pour waste down the drain, you need drains first, and they came as a result of demography. The population grew from about 40,000 in 1700 to 90,000 by 1800, and, including Leith, to 165,000 by 1841. By the end of the nineteenth century, the numbers living within the modern bounds of the city had reached 413,000. Growth slowed within the city, but not necessarily in the area round about, and by 2001 Edinburgh's population was 448,000. What had been viable as a way of disposing of waste in the eighteenth century became increasingly unworkable. As early as 1830 the countryside could hardly absorb the whole surplus any more, and in the following decades it was becoming more unwilling even to try. Fertilisers began to come in clean bags, nitrogen as imported guano, and phosphorus as manufactured bone-meal – why should farmers bother with human sewage unless it was made attractive in other ways? In rural areas, if it was cheap enough, it could still sometimes be sold or given away, at least until the end of the century, and sometimes beyond. In 1872, it was noted that the owners of some factories near Markinch in Fife still disposed of the sewage of their workforce, mixed with rag dust and ashes, to the local farmers, and even as late as 1935, in Stirlingshire, pollution of the River Avon from nearby mining villages was described as negligible, because 'the sanitary provisions are mainly middens or pail privies, the contents of which are periodically removed for manure'.[2] But for towns, especially Edinburgh, the expense of collection and the volume of production made such ancient and smelly frugality impracticable.

As the problem spiralled out of control, the traditional urban wells that supplied most of the city, not only proved inadequate for the greatly swollen numbers, but they became contaminated, and a wave of epidemic diseases struck, of a severity not seen since the bubonic plagues of the seventeenth century. Cholera arrived from India and created mayhem in 1832, 1848 and 1853 (there was a milder outbreak in 1866): it proved to be a disease carried in polluted well-water. Typhus struck especially in 1847, born by lice on unwashed clothes and bodies. Typhoid was a more constant killer, also water-borne. It was not a problem peculiar to Edinburgh or to Scotland, or even to the biggest towns, but Edwin Chadwick's great *Sanitary Report*

English Ideas of Scotland

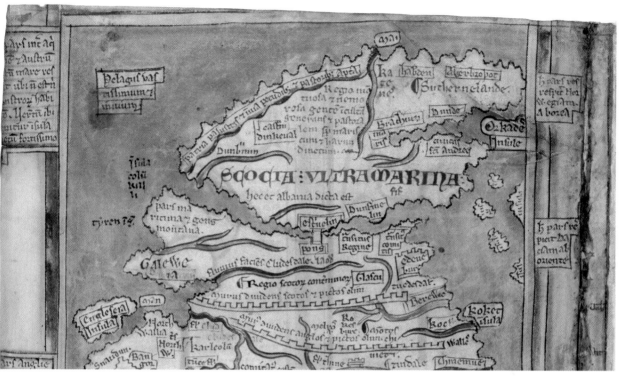

Plate 1. Matthew Paris's map of Britain, c.1250. Scotland begins somewhere between the two Roman walls and almost splits in two at Stirling Bridge. (© The British Museum Board: Cotton MS Claudius D.vi)

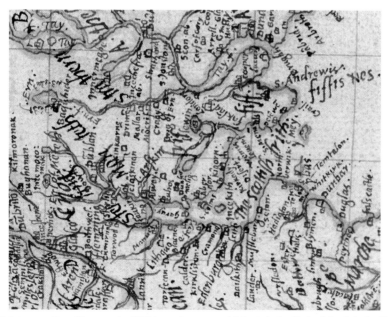

Plate 2. Laurence Nowell's map of c.1562 is much more sophisticated: the Firth of Forth is the 'Scottish Frith', with towns and islands that we can readily identify today. (© The British Museum Board: Cotton MS Domitian xviii)

A Firth of Contrasts

Plate 3. The inner Firth from the air in 2005: still only one crossing at Kincardine. Notice the land reclaimed from the mudflats by industrial ash, on the left beyond Longannet power station. (© Royal Commission on the Ancient and Historical Monuments of Scotland (Aerial Photography Collection). Licensor www.rcahms.gov.uk)

Plate 4. The Bass Rock in the outer Firth, seen from the air around 1990, with the colony of gannets beginning to encroach on the top surface. See also Plate 13 and the figures in Chapter 9. (© Colin Martin)

Sea Mammals in the Firth

Plate 5. A young grey seal swimming above the dense underwater kelp forests of laminaria, characteristic of the outer Firth and the Berwickshire coast. (© Alex Mustard/2020VISION/naturepl.com)

Plate 6. Blue whale, the biggest mammal ever to have lived on earth, washed ashore near Longniddry in 1869, drawn by Sam Bough. Only an occasional visitor, it was nevertheless first scientifically described by Sir Robert Sibbald from a specimen washed ashore in the Firth in 1692. (© National Museums of Scotland. Licensor www.scran.ac.uk)

The Changing Nature of Boats

Plate 7. Newhaven shore in the later nineteenth century, by Keeley Halswelle. The little yawls drawn up on the beach are characteristic of the small boats that caught oysters and fished locally. The figure coming down the beach with a train of ponies is bringing back mussels for baiting the sma' lines. (© Bridgeman Art Library)

Plate 8. Steam drifters in Anstruther in 1920, by James Gilmour. (© Scottish Fisheries Museum. Licensor www.scran.ac.uk)

The Triumphant Days of Big Sail

Plate 9. The *Reaper*, the only surviving Fifie, pride of the Scottish Fisheries Museum, Anstruther, one of the finest sailing drifters ever built. Sixty-one tons and 70 feet long, it was launched near Fraserburgh in 1902. (© Scottish Fisheries Museum Boat Club, courtesy of the crew of the Isabella Fortuna and the Wick Society)

Plate 10. The banner of the Cockenzie fishermen from c.1900 flaunts a Fifie to show their capacity, and a suitable motto to show their respectability. (© Port Seton Fishermen's Association. Licensor www.scran.ac.uk)

Reclamation and Back Again

Plate 11. Skinflats as it was, a reclamation from the sea for agriculture. (© Toby Wilson/Royal Society for the Protection of Birds)

Plate 12. Skinflats today, with the tide and the habitat for birds restored. (© Royal Society for the Protection of Birds)

The Bass and the Reefs

Plate 13. The Bass Rock in the late seventeenth century, by an unknown artist. Notice the grazed sloping top and the gannets around the cliffs. Compare with Plate 4 and the figures in Chapter 9. (© Sir Hew Hamilton Dalrymple and the Scottish Seabird Centre)

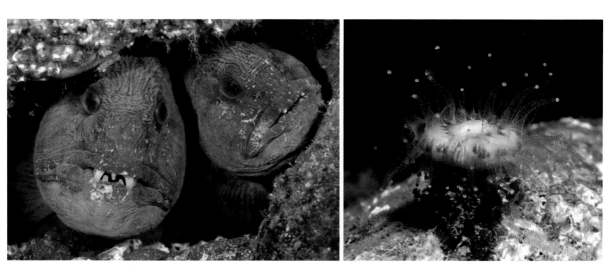

Plate 14. The underwater fauna of the Firth, especially off the Isle of May and St Abb's Head, is a mixture of north and south: left, wolf fish, characteristic of northern British waters. Right, Devonshire cup coral, characteristic of southern ones. (© J.J. Greenfield)

Iconic Seabirds

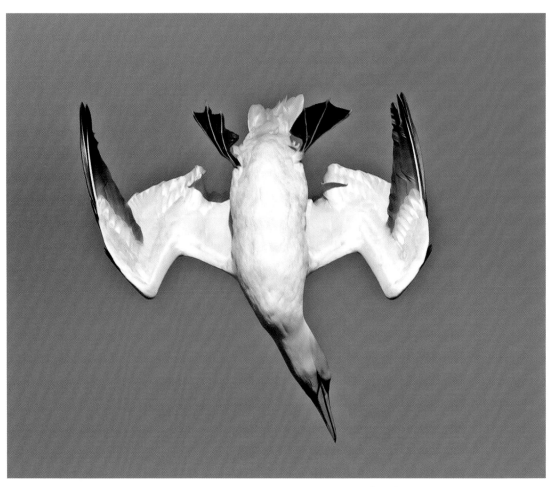

Plate 15. Diving Gannets from the Bass. (© John Anderson)

Plate 16. Puffins on the Isle of May. Left, with sandeels, their staple diet. (© Scottish Natural Heritage) Right, with pipefish, 2006, probably a fatal catch for the chicks. (© Harry Scott)

to Parliament in 1842 nevertheless concluded that the Scottish cities were the worst afflicted by 'fever' of any in Britain. His proposed solution was to close the urban wells, to pipe water from some pure and distant source into the homes in sufficient volume to allow for washing and flushing, and to make drains and sewers to carry human and industrial waste safely away.[3]

The town council and the ratepayers were shaken by the expense of Chadwick's notions. Perhaps, said the civic leaders, at the opening meeting of what became the Scottish Rights of Way Society in 1844, they could find an alternative, like providing the working classes with easier access to the countryside, which should make them healthier without any need for expensive sewers?[4] But they could not ignore much longer the logic of epidemical death. In 1848, an Edinburgh Police Act gave the local authority extensive sanitary powers, and soon a water company was supplying houses throughout Edinburgh and its satellite burghs of Leith and Portobello. The company tapped some 250 springs in the Pentlands, and, from 1879, Portmore Loch and Gladhouse Reservoir in the Moorfoots were developed as well. Provision per head per day went up from 12 gallons in 1844 to 42 gallons in 1886 despite the growing population. Still it proved too little, and around 1900 a new reservoir was built for Edinburgh at Tulla in the Tweed catchment, feeding water to the city down an aqueduct 35 miles long, followed in the 1960s by Fruid nearby, and subsequently by Megget Water also in the Tweed catchment, and West Water in the Pentlands.

From around the middle of the nineteenth century, therefore, the villas and tenements were attached to a network of drains and sewers built over some 30 years, and the need for a volume of water increased in the 1880s once the public took to the new flushable 'water closets', designed by Thomas Crapper and manufactured by firms like Shanks of Barrhead. The sewers, however, were at first designed only to reach the nearest natural watercourse. Other household refuse (such as the ashes from the coal fires) was collected by the council and consigned to landfill, and what could be burned was eventually put into incinerators, the first destructor, at Powderhall, being built in 1893.

Initially, the sewers on the west side of Edinburgh mostly discharged into the Water of Leith at Dalry, and quickly became the subject of complaint, as they fouled to an intolerable degree the mills that lined its banks and the harbour at its mouth. The situation was aggravated by a weakened flow in the river, caused by the abstraction from the Pentlands of water which would otherwise have found its way into the stream, so additional water had to be provided from reservoirs as compensation. New legislation in 1854, 1864 and 1889 gave the city powers to improve the situation. An interceptor sewer was built (and eventually extended as far as Balerno 20 miles upstream), to spare the river, and to take the waste directly to a marine outfall just beyond the harbour, where it was sent raw into the Firth of Forth.

Elsewhere in the city, the sewers flowed into small open streams which discharged into 'irrigation meadows'. The largest of the streams was the only too well-named Foul Burn, which fed about 250 acres of meadow on the edge of the sea at Lochend and Craigentinny (Fig. 7.2). A smaller stream at Lochrin watered about 60 acres, part of which is today Murrayfield international rugby ground.[5] The Figgate Burn (also known as the Pow Burn and the Braid Burn) in 1837 watered 30 acres of coastal dune recently laid out as meadow, and extravagantly described in the *Farmers Magazine* as 'one of the most beneficial agricultural improvements ever undertaken'. Later, it seems to have run seawards without any intervening meadows, taking the sewage of the expanding south side around Newington, Morningside and Liberton. It was particularly revolting, and Portobello in 1882 successfully took Edinburgh to court to compel the city to build covered sewers and a culvert to carry the waste to the sea – also raw, of course, and a good deal of it washed up again on Portobello's famous sands.[6]

This system of irrigation meadows had originated on a small scale around 1760, and in 1842 Queen Victoria elected to stay with the Duke of Buccleuch at Dalkeith, apparently to avoid the smell in Holyrood Palace from a meadow at the foot of the Canongate. But the *Farmers Magazine* applauded the way that 'the rich stuff . . . has made sand hillocks produce riches far superior to anything of the kind in the kingdom or in any other country'. The utilitarian mind of Edwin Chadwick was equally impressed, and he gave irrigation meadows the blessing of his publicity. For the next three decades, Craigentinny was internationally cited as an example of how the new liquid waste regime of sanitary disposal could be made to pay from recycling. The meadows were let by public auction in patches of a quarter of an acre, and produced grass cut four to six times a year, fed to several thousand cows kept in private stalls across the city. The grass was worth £24–30 an acre per year around 1870, and the city's medical officers of health declared the milk produced from it to be as safe as any other.[7] The farmer who lived in the middle with his six children told the *British Medical Journal* in 1873 that 'I have never felt the smell or effluvia doing myself or my family any harm'. No doubt Craigentinny was also a wonderful site for

Opposite: Fig. 7.2 – Sewage irrigation meadows at Craigentinny, 1847. They cover the entire area of modern residential Craigentinny as well as the golf course and the site of the Eastern General Hospital. From a plan for a proposed branch railway from Jock's Lodge to Leith.

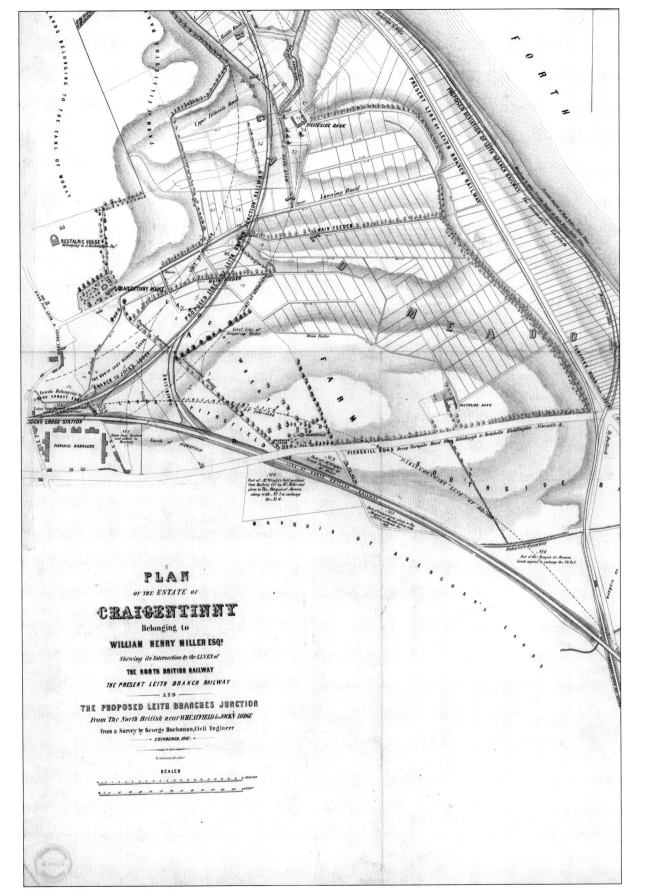

breeding and migrant birds, especially waders, ducks and gulls, but no early Scottish ornithologist seems to have been hardy enough to find out.

The downside, of course, was the disgusting smell that pervaded the vicinity of the burns and the meadows, and the fact that the filtering effect was evidently very slight, the sewage flowing out of the meadows and into the sea almost as nasty as it flowed in. By 1922, identification of the risk of tuberculosis in urban dairies had led to their closure. The value of the grass at Craigentinny fell, but the value of the land for development rose: the meadows were sold by the Earl of Moray for housing and for a golf course. The Foul Burn was covered over, and the entire sewage of Edinburgh from then until 1978 went into the sea with minimal filtration, through eight major outfalls spaced out along the shore.

Edinburgh was of course by far the largest single producer of the sewage that made its way into the Firth of Forth. Marine pollution was already so visible offshore in 1884 as to make the commander of the Granton fishery cruiser *Vigilant*, a witness to the Dalhousie Commission, wonder if the decline of fish was due to the discharge of sewage, rather than to the trawlers.[8] But everywhere the same story was to be told, of middens abandoned, the transfer of human waste to drains flushed by reservoir water, and thus the sewage running from drains to rivers and hence to the sea. Larger towns like Stirling, Falkirk, Dunfermline and Kirkcaldy, smaller old ones like Bo'ness, Anstruther and North Berwick, and smaller new ones like Cowdenbeath and Whitburn associated with mining developments, all went down the same path, and as they did so, they also coincidentally grew rapidly. What was not replicated outside Edinburgh was the experiment with the irrigation meadows. In most places, for the whole of the nineteenth century, and, at least where there was direct access to the sea, for most of the twentieth as well, the ordure just went off to the rivers and the sea pretty well as it came out of the population.

2. The Costs of Industry

Human sewage was not the only problem, nor in many places the most serious one. The Royal Commission on River Pollution in 1872 considered the Edinburgh neighbourhood excellent for study 'because of the curiously specific character of the foulness which the streams and running waters of Mid-Lothian severally experience'. While the streams draining from

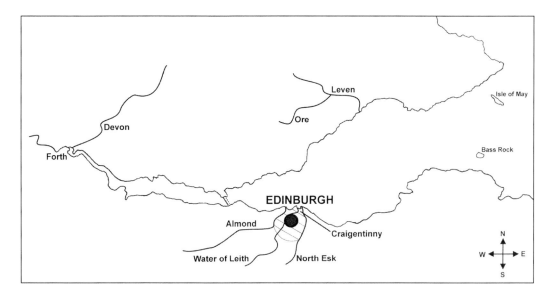

Fig. 7.3 – Main polluted rivers, c.1870

the capital carry the 'personal waste of the large population of that city' and 'may be said to represent almost strictly the nuisance due simply to town sewage', the North Esk and the Water of Leith (the latter having been mainly freed of sewage by then) 'exhibit almost exclusively the effect of paper manufacture on otherwise unpolluted streams'. The Almond, running in from West Lothian, showed in different parts the effects of flax steeping for linen manufacture, or drainage from whisky distilleries and of 'the distillation of oil-bearing shales'.[9]

The paper-mills of the North Esk had been a source of complaint since before 1841, when the Duke of Buccleuch, Lord Melville and Sir James Drummond of Hawthornden, all proprietors along its banks, combined to raise a civil process at law, complaining of nuisance against the manufacturers, 'stimulated by the horrible stench of a river which flows through their parks and under the very windows of their houses'. It took 25 years of legal wrangling to resolve, cost the Duke £6,000 in costs and the other parties 'very much more'. The 'Esk Case' went down in Scottish legal history as a demonstration of how complicated and expensive such litigation could become.

It involved three separate actions. The first resulted in a compromise agreement in 1843, whereby the paper-mills – nine mills were cited, involving seven sets of defenders – agreed to implement the recommendations of two scientists, Professor Christison of Edinburgh University and Dr Madden of Glasgow University, which would diminish the pollution entering the river. Although the manufacturers initially complied, significant increases of production, and the introduction of esparto grass, meant that the situation from the mid 1840s deteriorated again. The Duke and his co-pursuers therefore raised two further actions in 1864, which were not heard in court until 1866.

When it came before a jury at the Court of Session, the complexity of the case took up nearly two weeks of court time. The pursuers cited the 'intolerable nuisance' of the 'dirty and noxious refuse', such that neither human nor animal could now drink the water. Many witnesses were called to avow that the pollution had developed from around 1830, when trout still swam in the clear water, and it could be used for drinking and bathing. Thereafter, local inhabitants noticed a change in colour to grey and dirty brown, with froth and suds floating on the surface, and a 'loathsome soapy smell'. Cattle would not drink from it and the fish disappeared.

The defenders, the principal one being Alex Cowan and Sons of Valleyfield mill, suggested that the Esk had long been used for paper-making (since 1709) and that it had been a long time since it was the pure stream that the pursuers tried to make out. In fact, their lawyers suggested that the proprietors themselves were responsible for the pollution. This was not so unlikely a line as it seems – attack being the best form of defence – as both Melville and Buccleuch were themselves involved in riparian industries. Clearly other sources of pollution were also affecting the Esk, including seepage from coal works higher up the Esk, an iron foundry, a tan works, an estate sawmill emptying two tons per week into the river, and a carpet factory with dye-works owned by Lord Melville, which had 200 employees. In the past, he had also owned a paper-mill, although that had shut in 1828, some time before the industry had adopted more polluting methods of manufacture with esparto grass, and at a time when paper was still hand-made from rags collected in Edinburgh and surrounding towns. The Dean of Faculty, representing the mill-owners, suggested that 'the pursuers are bound to set their own house in order before they complain of the houses of others'. He used the defence of commercial expediency, as so many others have done ever since across the world: if the mills were forced to close because of the cost of expensive new treatment equipment, then the whole area would suffer. He laid out the importance of the mills to the local economy, claiming that the owners paid £80,000 in wages and employed 2,000 people. He added, 'such streams are running over golden sands and adding without limit to the wealth of the land'. It was a powerful plea to support the public good and never mind the environment.

The jury nevertheless gave a majority verdict, nine to three, in favour of the pursuers and the Lord Justice-General in his judgement gave a ringing endorsement to the principle of pure water:

> There could be no doubt that water had been sent for the use of man, and although he was to use it he was not to abuse it, and he considered it only right that man when using water should again return it to the river in as pure and wholesome a state as when took, so that his brother and neighbour below him might enjoy the same benefits by getting the water sent him in as pure a state as he got it.[10]

It still took another two years before the dispute was finally resolved between the Duke and the manufacturers.

The main impact of the Esk Case, however, was to demonstrate that the cost and length of such civil actions was so enormous as not to be an option for ordinary mortals trying to find protection in law against gross pollution, and it was to be many years before an effective statute filled the gap. Moreover, according to witnesses before the Royal Commission, the remedial action that the paper manufacturers were eventually forced to take, still left the Esk nearly as foul as the Irwell below Manchester, regarded as the most polluted river in England. The Water of Leith suffered in the same way and smelt even worse, as there was a weaker flow of water to carry the detritus away. The main cause of pollution in both rivers was the discharge of the liquor of the esparto grass imported from Spain to make the paper, an irony as the main harbour of import was Granton, which belonged privately to the Duke of Buccleuch.[11]

When the Royal Commission came to consider the pollution caused by the paraffin works along the tributaries of the Almond (and elsewhere), they found it even more serious. The manufacture rested on a process of distilling oil from shales, discovered and patented by James 'Paraffin' Young of Glasgow, who in 1851 built the first oil refinery in the world at Bathgate, and in 1866 started operations at a more ambitious second works at Addiewell by West Calder. There quickly came to be various other paraffin works on the shale deposits, and they all needed a lot of water, which they unhesitatingly discharged again as waste into the nearest stream. The Royal Commission

Fig. 7.4 – In 1959 the municipal water supply was disrupted in Cramond, and families drew water directly from the Almond as they had done in early Victorian times before it was polluted by paraffin oil works upstream.

in 1872 was told how at Linlithgow 'the oil floating on the stream could be ignited, burning with a large flame', and, even more dramatically by a witness from Broxburn, 'for curiosity I threw in a lighted piece of paper and it blazed up about twenty feet high'. The River Almond itself, 'previously a charm' and at Cramond Brig 'sweet and clean and much used for making tea' (Fig. 7.4), had become there 'a very dirty turbid river, smelling strongly of paraffin oil . . . useless for all purposes, constituting in summer a horribly offensive nuisance'. Even cattle would not drink from it.[12]

The Commission moved outside the immediate Edinburgh region, but they did not find things better. At the head of the estuary, where the Forth ran in from Stirling, there was terrible pollution, from town sewage, and effluent from textile mills, distilleries and an oil works. The River Devon below Dollar was a tributary of the Forth and particularly bad, stinking and rainbow-coloured, in summer a 'seething mass of polluted and disgusting corruption, deadly in its effects on the life of both man and fish'.[13] The River Leven in Fife was much the same, with pollution from untreated sewage, distillery waste, chlorine from paper-mills and bleaches from spinning mills, as well as water extraction from a brick factory and a foundry, and potential pollution from a cyanide factory, a creosote works, salt works, linseed oil mills and industrial laundries. It was later made particularly bad with the addition of a mass of coal dust from the washings of Fife Coal Company, established in 1877, and other local mines.[14]

Indignation about the state of the rivers ran high among many witnesses before the Royal Commission. Colonel Sir J.E. Alexander, a salmon proprietor on the Devon, is a good example. He expressed astonishment that 'in a professedly moral and religious community' mill-owners and distillers did not of their own accord take every means to keep the rivers clean. He assumed that 'hastening to be rich overcomes all scruples', and reported that offenders, when challenged, blamed their neighbours:

> It is not my stuff that hurts the fish, it's the chloride of lime from the paper mill, or the essential oil from the distillery, or the petroleum from the secret works.[15]

Manufacturers naturally had a quite different perspective. The papermakers, as in the Esk Case, were concerned lest 'blind interdicts' should hamper their competitive position, 'considering the paramount importance of manufactures in this country in regard to expense', and the owner of the Bathgate paraffin works observed that 'if manufacturers are hindered for the sake of the streams, it is a mistaken policy'. The factory owners were also well organised and close-knit, as well as complacent. When in 1869, for instance, the Forth River Salmon Board failed in an attempt to prosecute Robert White for discharging noxious waste into the river at Stirling, he wrote to the *Glasgow Herald* to publicise the conditions under which the case had been withdrawn, saying it was 'of importance to other woollen manufacturers throughout Scotland'. He added that if the Salmon Board were to bring another action, 'it will afford me an opportunity of proving to the satisfaction of all that the discharges from my works into the Forth have not the slightest prejudicial effect on the life of fish of any kind'. It was also easy for manufacturers to imply that protest against pollution was a class matter, of fish versus jobs, of proprietors versus the people, or of local selfishness versus national interest.[16]

In their discussions over the Royal Commission's report, Parliament largely sidestepped their findings, and attended more to the manufacturers' anxieties. By let-out clauses in the Rivers Pollution Prevention Act of 1876, manufacturers and others needed only to show they had taken 'the best practicable and reasonably available means' to avoid polluting a river.

Furthermore, then and in all subsequent legislation down to 1951, it was left to the local authorities to take action. Sometimes these did take a polluter to court, and they were occasionally effective, as against Pumpherston oil works on the Almond in 1903. But local authorities were themselves part of the problem. Elected by ratepayers, they were likely always to take the cheap option in sewage discharges. Dominated by local industrialists, they did not take kindly to curbs on their own businesses. If a civil action for nuisance was attempted by private aggrieved parties, this was expensive and protracted, with the likelihood of success very uncertain.

John Glaister, professor of forensic medicine and public health at St Mungo's College, in his address to the Glasgow Philosophical Society in 1897, showed how little progress had actually been made across Scotland by the end of the century. In the east, he once again picked out the River Almond as 'one of the worst polluted streams in Scotland, and that chiefly from refuse discharges of trade processes'. He noted that already in 1870 the Royal Commission had reported the stream as 'unpalatable and repulsive to taste . . . poisonous to fish life'. In 1891, the Chief Medical Officer of Health for Midlothian had demonstrated that it received pollution from 48 different sources, including sewage from many villages and towns, coal-washings from two collieries, the refuse of limestone mining and discharge from six ironstone pits ('which impart to the river and bed of the stream a yellowish-red or ochry appearance'), the chemical soakage from several old shale bings and, worst of all, the deliberately discharged waste from ten paraffin and other oil works. 'Fish cannot live in it: horses and cattle drink sparingly of it, if at all: and for industrial purposes it is almost useless, on account of the destructive effects upon boilers'. It was probably Glaister's report which led to the action against the Pumpherston works in 1903.

He reported that the situation was much the same elsewhere along the banks of the Firth of Forth. On the River Avon, 'the paraffin vapour rising from the water militates against the neighbourhood . . . as a residence', the River Carron was 'a most unsightly blot' polluted by a paper-works, a naphtha works and by the sewage of the town of Denny, and the River Forth and most of its tributaries were worst below Stirling where 'the beds of black and slimey ooze apparent on the banks amply testify to the gross pollution which it receives from that populous place'. In Fife, the River Leven between Loch Leven and the sea was, throughout its course, 'very highly polluted with sewage, paper works refuse, bleach works discharges, refuse discharges from woolen and spinning mills, and distillery waste'. The litany seemed endless, with only here and there a manufacturer trying to improve his performance, like a paper-maker on the Carron recovering their caustic soda and finding that the process paid, or a local authority like Fife County Council goaded to threaten action against one of the chief polluters of the Leven.[17]

Evidence from many sources pointed to the impotence of public authority faced with polluting private enterprise. The 1876 Act appointed an Inspector of Rivers for Scotland, with powers to issue certificates to manufacturers testifying that they had taken the 'best practicable means' to avoid pollution, which would exempt them from prosecution. By 1885, however, he had never issued a single certificate as the requirement was 'too absolute'. However, no action followed against polluters, because in other respects the Act was also too weak. Questions were asked in Parliament as to why a salary of £50 a year was paid to an official who could not carry out his duties, but he was spared the axe.[18] In 1898 the same powerless inspector observed that the main offenders along the River Devon were much as they had always been, three local authorities who poured untreated sewage into the rivers and 29 manufacturers. The river was still as turbid, discoloured and polluted as in Colonel Alexander's day, receiving tributaries 'rich in soapy water and the usual constituents of sewage', and on the main river 'bells of gas were to be seen breaking on the surface'.[19] It was all very well to have powers to name and shame, but as no one had much shame, it was quite ineffective to name.

In 1902 the Commissioners on the Salmon Fisheries summarised what had made the River Forth and its tributaries increasingly polluted over the last 40 years: the increase of sewage from settlements of all kinds, untreated organic effluents from breweries, distilleries, paper-works and tanneries (which deprived the water of oxygen), and toxic wastes such as chlorine from bleaching and paper-works, acids from steel and tin plate works, alkalis from wool washing, and gas lime, cyanides and carbolic acids from gas works. The only tributary where sewage was treated in any way before discharge was the Teith, presumably because the main pressures on local authorities in this very rural area were the landed salmon proprietors. On the Devon, the local authority of Clackmannanshire always deferred, said the Commission, to the problems of the manufacturers. The Forth District Salmon Fisheries Board, though set up under legislation from as far back as 1862 and 1868, remained powerless and unable to afford the litigation necessary to challenge either the councils or industry.[20]

One of the problems in the Forth system was that the natural flow had been much reduced by abstraction from the River Teith, the most powerful of all the streams in the area. At the point of junction, it actually carried more water than the River Forth itself, so strictly speaking the Forth was a tributary of the Teith. In 1855 Glasgow obtained a bill to empound Loch Katrine at the head of the Teith. Four years later the scheme was complete, 34 miles of lead pipe and brick tunnel carrying the water through the hills and moors of the Trossachs and down into the city. At the opening of the works by Queen Victoria, John La Trobe Bateman, who designed and built the scheme, called it, with no false modesty, 'a truly Roman work . . . work which surpasses the greatest of the nine famous aqueducts which fed the city of Rome', declaring that none of the other works of the city fathers would be counted 'more creditable to their wisdom'.[21] It initially took 50 million gallons a day out of the watershed.

The Admiralty had been nervous lest the diversion of water would deprive the Firth of Forth of its scouring ability, and the city council of Stirling feared for its fishing rentals. Glasgow satisfied the navy that its fears were overblown, paid a small annual sum to Stirling in lieu of any lost salmon revenue and built a compensation reservoir downstream at Loch Vennacher, complete with a fish ladder that did not work very well.

The Loch Katrine scheme was twice enlarged, and by the time of the First World War some 122 million gallons a day were being sent to the city, much of which would otherwise have gone to scour the Forth. The annual value of the fishings at Stirling fell from £1,300 around 1860 to £200 by 1910, though most of this fall was probably due to increased pollution on the other tributaries.[22] A further scheme by Glasgow to dam Loch Voil was opposed by the Scottish Fisheries Board on the grounds of the damage it would have done by removing the biggest remaining source of winter floods in the Forth system. The private bill necessary to authorise the scheme failed to pass the House of Lords, an assembly of gentlemen more likely to sympathise with the salmon and their fly fishermen than most.

3. Slack Water: The Interwar Years

In the interwar years, in other parts of Britain, advances in calibrating official standards recommended by the standing Royal Commission on Sewage Disposal, established between 1898 and 1915, and in establishing means for effective voluntary compliance, sometimes brought about considerable improvements, as along the Trent in Nottinghamshire.[23] Such was not the case, however, around the Firth of Forth, where attempts at improvement to river pollution were often either ineffectual, or entirely counterbalanced by new industrial and housing developments. The continued spread of modern sanitation, now generally replacing the old dry closets even in small burghs and rural areas, added to the problem: officials described how 'use of water carriage for sewage has undoubtedly increased the number of potential sources of trouble by adding to existing discharges and by creating the necessity for new outfalls'.[24] Although a Forth Conservancy Board had been established, by provisional order in 1919 and by Act of Parliament in 1921, with a remit for the 'improvement, control, protection, maintenance and conservancy' of the river and Firth of Forth, between Stirling Bridge to a point a little beyond the Forth Bridge, its interests were entirely in reclamation and navigation.[25] Despite a responsibility for the 'supply of water' it totally neglected water quality. Throughout the 1920s, the reports of the Scottish Fisheries Board in relation to salmon seemed to show the problems of pollution getting steadily worse, at least at the head of the estuary.

Hesitant steps towards improvement were nevertheless taken by central authority. In 1922, the Scottish Board of Health called for returns from county councils of the sources and extent of inland water pollution throughout Scotland, and put pressure on the most laggard to improve sewage treatment. A more significant step was the establishment in 1930 of a Scottish Advisory Committee on River Pollution, reporting to the Department of Health at the Scottish Office (successor to the Board of Health). It had no formal powers, but its detailed reports over the next decade underlined the intractable nature of the problems and the inadequacy of existing solutions, laying foundations for the reforms after the Second World War.

In 1935, its report dealt with the Almond, the Avon and the Grange Burn (which fed into the growing industrial complex of Grangemouth): it vividly illustrates what had and had not been done by that time. On the River Almond, most local authorities, prodded by the Scottish Board of Health, had indeed made some attempts to ensure that sewage entering the river was partially treated, either in septic tanks (a new invention since the start of the twentieth century) or in sewage filtration works. But several of these were inadequate to the task, and, at Whitburn, for example, raw sewage was still discharged straight into the river: 'the Town Council realised that they were committing an offence . . . [they] had in view that some form of purification would sooner or later be required'. The committee recommended eight schemes on the Almond (and eight on the other two waters) where new or improved treatment works were still needed. In the interval between the preparation of the report and its publication, the Commissioner for the Special Areas in Scotland, who channelled government assistance to depressed areas, made grants to four councils in Midlothian (and four more in Stirlingshire) to help them meet better standards. Piecemeal, but not comprehensively, small things could be done.

As for industrial pollution, of the five surviving oil works on the Almond, all had cleaned up their act, though one only did so following a visit from the committee: the 1903 legal proceedings against Pumpherston evidently concentrated minds. The biggest problem now came from 26 collieries, either working or abandoned, which discharged their effluent into the river: in the upper third of the Almond the ratio of mine water to rain water was 1:2. Below that point various shale works added their contribution, so that even well down towards the mouth 'the Almond was nearly black at times . . . the banks were coated with black coal dust or silt'. Coal discharge was not covered by the 1876 Act, and the committee considered that no progress could be made until it was brought under legislation. Finally, there was a single paper-works at Mid-Calder. The owner explained that the water they drew from the river was so dirty that they had to pass it through a series of four settling ponds before they could use it, though it was better than it used to be 20 years ago. He spoke of 'oil, yellow iron oxide, coatings of soda on the stones near the mill, and strong fumes of ammonia', once commonplace, as things of the past. Now, however, he considered that:

> the main impurities were coal washings and sewage and, in a spate, quantities of yellow material. His firm's position was that if they were to receive clean water from the river, they would be glad to purify their trade wastes, but that it would be a needless expense to purify their effluents, having regard to the present condition of the river.[26]

It certainly was not easy, in the middle of the great depression of the 1930s, for industry like this struggling paper-mill to help itself. Something more was needed from government than minor assistance and repeated investigations.

The impacts of all this pollution on biodiversity were most obvious on riverine fish. The Leven in Fife with its tributary the River Ore, for example, were considered among the foulest rivers in Scotland. It was reported by the advisory committee in 1930 that salmon had not been seen for 'very many years', sea trout had disappeared around the 1880s, and brown trout had died out 'recently'. The entire river, apart from a short distance below Loch Leven, was now 'completely devoid of all fish life'. Similar reports described the Esk and the Almond as equally lifeless, just as the Royal Commission had found them in 1872, and the Avon and the Carron were the same. Salmon continued to run up the River Forth and some of its tributaries, particularly the Teith, though in much reduced numbers: the drop in the value of the Stirling fisheries by over 80 per cent between 1860 and 1910 we have already referred to. Mass kills of fish repeatedly occurred in the tidal stretch of the River Forth, blamed by W.L. Calderwood, the Scottish Fishery Board inspector, as much on water abstraction as on pollution:

> In summer when the rivers run low the current is insufficient to carry off the deleterious matter, and when a strong stream tide sets in up the channel the sludge is stirred up, the poisonous gasses are given off, and large numbers of fish die.

He added that it got worse where the Devon joined the Forth, bearing a heavy load of distillery waste: 'the smell which arises from the mud at the mouth of this stream, at low tide, has seemed to me enough to kill a man, let alone a salmon'.[27]

The Fishery Board in 1923 called the death by poisoning or suffocation of numbers of salmon and sea trout in the river 'an annual occurrence', and in 1926 spoke of the large numbers of fish that died on the spring run 'in their attempt to penetrate the turbid waters of the estuary', with a similar problem for smolts on their return. In 1937, the Board warned of a disaster to salmon and sea trout fishing 'unless steps are taken to remedy the mixed domestic and industrial pollution of the upper tidal reaches'.[28] There is no sense at all in their reports of any gradual improvement taking place in the condition of the River Forth: the Board was alarmed that things were continuing to get worse, or at least had not changed since the end of the nineteenth century.

Up to this point, no one seems to have expressed much concern about what, if any, were the effects on tidal waters below the mouths of the various rivers. It was assumed that the sea could take it. Just as in the 1840s the main anxiety had been to get the waste out of the town and into the rivers, so up to the 1940s the main anxiety was to get the waste out of the rivers and into the sea.

4. The Road to Recovery

As with much else in British life, a new beginning was made after the Second World War. A report to government in 1950 advocated the replacement of the 1876 Act by a new, more effective and comprehensive statute, and the establishment of authorities with powers over entire drainage areas, with a small dedicated staff of qualified inspectors entirely independent of local government, to enforce recognised and measurable standards. Furthermore, commenting on 'the many and serious pollutions of tidal waters', they recommended for the first time that the remit of the new authorities should cover estuaries, and tidal waters throughout the Firth of Forth up to half a mile from ordinary low-water mark.[29] The upshot was the Rivers Purification (Scotland) Act of 1951, which imposed much more realistic penalties and standards, and made it obligatory to seek consent for all new discharges. It also created the Lothians River Purification Board, with responsibility for the catchments of the Tyne, the Esk and the Almond, and the Forth River Purification Board with responsibility for the entire River Forth catchment, and those of the Carron, Avon and Leven. The new boards did not initially have jurisdiction over the sea, but in 1960 their powers were extended by an Order of the Secretary of State in the way that had initially been recommended, to take in the estuary and coasts. In 1965, they obtained further statutory duties to impose consent conditions on all discharges, even those dating from before 1951. In 1974, a Control of Pollution Act established all the basic provisions for defining pollution offences. Recognising that the implications could be costly, it was designed to be implemented in phases. Each of these initiatives ratcheted up the power and effectiveness of the officials. In 1975 the two boards were amalgamated to form a single Forth River Purification Board. There was now a new sense of purpose, and a quite new scale to the resources available.

The changes came just in time. There was a period in the middle 1950s when the salmon and sea trout fishery of the River Forth was again considered by scientists to be 'in some danger of being wiped out' by the continuing heavy mortality of adult fish trying to get up the rivers to breed and smolts attempting to return to the sea.[30] The threat came from heavy biological pollution that removed dissolved oxygen from the water, creating 'oxygen sag', which, as Calderwood had noticed, was usually at its worst in summer when flow was low. Scientists, taking a new interest in the chemistry of the estuary itself, found a moving barrier of impassable water that floated from about five miles above Alloa at high tide to more than two miles below at low tide. It was in place for 18 days out of 25 sampled in 1961, with migrating fish lurking below it waiting for a chance to get through. After that, as the river purification boards began to get on top of the problem by controlling discharges, average levels of oxygen in the river improved, and by 1976 the estuary was only impassable on three days sampled out of 21.[31] The extent of the recovery within the estuary between Alloa and Queensferry by the end of the century was extraordinary: inputs of organic waste fell sevenfold between 1980 and 2000, the amount of mercury in fish and mussels fell between five- and tenfold over the same period, and the dissolved oxygen available to animal life about doubled between 1987 and 2007.[32]

Studying how the biodiversity of the mudflats of the estuary was affected by pollution also began in earnest in the post-war decades, in the 1970s and 1980s, with the work of Donald McLusky and others, many from the new University of Stirling which found an appropriate local focus in Forth studies. Several of these were concerned with the situation off Grangemouth, where in 1924 Scotland's first modern refinery had been built to process petroleum, greatly expanded after 1946 with the development by BP (and to a lesser extent by ICI) of linked refining and petrochemical operations. Inevitably there was pollution, even if not at all on the gross scale found in some non-European petrochemical complexes. The immediate vicinities of the outfalls were described even in the 1980s as abiotic, lifeless, but an area just beyond enjoyed very high biomass but limited biodiversity, that is, large numbers of few species, like the dense beds of filamentous blue-green algae near the effluent channels, and fauna 'dominated by opportunistic worms', especially annelids and oligochaetes. Great abundance of the latter was also associated with heavy sewage loadings in the estuary up towards Alloa and Stirling, and, indeed, wherever else there were outfalls carrying organic waste. Improvements in the discharges in the 1970s and 1980s were beneficial but could not tackle the 'large residue of historical contaminants' in the sediments, or completely remedy the impoverishment of the biodiversity. Nevertheless, the estuary remained rich enough to support big populations of intertidal organisms, and therefore of wading birds, such as red knot, and certain fish. The macro-invertebrate benthic fauna of the estuary, attached to the bottom below low-water mark, followed a similar pattern, with only two species surviving in 1987 close to the pollution at Grangemouth, but 82 in the area close to the bridges and furthest from the outfalls.[33]

Beyond the confines of Queensferry, the Firth of Forth had always been a much more saline and open marine habitat, with powerful currents helping to sweep pollution out to the open sea. Yet in a number of places along the outer coast there were serious problems in the 1970s, which were coming to be regarded as scandalous. By far the most pressing were caused on the Lothian shore by the sheer volume of raw (or roughly screened) sewage effluents and trade wastes from Edinburgh and its immediate surroundings, still pouring into the sea from a series of sewer outfalls between Cramond and East Lothian. A report to the Royal Society of Edinburgh in 1972 showed that the situation was not merely bad, but in some respects had continued to deteriorate since the first comprehensive survey of 1958, despite pressure from the Lothians River Purification Board. The latter had had some success in tackling purely riverine problems, such as the state of the Almond, which they had still found to be an 'open sewer' as late as 1955, and the Esk, also described as 'still very much an open sewer' in a Parliamentary debate of 1959. Here miles of river bed were still covered with esparto detritus, though the technology for dealing with this had been known since the 1870s.[34]

The problem in the open Firth off Edinburgh in the 1970s took several forms, some of them familiar from the estuary. In the vicinity of the main outfalls, there was significant oxygen sag, and water thick with suspended solids, caused by a mixture of sewage with industrial waste, notably paper-mill effluent at Musselburgh, spent gas liquor at Granton, gypsum silting from Scottish Agricultural Industries works at Leith, as well as large quantities of used grain and other waste entering the trunk sewers from the brewers and distillers of Edinburgh. There were characteristic sterile areas at some of the outfalls, surrounded by areas dominated by marine worms of large biomass but low biodiversity. There were also large beds of polluted horse mussels and common mussels unfit for human consumption,

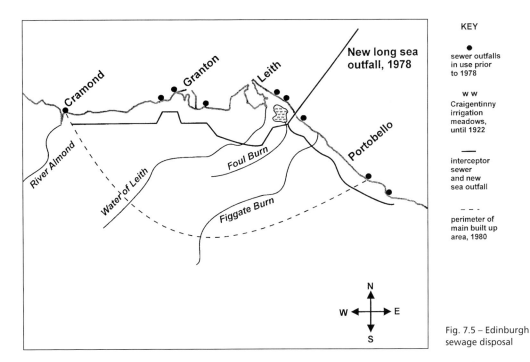

Fig. 7.5 – Edinburgh sewage disposal

very grossly polluted beaches, which bathers and walkers found revolting, and, worst of all to the official mind, heavy counts of *E. coli* bacteria in the sewage slicks, which, under certain circumstances of wind and tide could come ashore and present a risk to public health.[35]

Remedial action was eventually forthcoming when powers under the new legislation were brought to bear by the Lothians River Purification Board upon Edinburgh City Council and its successor, Lothian Regional Council. The upshot was the construction of a new Seafield treatment works near Leith, with a long interceptor trunk sewer necessary to bring the sewage to this single point for treatment, and subsequent discharge down a single outfall 2.8 kilometres long into relatively deeper water (Figs 7.5 and 7.6). The treatment plant and outfall were commissioned in 1978, costing around £52 million (equivalent to four times that in today's terms). The residual sludge was taken away in a boat (aptly named the *Gardyloo*) and discharged to offshore locations outside the Firth of Forth. This method was replaced by an even more expensive total onshore processing and secondary treatment before the end of the century, after European Union regulations forbade dumping at sea.

Fig. 7.6 – Workmen building one of the new sewers in Edinburgh, 1972. The complete scheme, with new treatment plant, was ready by 1978.

All this was a tremendous improvement. By the time of the next Royal Society of Edinburgh report in 1987, it was clear that the Firth was a rapidly recovering ecosystem, and the process has continued ever since. But it did not benefit everything. For one group of ducks, which had adapted to feeding on the marine worms and organic residues from the distilleries and breweries, the recovery in water quality was an unmitigated disaster. The scaup is a marine duck that breeds in Iceland and Scandinavia and winters in the Baltic, the North Sea and Atlantic coasts of Britain. The goldeneye has a rather similar distribution, though is not present in Iceland, and many winter inland. The pochard is much less marine and has a more southern breeding distribution. All three had taken to feeding in winter in large numbers off the outfalls of Edinburgh and similar places, the pochard resting by day on Duddingston Loch within the city and feeding on the sea at night, the others spending all day at sea. It is clear that the scaup had been rare in the nineteenth century: the minister of Prestonpans in the *New Statistical Account* of 1839 called it 'the most uncommon' of the 'duck tribe' in the Firth of Forth and mentioned that several had been shot after severe storms. In the early twentieth century, however, flocks a mile or more long began to be reported off Leith and fishermen would go out to catch them in nets. In the 1960s the scaup increased still more, and 30,000–40,000 were estimated off Leith in the winter of 1968–69 (Fig. 7.7). Up to 4,000 goldeneye were similarly present offshore at this period, and 5,000–8,000 pochard were counted in the 1960s on the loch. For all three species, Edinburgh was the prime British site, and birdwatchers came from all over to see the spectacle. As soon as water quality began to improve their numbers began to fall, and by 2010 fewer than 100 scaup were left in winter in the Firth of Forth. The decline in the other species was scarcely less remarkable, and the disappearance of the pochard flock from Duddingston may have reduced the number wintering in Britain by a third.[36]

The very last of the black spots to be cleaned up was in Fife. The sewage and trade effluents discharged into the River Leven (and its tributary the River Ore) were intercepted by a trunk sewer constructed by Fife County Council between 1949 and 1965, greatly to the benefit of riverine water, but discharged just offshore into the sea at Levenmouth, after only preliminary treatment. As late as 1999, the discolouration of the water made the sewage easily visible from the shore, and detritus was scattered down miles of beach. The rich inshore discharge also had the effect of increasing the numbers of scaup, goldeneye and pochard (the last roosting by day on Kilconquhar Loch), though on a smaller scale than at Edinburgh, and when in 2000 the Leven discharge was at last modernised and improved to European standards, all these duck except a few goldeneye, disappeared from here as well.[37] Even in quite small coastal communities, like Anstruther, the ending of raw sewage discharge has visibly changed what the birdwatcher is likely to see in winter on the coast: there is no chance of scaup now, and fewer goldeneye. Certain species of waders, like dunlin, turnstone and purple sandpipers have also become scarcer on the shore, and this effect, noticed elsewhere, has been related to cleaning up sewage effluent.[38] Cleansing the Firth of Forth may also have had adverse effects on other sea duck, as has been suspected in respect of eider, common and velvet scoter, and long-tailed duck, all of which have declined off the Fife coast since the early 1990s, but here the evidence is less clear cut: the decline is more likely to be due to other factors, such as damage done to other bottom-dwelling organisms by dredging for scallops or illegal electric fishing for razor-shells, or through increasing disturbance from recreation. It could also be related to climate change. At least in the case of velvet scoter, there has been a redistribution to the East Lothian shore, but eider and long-tailed duck have declined off Lothian as well.[39]

5. Conclusion

So the story of pollution in the Firth of Forth has a good ending, unless you are a scaup or an opportunistic worm. It tells of an environmental problem largely overcome. It has not of course been completely overcome. Old industrial processes leave relict sediments in the estuaries, and these can still be released by changing channels and strong tides as modern pollution. There are new concerns since the start of the later twentieth century, like 'diffuse pollution' by soil run-off from agriculture containing nutrients, pesticides and hormones, and by organic compounds and chemicals even in treated waste.[40] The sewage processing plant at Seafield is not perfect, despite the cost: its pervasive stink upsets local residents, and in 2007 the system failed and released a substantial quantity of raw sewage into the sea. There are occasional oil spills from boats using the Firth of Forth, but none so far have been catastrophic. There is a great deal of unsightly pollution along the coast in the form of washed up rubbish, some of it originating onshore but much of it from boats, including fishing boats. More than half is plastic that does not rot at sea like paper and wood: litter on British beaches almost doubled between 1994 and 2007.[41] On the other hand, the old view that the sea is an appropriate tip has gone. In the interwar years, many little towns on the coast,

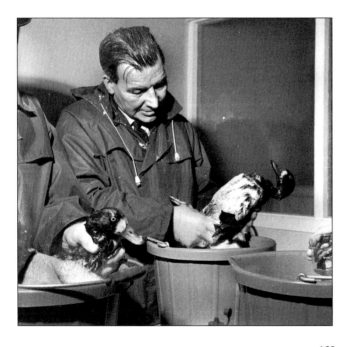

Fig. 7.7 – Scaup affected by an oil spill off Leith being cleaned up by SSPCA officers at Seafield Cat and Dog home. Most birds treated probably died later.

like Crail, kept a horse and cart to collect the community's rubbish, which they tipped over the cliff.

Generally speaking, the scene is transformed, the rivers again have fish and their waters run sweetly into a sea much less contaminated than for a century and a half. The regular return of bottle-nosed dolphins and some whales into the Firth is one sign of this transformation back to an earlier and better world. One memento at least remains: anyone diving in the clear water off Edinburgh today may encounter on the sea floor an indestructible scatter of tomato seeds. They mark the extent of the former plumes of raw sewage which the unenlightened capital once spewed forth.

How did this remarkable change come about? We have already listed the successive steps of effective legislation – the Act of 1951 establishing river purification boards with powers to license new discharges, followed in 1965 by extension of those powers to license all existing ones (no matter how long standing), and in 1974 by a comprehensive new Control of Pollution Act. The reorganisation of local government the following year gave the task of sewage treatment to the new regional councils, and also amalgamated the two local river purification boards which retained their supervisory and licensing roles. The bigger regional councils were more effective at laying sewers and installing better sewage treatment than the old town councils and county councils had been, but due to financial stringencies the water pollution clauses of the 1974 Act were only implemented in stages during the next decade, with some regional councils still struggling to finance larger schemes, like the Leven outfall in Fife. The next big step forward was the 1995 Environment Act, which created the Scottish Environment Protection Agency as a national supervisory and enforcement body, amalgamating all the River Purification Boards in Scotland, while at the same time actual sewage treatment in the area was taken over from local authorities by a new, separately funded, East of Scotland Water Board. In 2002 this was amalgamated with two other Scottish water boards into one national Scottish Water.[42]

Not least in importance as a prod to change was Britain joining the European Economic Community in 1973, which involved meeting an increasing number of demanding environmental directives by the end of the century, and beyond. The United Kingdom for more than a decade largely evaded and resisted (until threatened with fines) the Bathing Water Directive of 1976, which sought to impose standards of cleanliness that the government considered irrelevant to public health on British coasts. Mrs

Thatcher assumed that it would help British tourists on the dirty beaches of Italy and France, but could not imagine its application to Britain. No bathing waters were designated under the directive in Scotland until 1986. On the other hand Britain under John Major collaborated rapidly with the Urban Waste Water Treatment Directive of 1991 which halted dumping at sea, even though British scientists considered the way it was done off the Firth of Forth and elsewhere to be environmentally harmless.[43]

What is much more difficult to explain is why these steps to solve a long-standing environmental problem were taken by society at this particular time. What had been grumbled about but tolerated, even allowed to deteriorate, between 1860 and 1960 (or 1970 in some cases), was suddenly found to be intolerable. Why was heavy investment made at this point and solutions found within a generation?

Part of the explanation may lie in Oliver MacDonagh's celebrated thesis about the growth of government intervention in nineteenth-century Britain. Initially based on a study of the regulation of the conditions under which poor emigrants were shipped over the Atlantic, it stipulates that the concern of Parliament might be aroused by an abuse, find its first legislation ineffectual and be moved to create an inspectorate to enforce it. This new bureaucracy, initially weak and inexperienced in the face of vested interest, builds up over time a professional interest and expertise. By a feedback mechanism, it exercises pressure on its creator to act again, first with more enquiries to which the officials give evidence, then with more effective laws which strengthen the powers of the inspectorate, often in successive steps.[44]

The weakness of the thesis partly lies in its inability to account for variations in the timing of reform, which in the case of water pollution seems singularly long in coming to fruition. The first inspector of the state of the rivers came with the Salmon Act of 1868, which empowered the Scottish Fisheries Board to appoint one for the whole of Scotland: he could do little more than fulminate. Similarly, the Rivers Pollution Prevention Act of 1876 appointed another powerless inspector. Yet the slow growth of expertise, the involvement of health officials, the organisation of water and sewerage engineers into an international profession with journals, and their growing alliance with science to identify problems and solve them, must have been important. From the days of the standing Royal Commission on Sewage Disposal between 1898 and 1915 there was increasing attention to developing appropriate technology to treat industrial waste as well as domestic sewage. The seven reports of the Scottish Advisory Committee on

Rivers Pollution Prevention between 1931 and 1937, covering most polluted rivers in Scotland, built up a body of modern knowledge that especially informed advice to government in 1950. Critical in the latter advice were recommendations for the first time to define rigorous standards and appoint a proper inspectorate:

> qualified technical officers – i.e. river inspectors – who would advise on all technical matters and would keep in close touch with industries and local sewerage authorities in their areas and give advice . . . The qualifications we have in mind for river inspectors are Corporate Members of the Royal Institute of Chemistry and of the Institute of Sewage Purification, with considerable experience of the operation of sewage purification works.[45]

In fact, from 1951 onwards, the officials of the new River Purification Boards in Scotland often exercised the sort of pressure that MacDonagh described, protesting, for example against unhelpful medical orthodoxies about the relative harmlessness of beach pollution, and in 1979 planning to identify a series of clean bathing beaches under the European directive, until stopped from doing so by the Scottish Office.[46] Similarly, the diffusion of knowledge of appropriate technology by the officials encouraged a willingness by authority to spend money on the problem and, by its effectiveness in turn (by a virtuous circle) lowering cost and feeding a further willingness to spend.

Partly the initial delay had been due to the slowly developing character of the problem itself. Before modern sewerage, the towns may have been smelly and their wells dangerous, but pollution was contained within a narrow compass. Even in the 1850s the volume of sewage-bearing water was slight compared to what it rapidly became, and only in the later 1860s did fish disappear from most of the rivers. Chadwick himself assumed that private irrigation meadows would substantially cope with the problem in Edinburgh, and the scale of the new pollution that was to come from new and expanded industrial processes was still hardly imagined before about 1865, though the Esk was an exception. It took time for the shortcomings of the market in this kind of situation to be appreciated, and the need for much greater public intervention to be seen.[47] The first reaction was either that manufacturers were bound to clear up after themselves as they were gentlemen (this from the landowners) or that their deposits in the river either did no damage or were justified as being in the national interest (from the manufacturers themselves).

Solutions also came with the growing scale of governance. Victorian local government was far too local to tackle river pollution, which flowed through many jurisdictions, so the problem grew willy-nilly, whatever Parliament might decree. But when after 1951 the state imposed a supervisory authority as large as a catchment, then as Scotland, polluters could not so easily escape, and now that the European Union threatens even sovereign governments with sanctions, there is no escape for anyone. At the same time, the treatment authorities also progressively grew in scale to create and afford appropriate solutions. Big government was needed for such big problems.[48]

Underlying all this was a profound social and economic change in Western countries, which took the form of declining manufacture and a growing service sector. This was especially pronounced in Scotland in the Edinburgh area, where, for example, in the nineteenth century there had been 20 paper-mills on the Esk and the Water of Leith, but by 2000 only one was left and that was on the brink of collapse. In this case, the near demise of the paper industry occurred a decade or so after Britain joined the European Free Trade Association in 1960, exposing it to Scandinavian competition. In other instances the causes were less specific, as in the long-term decline of the textile and metal trades (Carron Company, harbinger of the Industrial Revolution in 1759, went out of business in 1982). The shale oil industry in West Lothian had been in decline since before the First World War, and the last plant ceased operation in 1962. In the case of the coal industry in Lothian and Fife, gradual exhaustion and decline over most of the twentieth century culminated in general pit closures following the failed miners' strike in 1986.

As deindustrialisation takes place, fewer vested interests survive to pollute the river, fewer jobs depend on them. The average citizen in the east of Scotland now works in computing, or for a bank, an insurance firm, in a shop, for government in the health or education sectors, or for the tourist industry. Citizens of advanced countries have more money to spend and more time to relate to 'post-material values' such as appreciation of the environment.[49] The defenders of the rivers and the beaches are now no longer the landed gentry but the working class and the middle class, concerned about their own health and their recreation. The consumer goods that they buy are manufactured cheaply abroad in China and India, where exactly the same arguments are made that pollution is essential to keep down costs as were made here a century ago – but it is not a pollution Scottish consumers need to see or suffer. Rising standards of living in the 1960s and early 1970s and again at the end of the century led not only to greater expectations, but also to a larger tax base for local or central government to afford expensive new processing plants and outfalls, and the accompanying bureaucracies of environmental management.

All these things were beginning to come to a critical threshold in the 1960s and 1970s, when at the same time events internationally were creating something like a 'green consciousness'. It focused round Rachel Carson and pesticide misuse in the US and Britain, around the rise of green political parties in Germany and elsewhere in Europe, and manifested itself by successful campaigns as various as those to restore the Rhine in Germany, to save old growth forest in America, to prevent another oil pollution disaster like the wreck of the *Torrey Canyon* in Cornwall in 1967, and in many European countries to stop new nuclear power stations. People could feel part of an international trend, and the cleansing of the Firth of Forth belongs very broadly in this context, though not at the forefront, and late in the 1970s rather than early in the 1960s, as befits a cautious national temperament.

The situation in Edinburgh was not marked by any sustained campaigns of local protest against the scandal of gross sewage disposal, though in other respects the citizens proved well able to stand up for themselves in the 1960s in defence of their property and amenity, being the first ever to defeat the planners over attempts to impose an inner-city ring road. The Scottish branch of Friends of the Earth was founded in 1970; its first relevant publication on water pollution (a one-page pamphlet on *Sand, Sea and Sewage*) arrived 21 years later. Even the British charity, the Coastal Anti-Pollution League, founded in 1958, maintained at this time that 'there was really no need for expensive treatment works' unless the discharge was 'in an estuary or on the shores of a confined stretch of water'. Official British medical opinion from 1959 until the late 1980s held that there was no danger from sewage in the sea unless the sea was visually so revolting that you would not want to swim in it anyway.[50]

Of course there were letters to the press in the 1960s and 1970s declaring the situation intolerable and a disgrace to the city, but then there had been such since the nineteenth century. Public pressure was of a more distant and general nature: feeling after the *Torrey Canyon* incident and pesticide scares, particularly in England, persuaded the Prime Minister, Edward Heath, in 1970, that it would be politically astute to amalgamate three existing departments into a new Department of the Environment at Westminster (even though for years afterwards it had little interest in the environment), and the Scottish Office was obliged to follow its lead. In Germany, France and the Netherlands, burgeoning green parties in the 1970s had more weight than in Britain, and the Continental system of proportional representation in regional and parliamentary elections leant them an importance to larger parties trying to form stable coalitions quite lacking in Britain. This translated into pressure on the European Commission to frame effective regulations. Thus public opinion and political bargaining in London and Berlin could count for more than opinion in Scotland.

Over two centuries, rivers running into the Forth have undergone several transformations. They are no longer in the least a source of power. First the steam engine and then the electric and diesel motors supplanted the force of water, and Scottish lowland rivers do not lend themselves to hydro schemes in the way that Highland rivers do. The rivers are not thereby necessarily made more 'natural', as beavers anciently built dams that had some of the ecological effects of the mills' weirs, dams and lades. The headwaters of rivers impounded into reservoirs have from 1850 onwards replaced wells and burns as sources of drinking water for city and town. And, most importantly of all, the rivers and the Firth of Forth itself went through a prolonged phase that lasted for over a century of being dumps for sewage and industrial waste, and the end of this misuse, beginning after 1950, marks a milestone victory for environmental responsibility.

These developments and transformations were brought about by people, yet they had important impacts on biodiversity. The removal of the mills made the river flow differently, to the detriment of dragonflies and frogs but to the benefit of other aquatic life. The fouling of the rivers destroyed their ecosystems and changed those of parts of the Firth of Forth itself: the cleansing restored the *status quo ante*, but some species which had adapted to the new conditions in the Firth lost out. Environmental history is never our history alone: it is always, in addition, the history of the organisms at our mercy.

CHAPTER 8
LAND CLAIM FROM THE SEA

1. Introduction

Today, the reaches of the Forth estuary between Alloa and Queensferry are relatively unknown to most Scots, and rarely viewed at close hand. Anyone travelling from Stirling to Edinburgh will probably take the M9 motorway, which skirts the Forth estuarine plain – the carse – running between the urban centres of Falkirk (with its satellite towns) and the refinery town of Grangemouth, before passing Linlithgow and the bing-studded landscape of West Lothian. The journey by train offers equally limited views of the estuary itself. The landscape from the carriage between Stirling and Edinburgh appears first as low, rolling hills, well cultivated and lined by old hedges. The first sign of industry is the chimney stack belching emissions from the chipboard factory at Cowie, while the massive tower of Longannet power station dominates the north shore of the Forth (Plate 3). Above all, the perpetually flaring stacks and steaming cooling towers of the Grangemouth petrochemical works provide an unmistakable landmark, visible across a wide area of the Scottish Lowlands (Fig. 8.1).

Much of this tangle of industrial apparatus sits on land which, less than a century ago, was part of a massive area of mudflats, some of which was infilled using refuse and mineral spoil to expand Grangemouth docks and to provide other land for industry. This twentieth-century land reclamation is, however, only the most recent episode to win land from the sea along the shores of the Forth. It has been estimated that between 45 and 55 per cent of the Forth's intertidal area has been lost over the last 400 years as a result of reclamation, or land claim.[1] Land was being taken from the sea perhaps as far back as the Middle Ages, and certainly over several centuries, both directly for agriculture and to protect the fertile carse lands behind from tidal encroachment. By the mid twentieth century, part of the

Fig. 8.1 – Grangemouth from the air, partially built on reclaimed land, with Longannet on the far shore: rich wildlife habitat between.

petrochemical plant, including the refinery power station, was erected on a concrete raft constructed on the mud east of the heavily engineered mouth of the River Carron.² On the north shore, the ground around Preston Island is also the product of land reclamation. It is no longer an island; the space between this rock in the Forth and the shore was infilled by fly ash from the nearby Longannet and Kincardine power stations over the last 50 years. The island itself – a little over one kilometre from the shore – was naturally a rocky outcrop which was encircled with dykes infilled in the early nineteenth century as part of an earlier attempt at winning coal and salt from nature's store.

The shores of the Forth estuary have constantly changed. Originally, much of the carse below Stirling was fresh-marsh, full of reeds and wildfowl, raised slightly from the sea only by isostatic uplift, the rebound of the land once the weight of the ice had been removed (see Chapter 1). In the drier parts, for example between Stirling and Kincardine and round the Hill of Airth, peat began to form, as it also did on an immense scale in the blanket bogs above Stirling. In the 1140s, when a grant of a salt pan was made to the monks of Newbattle, the scribes of the Norman royal court referred to 'blankelande' (white wasteland), possibly describing the white bog cotton flowers stretching across the surface.[3] Place-name evidence suggests that as early as the ninth century there began to be meadows and even arable land on the carse, and when in the twelfth century both kings and lay lords endowed the abbeys with land, Cambuskenneth was built on firm carse clay. Newbattle and Holyrood, particularly the latter, over the next few centuries led an ambitious programme to win more of this area for agricultural use. They were copied piecemeal by others, such as the peasants at Kinneil who extended their tenant holdings from the higher ground down to the salt flats. Such gains were at risk from flooding, particularly, but not only, in an exceptional storm surge and it seems highly likely that this required the construction of some form of defence from tidal waters. It is possible that a system of sea and river defences from Stirling to Grangemouth was created even in the Middle Ages. If so, the building and upkeep would have needed a level of co-ordination that only royal or monastic initiative could provide.[4] The sea dyke still visible at Queenshaugh, below Stirling castle, which was constructed to protect valuable agricultural land, may be a product of this system (Fig. 8.2). The dykes would be repaired again and again, depending perhaps on fluctuations in sea and river levels and on economic circumstances. This might explain why in 1533 the monks of Holyrood were disposing of land round a settlement called Little Saltcots because that was being regularly flooded by the sea and by 'the rivers known as Carron and Forth'.[5] This could have been due to neglect of the seawalls, or possibly because the exploitation of peat for their saltpans had lowered the surface of the land and undermined their stability.

Once reclaimed, however, the silts and clays of the carse below Stirling, like the polders of Holland and the fens of East Anglia, proved to be among the most fertile lands there were. By the end of the Middle Ages, there was a saying that 'a loop in the Forth is worth an earldom in the North'.[6] Robert Sibbald in the early eighteenth century said that 'the Carse grounds, for the best Grains, are equal to the fertilest land any where else.'[7] All this was long before the better known eighteenth-century improvers of Lord Kames's generation began to make such an impression on the carse above Stirling.

Fig. 8.2 – Sea dykes at Queenshaugh, Stirling, photographed from Abbey Craig; remnants of similar dykes run down to Clackmannan and Kinneil and might date to the sixteenth century or earlier.

Early industries also clustered along the shores of the Forth, such as salt-panning and coal mining (the developments at Culross in the sixteenth century are the best known). Several ports, including Bo'ness in the seventeenth century and Alloa and Grangemouth in the eighteenth, became transport hubs for trade with the Continent. More recent developments such as petrochemical works, power stations and oil transhipment points, gas plants and naval shipyards have made the estuary a focus for the modern industrial economy. Yet, despite all this, today much of the estuary seethes with wildlife. The intertidal habitats, including mudflat and salt marsh, are protected under national and European law through designation as Sites of Special Scientific Interest (SSSI) and as a Special Protection Area (SPA), providing food for internationally important numbers of wintering wildfowl and shore birds such as shelduck and redshank. Nevertheless, it is perhaps surprising that more of this vast expanse of mud and marsh was not reclaimed. Elsewhere, land claim has, for example, removed over 90 per cent of the intertidal area of the Tees estuary, and approximately 6,500 hectares of the Shannon estuary lowlands in Ireland are thought to have been reclaimed for agriculture and other purposes. Similar losses have been reported for other estuaries around the world.[8]

The story of land reclamation in the Forth estuary is long and eventful, if at times obscure. To H.M. Cadell at the start of the twentieth century it was a simple matter of using force in a good cause: 'War against the wild forces of nature is, from the point of view of humanity, a constructive, and in every way a useful, operation worthy of the attention of all statesmen'. Any reclamation was a human triumph: he was 'a hero and a patriot of the best kind, who uses his brains and money in laboriously overcoming the wild forces of nature, and snatching good land from the clutches of the stormy sea'.[9] Today we are not so sure. We value again the 'sleeches and slobs' of the intertidal zone as an oasis for biodiversity and consider reversing reclamation schemes.[10] Plans are now being formulated for large-scale habitat creation by controlled breaching of the sea walls, intended among other things to mitigate the damaging impacts of sea-level rise due to climate change. This chapter of environmental history is a story of changing human perceptions as well as of human effort.

2. The Intertidal Zone

The intertidal zone is where the land meets the sea. It is that no-man's land lying between the highest point to which the tide floods, commonly regarded as the high-water spring tide mark (HWS) and the lowest point to which the tide ebbs, or the low-water spring tide mark (LWS). Definition is complicated by the variation in tidal zones, with the average commonly referred to as the mean lines of highest and lowest tide (MHWS and MLWS), while recognition is also given to the more exceptional highest and lowest astronomical tide lines (HAT and LAT), which will not occur every year, and to extreme tides, which are associated with the storm surges, such as occurred most devastatingly in the North Sea in 1953. Biologically and historically, the variation in tides constantly creates conditions which shape and reshape the environment and therefore the precise loss of this zone is difficult to assess accurately.[11]

One of the most detailed assessments was undertaken as part of a British Geological Survey report on the engineering geology of the Forth, which estimated that 51 per cent of the intertidal zone had been lost since

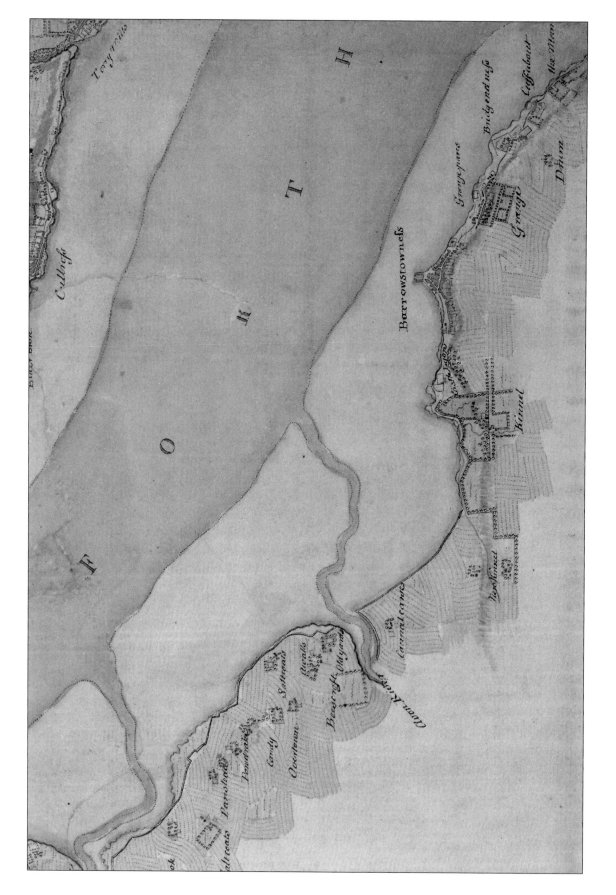

Opposite: Fig. 8.3 – Roy's Map of c.1750, depicting the Forth shore from the mouths of the Carron to the Avon. Note the angular double line likely to be a sea wall and the more natural irregular salt marsh to the seaward.

1600. This was based on the assumption that 4,000 years ago the line of so-called carse cliff, currently sitting three to five metres above sea level was the effective MHWS. However, this presupposes that this line, created by isostatic uplift between 6,000 and 4,000 years ago as the land bounced back from the enormous weight of ice from the last Ice Age, would effectively still have formed the line of MHWS until the seventeenth century. However, it has been suggested that natural forces acting since 2000 BC, such as continuing but more gradual isostatic uplift and the 'silting' up of sediments during more extreme conditions, would have continued to reshape the landscape.[12] In effect, we cannot be sure how much land has been taken from the intertidal zone as a consequence of human endeavour, though it must certainly have been around half and therefore very significant in terms of habitat loss. Moreover, we need to take into account the loss of salt marsh, which naturally would have extended from the mean high-water mark landwards for some distance, flooded only by the highest of spring tides. We also need to consider the loss of the meandering character of tidal reaches of rivers such as the Carron and even the upper Forth estuary below Alloa, changes which were often much smaller in scale but which resulted in straightening the river channels, with hydrological, sedimentary and ecological consequences.[13] One only has to glance at historic maps such as those produced by General Roy's military survey of Scotland around 1750, to see how much more naturally indented was the shoreline of the Forth estuary and some of its tributaries (Fig. 8.3). This is even more apparent if one examines large-scale maps of estates and farms of the eighteenth-century improvement era, many of which identify and depict the 'salt-greens' that acted as a buffer between the arable fields and the mudflats.[14]

Hydrographically, the Forth estuary comprises two distinct regions splitting at Kincardine (Fig. 8.4 and Plate 3). Between its tidal limit a little above Stirling, down to Alloa, the Forth is essentially just a tidal river, following a broad series of meanders, forming what are called 'the links of the Forth'.[15] The estuary proper, stretching downstream from Alloa to Queensferry, is characterised by a larger width-to-depth ratio and more complex bottom topography.[16] Along the seven-kilometre reach between Alloa and Kincardine, the estuary widens, and lateral water movements become important, resulting in the build-up of sand bars within the channel. The estuary narrows again at the Kincardine Bridge crossing, where it is constrained by raised banks of glacial deposits, resulting in turbulent flows and re-suspension of sediment during strong flood and ebb tides (Plate 3).

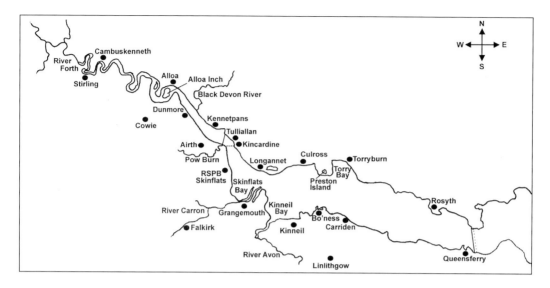

Fig. 8.4 – Location map: Upper Firth

Downstream of this point, the estuary rapidly increases in width and depth, until it is again constrained by the narrows at Queensferry. From Kincardine to Queensferry, a distance of 24 kilometres, it is flanked by large intertidal bays full of soft mud (Kinneil, Torry Bay and Skinflats). These constitute a big proportion of the area of the Forth that is flooded at high water and have been the particular focus for past land reclamation schemes. Webb and Metcalfe estimated that at MHWS these bays store 10 or 15 per cent of the volume of tidal water upstream of Queensferry and that the movement of this water on and off the embayments greatly affects currents in the main channel of the estuary.[17]

Within this zone, which, near Grangemouth, extends seaward for around two kilometres from the shore, inundation by salty water creates the special intertidal environment. Most particularly, this area of the Forth estuary is dominated by extensive mudflats and backed, nearest the land, by a relatively narrow strip of salt marsh. For humans, it is a hostile environment where the mud cannot be easily crossed and the tides can quickly swamp any human endeavour, but the intertidal zone is rich in biodiversity; its ecological niches provide important nursery grounds for young fish and a feeding ground for birds, both depending upon a range of molluscs and invertebrates which live in the mud. It has recently been estimated that the loss of mudflats on the Forth as a result of land reclamation has removed 24 per cent of the natural fish habitats, which equates to a 40 per cent reduction of their food supply.[18] In the past, the estuary offered rich resources of fish and shellfish, as explained in Chapter 6. Seaweed, associated with the rockier shorelines, was another product of the intertidal zone, exploited as fertiliser (see Chapter 1). The salt marshes were valued as good grazing land, providing strips of intermittently inundated swards of salt-tolerating grasses and herbs adjacent to the rich cultivated carse lands, ideal natural feeding for plough animals, and also for sheep in an agricultural system which traditionally focused on arable cropping. The ancient use of these fringing marshes is demonstrated by the frequency of place names like Saltgreen and Seagreen.

Similarly, the memory of one of the earliest industries to develop in the estuary – salt-panning – is also enshrined in the names of farms and villages along the shore, such as Kennetpans, Grangepans and Saltcoats. Richard Oram has recently demonstrated how salt extraction, so critical for preserving meat and fish, was important in the estuary as early as the first half of the twelfth century.[19] Both monasteries and lay proprietors were involved in 'sleeching', a method by which the salt was extracted by rinsing dried silts with sea water and heating the resulting brine in iron pans. Initially, wood was used as fuel, but, as the local woods round Falkirk, Airth and Kincardine came under pressure after about 1150, it was increasingly replaced by peat. This was also needed in large quantities and the salt industry may be partly responsible for the loss of peat cover from parts of the area by the sixteenth century, even outside the carse itself, as at Blairhall. By the close of the Middle Ages, though sleeching continued in the estuary for several more centuries, a more efficient means of winning salt was found by directly boiling seawater in pans over coal fires, and the centre of gravity of the industry moved downstream beyond the estuary, towards Prestonpans in East Lothian and Dysart in Fife, where there was outcropping coal. Increasing shortage of outcrops led in the sixteenth century to improved mine drainage using horse-powered bucket wheels or 'gins', so that coal could be dug by following the seams below ground. Sir George Bruce of Carnock was the major pioneer on his estate at Culross, where, in 1575, he constructed a mine that ran beneath the Forth and came out in an artificial island offshore, so that the coal could be loaded on to ships.[20] Culross became, for a time, possibly the most important salt producer in Scotland, until 1625 when the pit was wrecked by a storm. So the manufacture of salt around the intertidal zone led on shore both to landscape change and to technological advance.

Place-name evidence can help to confirm the location of the earliest coastal settlements, their likely period of origin and their position relative to the current shoreline. Such evidence is to be found in recent comprehensive research published by Reid on Falkirk and east Stirlingshire and Taylor on Fife.[21] Used alongside cartographic, archaeological and documentary sources, such as early charters, it is possible, at least partially, to reconstruct the line of the shore. Reid, for example, identifies a fourteenth-century reference to 'Isle of Erth', or Isle of Airth, which was large enough to pasture six horses.[22] The Roy map of around 1750 also depicts an island – at the mouth of South Pow, just south of Airth, near Higgins Neuk – which is not shown on any other extant maps, but which conceivably is the same as the one referred to as the Isle of Erth. The Blaeu map (originating in Pont's survey of around 1590) appears to show an island off Kinneil at the mouth of the Avon. This feature does not, however, appear on any other map and there is no mention elsewhere of an island at this locality, so it may instead represent a prominent mud bank, perhaps the Ladies Scaup, which Sibbald, in 1710, described as a shellfish bed associated with this locality, similar to those at the mouth of the Carron.[23]

Whether these map-makers depicted such features accurately or not, it is abundantly clear from old maps that the shoreline of the Forth, its meanders upstream from Kincardine, its tributaries, and the characteristic, tidal-influenced pows (streams), which flow into the estuary, were considerably more irregular and meandering than they are today. Not all of the change was the result of human endeavour, and the River Carron at least was known to have naturally altered course, as rivers do. The *Statistical Account* for Bothkennar, on the north side of the Carron, describes the Carron 'having changed its course, now intersects both the parishes of Bothkennar and Falkirk, leaving part of the former on the south, and a small part of the latter upon the north side of it'.[24]

3. From Sea Defence to Land Claim

Since the land which flanked the estuary was so important at an early date both for agriculture and for industry, it must always have been worth defending from natural estuarine processes. There is evidence for sea-wall construction along the Forth from the early seventeenth century, and, since building these structures would have required considerable co-ordination, their origin (as we have seen) may lie in the Middle Ages, associated either

LAND CLAIM FROM THE SEA

with monastic houses or royal possessions. By the early modern period it is likely that the coastal fringe was already altered by human activity, even if only in a relatively minor way, for the defence of the rich carse lands.[25] Early sea walls were low and their function was to prevent all but the very highest, surge tides flooding the land – not total prevention.

Some of the earliest reclamations of the intertidal zone for agricultural purposes, building on extant sea defences, may have taken place around Airth (Figs 8.5 and 8.6), which, by the sixteenth century, was also a busy port, trading in coal.[26] Historically, the lands of Airth were synonymous with the family of Bruce of Airth and Stenhouse, who gradually began to accumulate land in and around the settlement from the fifteenth century. In 1597, John Bruce fell heir to the lands of Airth. By this time, the family's expansionist policy had created a large and complex estate, illustrated by the charter granted to John Bruce on his accession. As well as rents derived from farming, there

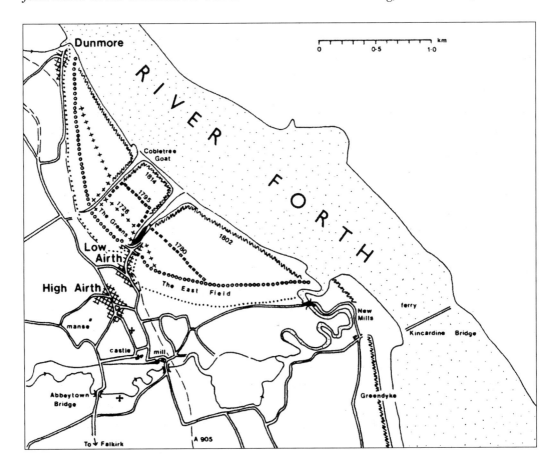

Fig. 8.5 – The shoreline at Airth showing successive lines of reclamation in the eighteenth century.

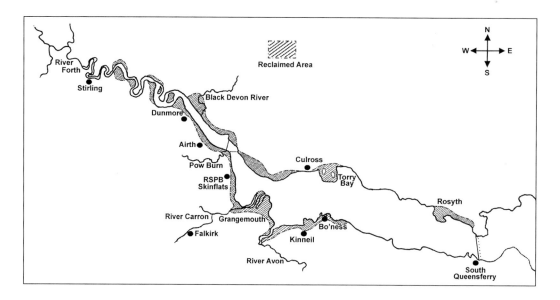

Fig. 8.6 – Land reclamation on the Forth Estuary

were revenues from coal, fishing, salt-making and milling activities, and also from a ferry. Over and above these, the family had control of the port of Airth.[27] Evidence suggests that the Bruces vigorously exploited all these assets. The carse land between Stirling and Bo'ness was economically and agriculturally advanced, where crops were not grown simply for subsistence, but also for the market, and they included wheat, regarded as a luxury food, which was sold to the towns to make bread. Wheat was known to have been grown in Bothkennar parish from the late thirteenth century,[28] with other evidence suggesting that elsewhere in the carse it was intensively used for arable farming from at least this time.[29] John Harrison has uncovered documentary evidence dating from the early seventeenth century for long-standing tenurial obligations to maintain and repair sea dykes at Ferriton, near Clackmannan, where a tenant was required to see to 'reparying and re-edifieing of the hale sea dykes demolished by Invadations of Watter and sall menteyne the same in as gud estait as they have bene they yeris bypast.'[30] It would seem entirely possible that similar constructions were put in place by the Bruce lairds of Airth. It is not a great leap from defensively building dykes against floods to aggressively building dykes to claim land from salt marsh and mudflat, and so adding arable land to the farms.

Sea defences against flooding appear to have been an important element of the eighteenth-century landscape. In 1710, Sibbald described the continuing use of dykes to protect farm land along the Carron:

> This part of the Countrey also, much of it in ancient times hath been covered with the Sea, and oweth its Fertility to the Slime and Earth brought down by the Spates of Carron Water, which some time carries off large Parcels of Ground, and lays it sometimes on the one side, sometimes on the other, upon which account the Inhabitants are obliged to make large Dykes, which in few years must be altered and placed elsewhere.[31]

That such dykes were still being used as sea defences in the eighteenth century is graphically illustrated by Roy's depiction of a solid double line, running roughly parallel inland of the more indented true shoreline, between the mouths of the Avon and Carron, and continuing eastward as far as Kinneil (Fig. 8.3). Seaward of this line, between the mouths of the Carron and the Avon, there appears to be a more natural shoreline, which may indicate the extent of salt marsh in this period before the major infill and realignment of this stretch of the estuary in the nineteenth and twentieth centuries.

By the mid eighteenth century, the use of embankments or dykes as part of schemes to create arable land from salt marsh and mudflat is well documented, particularly around Airth, so it is feasible that the sea walls depicted on Roy were intended to extend the arable acreage as well as having a defensive function. The sea wall on the Roy map continues eastward of the mouth of the Avon towards Kinneil, although without the same fringe of land on the seaward side. In 1774, the Duke of Hamilton, who owned Kinneil, instigated the building of a large embankment, nearly 2½ kilometres long, which, it was said, was not to gain ground but to defend what lay behind from the sea, perhaps replacing one depicted by Roy.[32] It is known that there was an early reclamation at Kinneil, which must be one of the first extant references to such attempts at claiming land from the sea. This is alluded to in a Kinneil charter of 1540 as 'terras infra mare'.[33] The earliest embankments must have been built on salt marsh subject to flooding at spring tides, but once defended, the marsh could be converted to cultivatable land by applying lime, which counteracts the salt, thus extending the highly fertile, but still vulnerable, carse lands.[34] Once salt marsh is reclaimed and protected from future tidal intrusions, if the conditions are right (i.e. substrate, depth and exposure) then new salt marsh will recolonise seaward of the embankment; as can be seen today along the 16-kilometre stretch of coastline from Kincardine Bridge to the mouth of the Carron. On the interior of Alloa Inch, an island which was periodically submerged by the tide until it was protected by a huge bank in the early nineteenth century, salt marsh developed after 1983 following the partial collapse of this bank.

In 1710, Sibbald remarked (without giving details) that:

> the Dutch did offer some time ago to make all that Scape, good arable ground and Meadow, and to make Harbours and Towns there, in convenient places, upon certain conditions which were not accepted.[35]

If the Dutch, who were uniquely experienced in gaining much of their own land from the sea, did indeed make such an offer, we can only speculate on the conditions and why they were refused. Possibly they threatened too many local interests by demanding control. The future lay not in any grand or all-embracing scheme of this kind, but on piecemeal decisions in a complicated legal framework.

In the eighteenth century, control over the coastal fringes, particularly the salt marshes or 'salt-greens', was frequently contested between landowners. Many were originally 'commonties', a form of tenure with ancient and strong traditions of common use built into the annual cycles and subsistence economies of rural communities, most frequently relating to grazing and to provision of peat, turf and wood.[36] Salt marsh provided nutritious pasture for farm stock (as it still does today), allowing the carse lands themselves to become the focus for crop growing, and some were a good source of turf which was necessary for so many aspects of pre-improvement farming life such as housing, roofing and dyking.

By the end of the seventeenth century, commonties throughout Scotland were becoming regarded as an impediment to improvement because they were shared and were unenclosed, therefore could not be exploited to their full potential by individual owners. They were said to prevent a landowner from profiting from these areas without the agreement of neighbours, some of whom may not have been of the same 'improving' nature. In the case of the salt-greens, this might prevent them being reclaimed and added to the arable carse. In 1695 the Scottish Parliament passed an important Act allowing division of commonties between neighbouring landowners, which was relatively simple, quick and cheap (though not necessarily trouble free), compared with the procedure in England, where each enclosure might require an Act of Parliament. This single Act in Scotland enabled all commonties to be divided through the courts.

Court records from the eighteenth and nineteenth centuries are replete with instances of attempts, for the most part successful, to do away with commonties, although the process of division often led to disputes between landowners, resulting in a long legal process to achieve satisfactory division. We know such disputes took place on the Forth, which related to salt-marsh commonties. A map of the Haughs of Airth, lying between Higgin's Neuk and Powfowlis on the south shore of the Forth, in 1784, depicts a pattern of fields like earlier runrig, with strips of fields allocated to individual tenants, but which in this case represented the intricate breakdown of land ownership itself.[37] This marvellously detailed map identifies sea greens along the shoreline, marked as 'com'y', and is apparently associated with a disputed division that took place from 1771 to 1789, one landowner embanking land from the sea without waiting for a formal division.

The position was further complicated by the intricacies of foreshore rights. Ownership of the intertidal zone for the most part rests with the Crown, today managed 'in trust for the people of Scotland', by the Commissioners for the Crown Estates. When an area of mudflat is proposed for reclamation, ownership naturally becomes a little nebulous, an issue which continues to cause some problems for those proposing to breach sea defences on the Forth and to restore intertidal habitats on land which for centuries has been beyond the foreshore. This complex set of factors may well have impeded reclamation schemes as early as the apparent Dutch offer, and probably also led to the failure of one extremely ambitious proposal in 1875 to embank and reclaim some 3,600 acres between Bo'ness and Grangemouth, which, we are told, failed because of opposition from 'numerous riparian proprietors'.[38] In this last case, the scheme's proposer, Mr Livingston Learmonth, did not actually own any of the shoreline, whereas all other successful nineteenth-century schemes were undertaken by the riparian proprietors.

That land reclamation was profitable in the eighteenth century was demonstrated by H.M. Cadell, landowner, industrialist and geologist who trained in the 1870s under Alexander Geikie at Edinburgh University. As we have seen, he was an avid supporter of reclamation and wrote a detailed historical account of it in the Forth estuary up until that time, describing the methods adopted, acreage claimed, and the cost and consequent value of the land obtained. For example, Lord Zetland, founder of Grangemouth, reclaimed 200 acres about 1784 near the mouth of the Carron in Bothkennar parish. The traditional method used to convert mudflat to agricultural land then and earlier, was called 'warping'. This involved driving wooden stakes around seven-foot long into the mud, at intervals along the line of the eventual embankment. These would then be interlaced, or 'warped', by wattles of brushwood, creating a cage to trap the silts suspended on each incoming tide. This was highly effective, and it was recorded that as much as three feet of mud had been trapped by this device after just a year in operation. Further lines of warping were then constructed within the initial line, and a year or two later the level rose equal to the adjacent land. A year or two after this, crops could be grown and the value of the new ground was comparable with adjacent farm land. Cadell records that 200 acres were reclaimed in this manner by Lord Dundas at West Kerse of Bothkennar at the end of the eighteenth century, costing £200, which, five years after the process began, realised £4 per acre for the reclaimed land.[39] By the 1820s, it was estimated that over 800 acres of land had been reclaimed or protected from inundation in the previous century.

During the nineteenth century, several hundred more acres were added to this tally, involving some big engineering schemes. Among these were the so-called Tulliallan reclamations, instigated by George Elphinstone, first Viscount Keith, on foreshore adjacent to his lands. These were begun in 1821, initially between Kincardine and Kennetpans to the west (Fig. 8.7), which was completed by 1824 enclosing 152 acres, followed by a scheme which enclosed 214 acres of foreshore between Kincardine and Longannet Point to the east (Figs 8.7 and 8.8). The latter was begun in 1829, but was more difficult to accomplish, though it was largely completed ten years later. However, the cost of these schemes far exceeded the £1 per acre for the scheme at Bothkennar in 1784. The first Tulliallan scheme cost £40 per acre and the second £60 per acre. The increased cost of the latter was accounted for by the length of embankment – warping with stakes does not appear to have been the method used. It stretched over 3,000 yards, and required considerable amounts of quarried stone to shore up the bank. At the same time the trustees of Viscount Keith, who had died in 1823, began a reclamation scheme on the opposite shore at Higgin's Neuk, utilising the newly extended ferry pier. This venture successfully used warping, resulting in around 30 acres being partially reclaimed, but it was never completed because a sea dyke was not built to enclose the seven feet of mud which had built up behind the warping. The result was that, as Cadell states, 'the outer part of the warped ground at Higgins Neuk is still a salt marsh' (Fig. 8.8).[40] Almost 100 years after Cadell described the results of this small failed reclamation scheme, this area remains a salt marsh, one of the most extensive in the upper Forth estuary, and the piles driven into the mud nearly 200 years ago are still a feature of the landscape (Plate 3).

The Tulliallan schemes were among the last undertaken to gain land for agriculture. Thereafter, effort was focused on creating land for industrial developments, often associated with the expansion of port activities and, in the twentieth century, also with the growth of the petrochemical industry and power generation along the estuary. There are several reasons why. The earliest schemes were instigated at a time when agriculture was the main occupation and source of national wealth, so any additional acreage was highly desirable, especially close to major urban centres such as Edinburgh and Stirling. The method was labour intensive, but there were still good numbers of people working the land. It required only stakes and brushwood, which could be sourced from scattered woods in the hinterland like those at Callendar, Airth and Clackmannan, and from new plantations that proliferated on either side of the Forth. Warping was best employed where

Fig. 8.7 – The site of Longannet power station in 1965: it was to be built on land reclaimed from the sea. The power station in the background is Kincardine, now demolished but also built on reclaimed land.

GB 22 13 bc
Geheim

KINCARDINE
Munitionslager

Karte 1:100 000 Blatt Sch. 23.
1:63 360 Blatt 39.

Kriegsaufnahme: 596 R 78

Länge (westl.Greenw.): 3° 41′ 0″ Breite: 56° 4′ 0″ (Bildmitte)
Mißweisung 13° 48′ (Mitte 1938)

Maßstab etwa 1 : 15 600   (1cm ⊢ 156 m)

Nachtrage: 2.10.39.

Ⓐ GB 22 13 Munitionslager

1) 17 Bunker   etwa 12 500 qm
2) Lager- und Verwaltungsgebäude   etwa 1 000 qm
3) Gleisanschluß
4) Straßenanschluß
5) Baustelle
   bebaute Fläche (Schwerpunkte)   etwa 14 500 qm
   Gesamtausdehnung   etwa 154 500 qm
   Gleisanschluß vorhanden

Ⓑ GB   Straßenbrücke

1) Straßenbrücke mit 17 Pfeilern (Eisen)
   Länge etwa 850 m, Breite etwa 13,5 m
2) ausschwenkbarer Teil (drehbar)

Opposite: Fig. 8.8 – A Luftwaffe photograph of October 1939, showing Kincardine Bridge and Longannet ammunition dump on reclaimed land. Note the wide saltings on the opposite shore.

the land was at least partially embayed, so that the length of enclosure was limited in relation to the acreage reclaimed, or where siltation was likely to be profuse. In the 1820s and 1830s, when the Tulliallan schemes were under construction, agriculture was in transition and industrialisation about to take off. From this point onwards, reclamation for agriculture was not regarded as financially viable in Scotland, though in England, at least around the Wash, it continued until well into the twentieth century. For example, near Boston a scheme using 'Borstal boys' as labour reclaimed over 900 acres from the North Sea between 1935 and 1978. In Ireland, where a different set of socio-economic conditions prevailed, land claim became a popular priority for government support, and was also primarily for agriculture development.[41]

4. Land for Modern Industry

Rapid Scottish industrial and commercial expansion in the nineteenth and twentieth centuries demanded land on the foreshore. On the Forth, the emphasis from the mid nineteenth century was on claiming land for trade, like the project at Bo'ness, focused on the importation of pit-props for the burgeoning mining industry. A new dock was built here in 1881 by the North British Railway Company, involving in its heyday 120 acres of storage yards served by ten miles of railway sidings employing about 1,000 people.[42] The alteration to the shore at Bo'ness is evident from a comparison of the 1856 and 1895 editions of the Ordnance Survey maps, the topography of the harbour area being completely altered, and the piers shown on the earlier Roy map obliterated by nineteenth-century improvements.

At much the same time, Grangemouth was also transformed. The growth of the town had been stimulated by the success of Carron ironworks, the canalisation and redirection of the Carron river between 1768 and 1785, and the completion of the Forth–Clyde canal in 1790. By the time of the first edition of the OS map in 1855, it had already basically altered the landscape of this stretch of the Forth. The railway arrived in 1860. A large dock was excavated in 1882, and added to in 1904, when it was estimated that the area reclaimed, including the docks, formed a promontory 260 acres in extent, extending a mile out from the shore to the low-water mark, something which remains a distinctive feature of the Forth shoreline (Fig.8.6).[43]

A similar process of dock development, but for the military, took place at Rosyth, after the Admiralty decided to site a naval base and dockyard there, taking advantage of the deep tidal water off St Margaret's Hope (Fig. 0.4). Work started in 1909 and was completed in 1916. Today, the dockyard is almost 1,300 acres in size, a large proportion of which was reclaimed during construction.[44]

Not all proposals were successful. The failure of Thomas Livingston Learmonth's 1875 reclamation scheme we have already mentioned. Five years later, there was an unsuccessful attempt to set up a Forth Conservancy Board, the intention of which was to dredge and improve the estuary, and at the same time to reclaim foreshore at a scale only marginally smaller (at 2,950 acres) than the 3,600 acres aspired to by Livingston Learmonth. The Forth Conservancy, which would have required a private Act of Parliament to create a body with appropriate powers, was proposed by a firm of chartered engineers, Messrs Thomas Meik and Sons. The legacy of this scheme is a set of detailed plans (RHP80890), elaborating the line of embankments and associated engineering works that would have been constructed, enveloping most of the mudflats from Bo'ness to Higgin's Neuk. Why it did not proceed is unclear. According to Cadell, it failed because of the opposition of shipowners, who threatened to refuse payment of the Conservancy dues which would be required to finance the scheme. He blamed 'the short-sighted old grumblers' who prevented the board from being created, but at the time plans were also afoot for the first crossing of the Forth below Stirling – the Forth rail bridge – the construction of which was begun in 1883 and completed in 1890 (Fig. 0.2). There was much debate and negotiation concerning the bridge and it is possible that a Forth Conservancy might have been regarded as an additional obstacle.

Whatever the reasons for the nineteenth-century failure, Cadell was delighted to report in 1928 that 'a more enterprising generation' had succeeded where the 'old grumblers' failed. The Forth Conservancy Board was finally constituted by an Act of Parliament in 1921, chaired by Lord Elgin, with powers to improve the Forth between the rail bridge and Stirling, and to reclaim the foreshore where considered advantageous.[45]

Cadell himself was a power behind the new Conservancy. He had long felt that the dredgings from harbours along the Forth above the bridge were wasted when taken by barge and deposited in deep water beyond the bridge, not least because he guessed that much of the mud would be swept back up the estuary and re-deposited. He was adamant that the best use of this mud was to re-employ it as infilling within new embankments as a supplement to natural accretion. It was a method he had employed on his own reclamation schemes to the east of Bridgeness harbour, which he owned. In his first scheme, begun in 1889, he paid the tugmaster who dredged Bo'ness and Bridgeness harbours to deposit his load behind a low bank, enclosing 30 acres of foreshore. Warping was not employed on this scheme, Cadell later explained: 'the idea of a dyke of stakes and wattles did not commend itself to me, as there was available plenty of heavy rubbish from the colliery'. Instead, the banks were composed of his own Bridgeness colliery waste (redds) and ash procured free from a railway company.[46] The ashes were emptied cheaply by employing 'old miners . . . and a muscular old woman of a bygone race'.[47] In 1912, Cadell was finally able to rent out the reclaimed land as woodyards, and although it had taken 23 years, he was satisfied, and he had already embarked on another small reclamation on neighbouring land. Between his own efforts and those of the North British Railway Company, 81 acres were reclaimed before 1921.

The bill to establish a Forth Conservancy Board was also successful that year, although only after an amendment allowing Inverkeithing paper-mill to continue to discharge into the bay 114 different kinds of liquid wastes, which illustrates how little anyone cared at that stage about polluting the estuary.[48] The Board was quick to start planning for future foreshore reclamation schemes, with Cadell clearly in the driving seat, although inevitably there was much deliberation about how to proceed. In 1922, Kinneil estate, which had been in the hands of the Dukes of Hamilton since 1326, was broken up and sold. The Board took the opportunity to buy most of the foreshore of the old estate (1,000 acres of mudflat), while Scottish Oils Ltd (later BP) acquired land west of the mouth of the Avon, effectively establishing the future utilisation of this land as industrial land (Fig. 8.9). The Board's initial plans were relatively cautious. Operations began in 1926 to reclaim 310 acres within a bank some 2,370 yards long, linking an old slag reclamation site a mile from Kinneil colliery and the old Kinneil sea dyke of 1774. The line of the dyke extended only as far as the channel of the River Avon, the crossing of which would have been technically difficult and costly. After some setbacks the dyke was completed in 1928 and the lagoon was reported by Cadell to have begun to silt up to his satisfaction. At the east end, colliery waste was dumped by one of the railway companies, which quickly infilled about three acres of land by early 1929 (Fig. 8.10).

The reclamation schemes of previous centuries had used natural

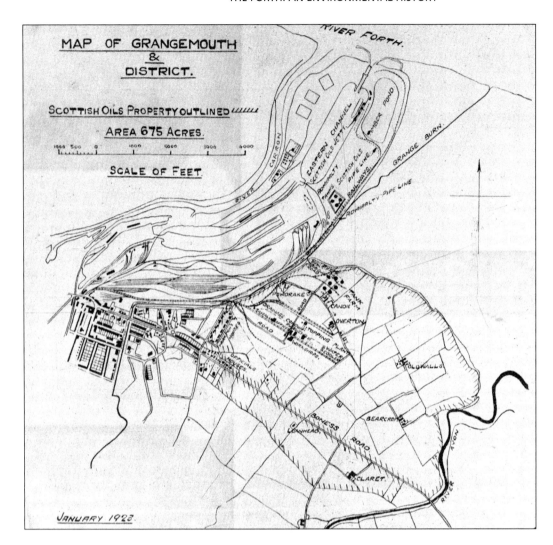

Fig. 8.9 – The property of Scottish Oils in 1923. Next year they built Grangemouth refinery on agricultural land at West Saltoun farm and in due course BP took over the entire property.

sedimentation processes to create the land, but with a continuous supply of waste from local industries, particularly from coal mining, infilling embanked lagoons with such waste became the norm. Indeed the shore became one long disposal tip for domestic and industrial detritus. In 1900 there were 50 pits within a mile of Bo'ness, excavating huge quantities of spoil, and foundries, as well as potteries, salt pans, railways, many trades and dwelling houses, all converting coal to ash. Industrial land was at a premium and waste found a ready use extending riparian property to seaward, a process that continued into the twenty-first century.

LAND CLAIM FROM THE SEA

Fig. 8.10 – The Kinneil shoreline in 1929, two years after the Forth Conservancy set about embanking and infilling the mudflats. Note the small-gauge railway being used to build the embankment.

Cadell died in 1934, and with his passing the impetus was at least temporarily lost. Aerial photographs of Kinneil from the 1940s clearly show the line of the 1928 embankment, with perhaps a quarter of the lagoon infilled within a line running north–south on the Bo'ness side of the site, which consolidated the rather uneven slag-created reclamation mentioned above. Infilling artificial coastal lagoons with waste material from industrial processes nevertheless continues to the present day, Kinneil Kerse having also been used for domestic and commercial landfill since the 1950s. Today, all but one large lagoon of about ten hectares, known as Kinneil Pan, along with a series of smaller brackish pools, have been infilled. However, Kinneil Kerse is also among the most important wildlife sites in the area. Parts have been designated as an SSSI, and it is a component of the Firth of Forth Special Protection Area on account of its attraction to wading birds and ducks.[49]

After the Second World War, the Forth became a focus for electricity generation, now required both to power heavy industry and to meet the needs of a population, which, after the bleak years of war, was eager to acquire new household appliances and enjoy light at the flick of a switch. The

Fife and Lothian coalfields provided a ready supply of fuel, and the shores of the Forth offered an optimal location, taking advantage of improvements in steam turbine and generator technology and of the cooling powers of the Forth waters. Kincardine power station was commissioned in 1954 and opened in the early 1960s, followed soon after by Cockenzie which began producing electricity by 1968. Longannet, the third and largest of the Forth's coal-fired power stations, was constructed downstream of Kincardine on land reclaimed from the estuary, using ash from the Kincardine station, which was itself built on land originally reclaimed in 1824 (Fig. 8.7). When Longannet reached full operation in 1973 it was the largest coal-fired station in Europe.

With this level of electricity production and concomitant waste, disposal of the ash was focused on Torry Bay and the vicinity of Preston Island, where a series of settling lagoons were created on the mudflats, serving the dual purpose of land reclamation and disposal of the slurry pumped, day and night, from Longannet. Further down the Forth, slurried ash from the Cockenzie power station was piped along the coast to settling ponds behind a sea wall constructed from Prestongrange west to the mouth of the River Esk. This created an important bird reserve much visited from Edinburgh, the Musselburgh pools. They have been a magnet to a remarkable number of rarities, ranging from a western sandpiper and a royal tern from America to a lesser crested tern from Africa. This locality has rivalled the natural inlets on the East Lothian coast, Aberlady Bay and Tynninghame Bay, for its attractions to the dedicated twitcher.

5. Conclusion: Re-naturing the Forth Estuary

The impact of modernity on the Forth's ecology goes beyond the reshaping of the shoreline and infilling of the intertidal zone. Estuarine ecology, as we have seen, is complex and ever-changing, the result of natural and anthropogenic factors impacting on a constantly fluctuating ecosystem. As we saw in Chapter 7, discharges of sewage and industrial waste into the Firth of Forth have had a significant historic effect on the invertebrate fauna of the mud and marshes, as well as on the birds and fish utilising these areas as nurseries and for other seasonal activities. Studies have demonstrated a loss of biomass and biological production proportionately greater than the loss of intertidal area itself, but this is complicated by enrichment of estuarine waters.⁵⁰ In the absence of some forms of organic pollution, especially sewage, the intertidal zone might not have remained as productive as it did, and the loss of food for fish might have been even greater. The same sort of argument must also apply to birds feeding in the estuary – the sewage was enriching their habitat just as land reclamation was eroding it, so two aspects of modernity worked in different directions.

After industrial discharges had been reduced through a combination of plant closure and modernisation, and sewage effluent treatment works installed from the 1970s, there were again changes, and increases in marine biodiversity. Nevertheless, within the intertidal and sub-tidal silts the presence of hydrocarbon and heavy-metal contaminants may yet have long-term effects on the ecology of the estuary. It has been reported, for example, that a particular combination of tide and weather conditions can unsettle these contaminated substrates causing episodes of fish mortality.⁵¹ This is a potential toxic time bomb which could be made unstable by predicted increases in storminess, precipitation and peaks in river flows.

The Forth estuary holds some remarkable and unexpected wildlife sites which owe their interest to human manipulation. One such is Alloa Inch, an island now owned by the Scottish Wildlife Trust, which was transformed by a storm breach of the sea wall: the farmland grass of the interior was replaced by the botanical communities of the salt marsh, including, in autumn, mauve sheets of sea aster. This same plant is found in unexpected localities; such as around the docks and industrial installations of Grangemouth, where there are extraordinary examples of natural diversity appearing within man-made alien landscapes. Botanists are also fascinated by a new botanical community colonising reclaimed land at Valleyfield Lagoons in Torry Bay, Kinneil Kerse and Bo'ness Island: it comprises a range of non-local pioneer species creating a grassland type not previously recorded, one including tall melilot, a member of the pea family.⁵² The common factor appears to be the unnatural substrate, which, in the case of Bo'ness Island, may also be composed of ships' ballast.

There has been a long history of use and alteration of the Forth estuary and its shores, at first very light and gradual, mirroring the pace of economic progress here and elsewhere in Western Europe, but quickening from the eighteenth century into the industrial age, when much of the semi-natural estuarine landscape was almost obliterated. Today that stretch of the estuary where the land meets the sea, has been pushed seaward, and the sea greens, sleeches and slobs of the past are now occupied by a new 'blankelande' of industry, commerce and settlement. This is something that is unlikely to change even in Scotland's post-industrial age, where the petrochemical and power-generating industries are constantly striving to renew their technologies to meet new demands. However, biodiversity has by no means been wiped out. With the emergence of the environmental movement, and the will and wherewithal to 're-nature' certain land not valued as highly as before, new schemes are being devised to create wildlife habitats with the support of industry and commerce.

The rise in global sea levels and an increased risk of storm surges has placed coastal infrastructure and habitats under increasing threat. Erosion rates may quicken, the incidence and severity of flooding events may increase, and intertidal profiles may alter. All this may well increase demand for coastal defences. Grangemouth is held fast within a concrete bulwark, which should endure the most overwhelming of storm surges, but the adjacent carse is considered by the Scottish Environment Protection Agency to be at risk.[53] Grangemouth will not be sacrificed, but other areas along the shores of the estuary have been earmarked for 'managed retreat', where the sea is being allowed back in under strict conditions.[54]

In the late 1990s, a controlled breach of the flood embankments of the Black Devon was undertaken to allow the reclaimed salt marsh adjacent to the tidal part of the river to be inundated, an area of some 18 acres. This was followed by the restoration of the neighbouring landfill, which resulted in more wetland being created. Similarly, with the construction of the Clackmannan Bridge over the Forth in 2008, a section of sea wall was breached to create new intertidal habitat. In 2010, at the RSPB's Skinflats reserve a different approach to managed retreat was instigated, using a technique called 'regulated tidal exchange', where a pipe was inserted through the sea wall and a series of sluices created to control the tidal flow into an area of former reclaimed agricultural land, within which a lagoon was also excavated (see Plates 11 and 12). This is more engineered 're-naturing', necessitating a new flood embankment at the back of the site to contain the water within the reserve and ensure that no other properties are affected. The RSPB, in alliance with a range of other organisations, now have plans for a much grander, landscape-scale project, with managed retreat at the core of its aims, which if successful in securing funding, may result in several hundred acres of the shores of the Forth estuary being given over once more to salt marsh, slobs and sleeches.

CHAPTER 9
THE BASS AND ITS GANNETS

1. Introduction

Scotland's seabird colonies are considered the most important in Europe, for their number, the size of their populations and the diversity of their species. None is more dramatic than the immense gannetry on the Bass Rock at the mouth of the Firth of Forth. To an approaching boat, the island appears to be swathed in flying and resting birds: it is the core of an ancient volcano, 365 feet high, a mile in circumference, and it looks crystalline, a sugar lump, shining in contrast to the greys and greens of the neighbouring islands and shores. As the boat draws closer, it becomes clear that the white is not merely guano, but the living bodies of over 100,000 gannets arranged on the sloping surfaces and cliffs as closely and carefully spaced as the dots on a pointillist painting by Georges Seurat. The fact that the Bass also holds large colonies of auks and gulls is eclipsed for the observer by the sheer biomass of gannets. The rock holds 11.5 per cent of the world's population. They are birds with a six-foot wingspan and a four-inch spear of a beak (Plate 15). They come all around the approaching boat, flying, diving, staring at the watcher with pale grey eyes like Swedish girls. Gannets arrive on the Bass each year in March, and depart at the end of September when the young have fledged. They range in summer to feed over the whole North Sea, sometimes flying hundreds of miles before returning; flocks and individuals will seek out shoals of herring and mackerel, or sandeels, and plunge vertically from a height of 60 feet or more, or at certain times of the year they will follow returning trawlers and feed on the discards thrown over the side. They are long-lived – 16 years or more on average. They have no predators except man: they are too big. They cannot be ignored, and never could. The Anglo-Saxon poem 'Beowulf' referred to the North Sea as the 'gannet bath', as we might call the Atlantic 'the pond'.

The origins of the association between the gannets and the Bass are lost in time. The Gough map of Britain of *c.*1360 clearly shows both the Bass and the Isle of May as landmarks for sailors at the entry to the Forth. The gannet in French is 'Fou de Bassan', in Dutch 'Basaangans', in German 'Basstölpel', testifying to an ancient link between the rock and the bird in the minds of those Europeans most likely to sail to Scotland. Linnaeus in 1758 named it *Pelecanus bassanus*, from which the modern scientific name *Morus bassanus* is derived. Buffon around the same time explained that 'L'isle de Bass ou Bassan, dans le petit golfe d'Edimbourg' gave its name to the species because it was generally thought to breed nowhere else, but he cited authorities including the earlier Scottish naturalist Sir Robert Sibbald to explain that it also nested on Ailsa Craig in the Firth of Clyde, in the Hebrides and in the Faeroes.[1] The Bass, however, was still the only colony in the North Sea at that time, and they did not breed in France until 1930.

In 1966 the British ornithologist James Fisher made the interesting suggestion that the Anglo-Saxon spiritual poem 'The Seafarer' referred in one passage to the Bass. The poet goes voyaging, but it is unclear at times whether he is on sea or land: at a place 'where on stone cliffs the seas came crashing', he speaks of the swan calling (plainly a whooper swan), of the gannet, the gull, the tern, the curlew (or perhaps the whimbrel), and the sea eagle, all making appropriate calls, and Fisher ingeniously suggests that only on the Bass in late April would all these occur together. The poet speaks of 'the gannetes hlothor' (in one translation, 'the gannet's blether', in another 'the gannet's language'): outside the breeding colonies gannets are largely silent, and the Bass is the only colony known within Anglo-Saxon territory.[2] The Scottish historian Alex Woolf has made the further suggestion to us that the author may have been St Baldred, an eighth-century hermit who lived on the Bass and used it as a base to evangelise the adjacent coast. The naming of his stone 'boat', which, along with his 'cradle', is one of the geological features of the nearby Lothian shore, may reflect some memory of his reputation as a seafarer.

All this is of course conjecture. The first unequivocal reference to gannets on the Bass comes from Walter Bower's *Scotichronicon*, in 1447, where the writer, abbot of Inchcolm, speaks of the Bass 'where solans nest in great numbers'. Solan or solan goose (or soland goose) was the traditional Scots term for gannet, allied to the Norse 'sule'.[3] From the early sixteenth century onwards, every notable topographical writer mentions the Bass and its birds: John Major in 1521, Hector Boece in 1526, John Lesley in 1578 were early examples.[4] From the seventeenth century onwards, in addition to topographers and travellers there also came a remarkable sequence of scientists and naturalists, including William Harvey the discoverer of the circulation of the blood, John Ray the renowned naturalist of the Royal Society, with his friend Francis Willughby, in the seventeenth century, Thomas Pennant and the Revd Professor John Walker in the eighteenth century, William MacGillivray with the great American bird artist John James Audubon in the earlier nineteenth century, followed by H.F. Witherby and J.H. Gurney later. Gurney's monograph on *The Gannet: a Bird with a History* (1913) was both an early environmental history to which this account is very much indebted and a remarkable scientific treatise which formed the take-off platform for subsequent gannet studies. In the twentieth century the roll-call continued – James Fisher, Nikko Tinbergen, Roger Tory Peterson. In the 1960s, Bryan Nelson made his home on the Bass as St Baldred had done a millennium before, and in 1978 published his classic study of the biology of the gannet based on these field studies.[5] J.M.W. Turner painted the Bass; Hugh Miller and R.L. Stevenson wrote about it; George Washington Wilson and Robert M. Adam photographed it. No other seabird colony in the world has attracted such study and interest over so long a historical period.

2. The Many Ways of Using Gannets

Part of the reason for the early interest was that the gannet was a valuable economic resource, providing a range of products and services at different times: grease, feathers, meat (mainly from the young captured just before fledging), recreational killing (alias sport) and latterly recreational observation. Its value triggered one of the first pieces of conservation legislation in Britain, an Act of the Scottish Privy Council in favour of George Lauder of the Bass in 1584, ratified and turned into a statute by the Scottish Parliament of 1592.[6] This laid out how, through the providence of God, 'the solane geise and utheris fowlis quhikis hantis, reparis and biggis within the Isle of Basse yeirlie' were profitable to the 'commoun weill of his realme', and related that the adult gannets, though themselves inedible and of no value except for their feathers, were being slain by fishermen and sailors with barbed hooks and nets. Thus they did not breed and the commonweal was deprived of their much more useful young. Killing them was therefore forbidden under various penalties. Very probably, the fishermen, as well as using their feathers for bedding material, also used the flesh of the adult to bait fishing lines and crab pots and, despite what Parliament believed, ate them themselves in an emergency, as the poor fishermen of Dunbar still did in the nineteenth century.[7] One way to catch them was to tow behind a boat a herring tied to a sunken board, upon which the gannet plunged from 60 feet and broke its neck: some doubted if this was more than a tall story, but Robert Gray in 1864 actually saw it done.[8]

The statute of 1592 was ordered to be read from the market crosses of every coastal burgh from Montrose to Inverkeithing on the north shore of the Forth and its approaches and every coastal burgh from South Queensferry to Eyemouth on the south, which gives an indication of how widespread this problem of gannet luring was believed to be. The Lord of the Bass, George Lauder, was empowered to hold baron courts to try the offenders, sharing the fines with the Crown. This law was probably not widely or long obeyed, but it does demonstrate an attitude of mind concerned with maintaining a sustainable resource.

Modern conservation began for the Bass with the Sea Birds Preservation Act of 1869, which imposed a close season on seabirds between 1 April and 1 August, and the comparable Act of 1880, which extended the season back to 1 March. But the protected season ended too soon, and in 1893 a writer to the *Field* told of finding a heap of empty cartridges and the bodies of old gannets lying about the rock on 12 August. The close season was eventually extended in 1905 under bylaw powers by Haddington County Council from 1 August to 1 November, and thereafter things improved. The motivation behind this legislation was quite different from that behind the Act of 1592 – the Victorians became very concerned about the cruelty of shooting up seabird colonies in the breeding season, after centuries when it had been assumed to be harmless fun, and even the sport of kings.[9]

The first insight that we have as to the economic value of the gannet comes from a legal dispute of 1493. In that year the nuns of North Berwick went to the considerable trouble and expense of petitioning the Pope in Rome, complaining that the Lord of the Bass had withheld from the convent the customary tithe of barrels of seabird grease, and the Pope instructed the prior and archdeacon of St Andrews to sort out the problem.[10] This has to be seen in the context of the political atmosphere at the time: Norman Macdougall pointed out to us that Lauder of the Bass was out of favour with the king, who no doubt backed the prioress in her suit. Even so it seems at first sight a fuss about a trifle.

It was not. In 1521, John Major, at that point Principal of the University of Glasgow, wrote his *Historia Majoris Britanniae* and gave a detailed account of the Bass, which deserves credibility as he was born and brought up near to the East Lothian coast within sight of the island. He described how the gannets arrived in spring and circled the rock for two or three days without landing (they still do), 'during which time the dwellers on the Rock are careful to make no disturbing noise', how the adult brought fish to the young which the dwellers sometimes intercepted and used themselves (presumably by making the gannets disgorge a half-digested mackerel or herring), and how the fat young birds were sold in the neighbouring country: 'if you will eat of them twice or thrice you will find them very savory'. Only the lean parts were eaten, and he explained significantly, 'these birds are extremely fat, and when the fat is skilfully extracted, it is very serviceable in the preparation of drugs'.

But what drugs? His fellow principal at Aberdeen University, Hector Boece, gave another account of the Bass, repeating much of what Major had said, but was much more explicit about the drugs. The young gannets possessed, he said in a near-contemporary Scots translation, 'ane fatness of singulaire medicine, for it helis mony infirmeties, speciallie sik as cumis be gut and cater disceding in hanches or lethes of men and women'. The exact meaning of some of these difficult words has been misunderstood: 'gut' is certainly 'gout', but 'cater' is not 'catarrh', as Gurney supposed, but rheumatism, and 'lethes' are not 'groins' but 'joints'.[11] So the grease extracted from young gannets was used against gout and rheumatism, presumably as an unguent rubbed on where it hurt. It is easy to believe that the nuns would have dispensed it as part of their mission to the sick. William Turner, the pioneering English ornithologist, never visited Scotland, but in 1544 he too stated in the context of the Bass that a 'salve' was made from gannet fat, which 'may deservedly rival the Commogenum vaunted much by Pliny in its virtue and number of cures'. This could be copied from Boece, but English medical literature of the time also independently refers to the virtue of gannet grease. Then in 1578 Bishop Lesley again mentioned it as a produce of the Bass used against gout and other 'dolouris of the body', and said it was worth a great price. Finally, in 1585 the young German Lupold von Wedel visited the Bass to see the gannets, describing how a boy was let down the cliffs by a rope to kill the young ones with a cudgel, throwing the bodies into the water to be picked up by boat. He observed that gannet fat 'is very good in cases of paralysis, the limbs being rubbed with it'.[12]

At some point thereafter, however, the medical reputation of gannet fat evaporated in the mind of mainstream physicians, and no later visitor or commentator on the Bass, even the physician William Harvey, mentions it in this context. Its reputation as an unguent persisted as folk medicine in the Hebrides, and in the 1960s Bryan Nelson found from personal experience that it was efficacious against bruises. In John Walker's valuation of the produce of the Bass around 1765 the oil was worth only £2 13s 5d compared to £108 0s 0d from the produce of the meat and £5 0s 0d from the feathers. The oil sold at 8d a pint then. A century later, when the value of money was a third lower, it was worth only 6d a pint or less. By the Victorian period it was being used only for oiling boots and greasing farm carts and threshing machines. Sometime around 1875 its sale ceased altogether.[13]

The young birds were stripped of their feathers and their fat at Canty Bay, on the shore near Tantallon Castle opposite the Bass, and the fat boiled down in great vats. E.T. Booth in the 1880s described the little settlement permeated by a revolting smell, and half a dozen old women employed at 1s 6d a day to pluck the birds. They sat gaunt and grim in a circle of 'ruined and gloomy plucking sheds, their heads wrapped around with cloths, half smothered in dirt and feathers', so that 'a view of the proceedings almost conveys to an observer that he is gazing at an assemblage of witches'.[14] The feathers, both from young birds and old, had always been another valued by-product, primarily used for bedding after they had been baked to remove the smell. They held their value better than the oil: 10s a stone in 1765, but 15–18s a century later. Eighty to a hundred gannets were needed for a 'stone' of feathers, of 24 lb, and three stone for a moderate sized bed. However, the value of the feathers was always slight compared to that of the flesh of the young birds: for instance, around 1765, £5 for the feathers compared to £108 for the meat.[15] So from the seventeenth century until the mid nineteenth, the product of the Bass that was prized the most was the young gannet meat: in the absence of earlier prices for the oil, when it was of most value for medical purposes, we cannot be sure if that was true in the sixteenth century. The birds were killed in August just before they left the nest, when they were at their fattest: at about nine pounds, they were heavier than their parents, and the weight of a large human baby.

The birds from the Bass were prepared and eaten in a different way from those taken in the west of Scotland. On St Kilda, gannets were part of the staple diet of the islanders along with fulmars, puffins and other seabirds, preserved by drying, until in the nineteenth century sufficient salt became available for them to be barrelled and packed like fish. On Lewis, the inhabitants of Ness are still allowed (by dispensation from British and European bird protection law) a yearly cull of 2,000 young gannets from Sula Sgeir, on the grounds that this is a 'cultural tradition', analogous to the Inuit catching whales. The birds are pickled in salt in barrels at the colony, fat and all, and eaten by crofters and others in Lewis. They are cooked after being immersed in washing soda and boiled in a pot for 20 minutes: only when this has been done three times are they edible, having a salty fishy taste but the texture of meat. They should be laid down in the salt for at least three days to get the best flavour, and they will keep in this condition throughout the winter. Understandably, perhaps, not everyone likes them.[16]

The gannets of Ailsa Craig in Ayrshire were eaten in a different way again, sometimes fresh roasted rather than dried, pickled or boiled. Sir William Brereton in 1636 described a feast of solan geese at the home of the Earl of Cassillis, owner of Ailsa:

> So extreme fat are the young as that when they eat them, they are placed in the middle of the room, so as all may have access about it; their arms stripped up and linen cloaths placed before their clothes, to secure them from being defiled with the fat thereof, which doth besprinkle and besmear all that near unto it.[17]

In the eighteenth century, the Earl distributed young gannets as presents to local lairds and other friends, and also sold some to poulterers between 1764 and 1781 at the low price of 6–9s a dozen – less than half the price of a Bass gannet. He also kept some to eat dried in winter. They were not here a food exclusively for the rich, any more than they were in the Hebrides. In 1696, William Abercrumbie reported 'that the very poorest of the people eat of them in their season at easie rates'. As late as 1822 Ailsa gannets were reported as being salted to serve as winter food.[18]

The Bass gannets were eaten in a quite different way. Firstly the birds, stripped of layers of fat at Canty Bay, were eaten roasted and served exclusively as an elite dish. The English poet John Taylor in 1618 found that they were served at a sideboard before dinner 'unsanctified, without grace' and needed to be 'washed down with two or three good rowses of sherrie or Canary sack'. Because the fat had been removed as far as possible (they must still have been greasy enough) there was no need for the dressing up palaver that went on in Ayrshire.[19] In the west, the gannet was part of peasant subsistence, mainly eaten pickled and boiled or dried. In the case of Sula Sgeir, 40 miles offshore, the remoteness of the island and the length of time needed to complete the cull and return, also meant that it was best for the catch to be preserved and salted down on the spot. On the Bass, so close to the shore, birds could be caught and processed in a day, and the rock visited daily in August to ensure a fresh catch. The fat was stripped because it was seen as having its own medicinal value (perhaps considerable in the late Middle Ages) and the meat, enhanced by this process, was destined for immediate use by the well-to-do. The supplier had a monopoly – in the eighteenth century the tenant was usually a single dealer with an outlet in the Edinburgh poultry market – and the price was correspondingly high.

Even Bass gannet was not to everyone's taste. John Major, for whom it was a local delicacy, implied that it might take two or three attempts before 'you will find them very savory'. Taylor thought it a 'most delicate fowl', but his compatriot John Jonston five years later found the flesh hard, requiring wine and smelling of herring, and John Ray in 1662 wrote that it 'smells and tastes strong of mackerel and herring'. Charles II, sent some in London by the owner of the Bass, allegedly remarked that the two things he disliked most about Scotland was the Solemn League and Covenant and the taste of solan goose.[20] On the other hand, most elite Scots evidently found it delicious, at least up to the nineteenth century. Sir Robert Sibbald in 1684 averred that the art of cookery held no finer dish than roasted solan goose, 'of such delicate flavour, and combining the tastes of fish and flesh'. Gannet feasts had appeal far into the eighteenth century: thus, for example, Alexander Innes at the Royal Bank at Edinburgh received an invitation from Adam Ewart, written at Newbigging outside Musselburgh, on 8 August 1762:

> Our Solen or Solon Geese feest is to be holden Friday first the current and by your burgess ticket you have right to all the priviledges of this toun and your presence will be very agreeable to the Company.

Cooking it could be a problem – Sir Walter Scott said the smell was so bad that it should take place outside the house.[21]

Harvesting the Bass was a valued concession. In the eighteenth century William Mitchell, poultryman, bought the tenancy of the island from Sir Hew Dalrymple the owner, and from the 1740s to 1768 at least, Margaret Watson and then John Watson junior, poultry dealers, had the rights. The

	Price (pence per young bird)		
1511	12 [Scots]	–	James IV Household Accounts
1598	144 [Scots]	–	James IV Household Accounts
1624	10–12	–	Howard of Naworth Accounts, Cumbria
1652	30	–	Earl of Hartfell, Dumfriesshire
1654	25	–	Gordon of Straloch, in Blaeu, Edinburgh
1662	20	–	John Ray, Dunbar
1674–75	17–17½	–	Bass accounts, Maitland, wholesale
1675	29	–	Lauder of Fountainhall, Edinburgh
1688	24	–	Edinburgh Burgh Council
1701	24	–	Foulis of Ravelston, Edinburgh
1704	26	–	Foulis of Ravelston, Edinburgh
1710	24	–	Robert Sibbald
1755	20	–	Breadalbane Muniments, Perthshire
1764–67	20	–	Bass accounts, Walker, wholesale
1769	20	–	Thomas Pennant
1806	20	–	Buccleuch Muniments, Edinburgh
1833	20	–	Selby, Ornithology
1848	6–9	–	McCrie
1848	12	–	Wolley
1852	9–12	–	MacGillivray, British Birds
1860	9	–	Harvie Brown, Bass
1880	8–10	–	Booth, Rough Notes: English towns

Tables 9.1 – Bass gannet prices. Sources: Gurney, *The Gannet*; *Dictionary of the Older Scottish Tongue*; NAS: GD 112/15/337/12; GD 224/351/120; NRAS: 2171 bundle 60/1; Fleming in McCrie (ed.) *The Bass Rock*; MacGillivray, *British Birds*; Selby, *British Ornithology*; Booth, *Rough Notes*.

A note on the value of money: Inflation in the course of the sixteenth century was approximately twelvefold, which means that in real terms a young gannet would have cost about as much at the start of the century as at the end. In 1603 the value of the £ Scots was fixed at 1/12 of the £ sterling and all prices thereafter are quoted in sterling. In 1598 it would have been more like 1/9 of the £ sterling, but inflation lessened thereafter. In rough terms the value of money was approximately constant from 1620 to 1760, but price levels doubled between the 1760s and 1806, falling back by the 1860s to 50% higher than a century earlier. So a gannet bought in 1806 cost half what it had in 1765, in real terms, and in 1860 about one third.

	Bass valuations		
1535	400 'gold pieces'	–	value to the commander of the feathers and fishes brought ashore by the gannets: Peter Sware, Danish visitor.
1618	£200	–	profit to the Laird of the Bass: John Taylor, English visitor.
1635	£200	–	value of the birds as sold: William Brereton, English visitor.
1662	£130	–	yearly profit to the laird: John Ray, English visitor.
1678	£75	–	rental agreement with Charles Maitland of right to birds.
1715	£33	–	subtack to Edinburgh poultryman of right to birds.
1764–7	£46 13s 4d	–	rent of the Bass – Walker.
	£120 13s 5d	–	'produce' of the Bass – Walker.
1820	£35	–	rent of the Bass – E.G. Fleischer.
1833	£60–70	–	rent of the Bass – P.J. Selby.
1841	£30	–	rent of the Bass – Statistical Account.
1876	£20	–	rent of the Bass – Gurney.
1913	£17	–	rent of the Bass, recouped by visitors' fees – Gurney.

Table 9.2 – Bass valuations. Sources: Gurney, *The Gannet*, NAS: GD 110/735.

Notes: The first quotation can be disregarded as vague and incredible. The next three are also only opinions given by travellers, and refer to the gross yield of the Bass; as the more detailed account of 1764 shows, there was a substantial difference between the rent paid to the laird and the gross produce of the island. There is something odd about the 1678 rental, as the rent was roughly equal to the gross yield of the birds in the previous five years and makes no allowance either for profit or for expenses. And the discrepancy between Fleischer (1820) and Selby (1835) makes it unlikely both were correct.

The most significant point is that the rent in 1876 was less than a third in real terms of what it had been in 1764–67.

combination of elite fancy and monopoly supplier kept the price high for centuries (Tables 9.1 and 9.2 and their associated notes). This can be seen from the very first price quotation for supplies to the royal court in 1511. Around 1700, when a young solan goose from the Bass was selling for 2s, a farmyard goose could be bought in and around Edinburgh for 10d, a capon for 8d and wild duck from 4d to 6d.

Access from the landing place on the Bass was protected by a wall and a locked gate, but thefts occurred. In 1723, a group of men from Fife, including a skipper from Pittenweem, a shoemaker in Cellardyke, the ground officer at Innergellie House and others, allegedly broke through the defence, stole £50 sterling value of young gannets by going up the cliffs on ropes and throwing the dead birds down into the water.[22] These were said then to be worth 1s 8d each. But values changed: by the 1790s when a young gannet was still worth about 1s 8d, a fat domestic goose in the east of Scotland cost at least 2s 6d. Pennant remarked that solan goose was one of the few provisions the price of which had not gone up in 100 years, and its relative value was clearly already falling.[23]

As the tables show, in the nineteenth century the popularity of young gannet meat suddenly plummeted. By 1806 it was worth in real terms only half of what it had been 30 years before, and by mid century less than a third. Tastes were changing, and the elite were less enamoured by what Henry Cockburn termed 'Scotch peculiarities'. As the bottom fell out of that market, the tenants of the Bass took advantage of the new railway network to send gannets pre-cooked at Canty Bay to London, Newcastle, Sheffield, Birmingham and Manchester, where they 'were consumed at the commoner eating houses'. They also continued to sell them in Edinburgh and to hawk them round the countryside in carts. Local farmers bought them as a cheap treat for itinerant Irish harvest labourers: 200 gannets 'landed in boats from the Bass and afterwards cooked in brick ovens are said to have been sometimes consumed at a single feast'. It was to no avail, and after 1885 it was judged no longer worthwhile to collect the birds, and the trade of centuries came to an end.[24]

As the sale of young birds declined, there was an effort to compensate by selling the eggs instead. In the past this had been discouraged. As an Edinburgh lady, Marion Trotter, explained in a letter to her friend Jane Innes in 1822, accompanying the gift of two hard-boiled gannet eggs, the tenant of the Bass did not usually allow their taking 'as that always makes fewer geese to sell'.[25] By 1860, however, the tenant of the Bass was selling them at 6d each, advertising them with a flyer that reproduced a letter from Buckingham Palace, following a gift to the Queen.

> The eggs . . . were as duly handed round the table, universally tasted, and admitted to be indistinguishable from plover eggs. Royal thanks are therefore your due, the expression of which I am commanded to convey.

Queen Victoria or her officials seem to have been more tactful than Charles II. There continued for a time to be some market for eggs. Some 2,000 were taken in 1880, and 1,800 in 1885, but very few by 1913 when it was said that 'nobody eats them'.[26]

When the market for gannet produce began to disappear, a much more serious threat to the colony emerged in the form of 'sport'. Shooting nesting seabirds at their colonies for pleasure had a long history. James IV on a trip to the Isle of May in 1508 with local lairds and clergy from the priory of Pittenweem, shot at seabirds with culverins, and it has been surmised that his trip to the Bass in 1497 had also been for the sake of sport.[27] However, lesser mortals certainly did not get to shoot at the gannets or anything else. In the fifteenth and sixteenth centuries and much of the seventeenth century, the Bass was a garrisoned fortress guarding the entrance to the Firth of Forth, under the control of Lauder of the Bass who was also zealous to protect his valuable birds. After the Restoration it became a state prison to incarcerate Covenanters and other dissidents. When the political wheel of fortune turned, it became a prison for Jacobites, who turned the tables on their captors and held it in the name of James VII for three years, no doubt eating gannets when under siege. Though in 1706 it ceased to be fortified and passed to the family of Sir Hew Dalrymple who own it to this day, it continued to be strictly protected for the value of its produce. When Thomas Pennant visited in 1769, he explained that it was not permitted to shoot the gannets – 'the place being farmed principally on account of the profit arising from the sale of the young of these birds, and of the kittiwake'. And P.J. Selby as late as 1833 said that great care was taken to stop the adults being shot or trapped within 'a certain limited distance of the island', though naturalists were sometimes allowed to take specimens.[28]

Within a generation came a basic change. The price of young gannets had fallen even further, the sale of eggs, feathers and young far from made up the gap, the rent had halved, and there was little incentive for tenant

Opposite: Fig. 9.1 – Gentlemen shooting seabirds on the Bass in 1876. Notice the wine in the picnic basket.

or landlord to continue to protect the birds. Cheap shotguns, firing cheap cartridges, were available first as a novelty, and then as an essential tool for every would-be sportsman. This created a revolution in the affordability of recreational killing, and ordinary tourists on steamers, or fishermen in their sailing boats, were able to take part in shooting trips to seabird colonies, which they did all over Britain from the Yorkshire coast to the Hebrides.

For the elite, too, shooting at nesting seabirds had a particular cachet (Fig. 9.1). This was well expressed by John Colquhoun, whose *Moor and the Loch* had run to four editions by 1878. In a chapter entitled 'Sea-fowl shooting in the Firth of Forth' (despite the title of the book) he gave an ecstatic account of the pleasures of shooting up Fidra, the Isle of May and the Bass, in the breeding season. He extols the superiority of the discriminating sportsman seeking specimens in wild places all year round, over the 'mere shooting-machines, whose delight is a partridge or grouse *drive*'. He explains how wonderful it is to take a fishing-boat and go out to the nesting cliffs:

> Should he secure the coveted object of pursuit, it is not alone its rarity or beauty which makes his heart to dance like his little shallop over the waves, but the associations sure to cling to it in future and far-distant years. To him each of those seabirds that grace his museum suggests its own wild tale of grandeur or beauty. The beetling precipice, the gleaming, tranquil sea, the jutting headland or looming, boundless old ocean, rise to his mind's eye, fresh and glorious, even by a passing glance at that little denizen of the deep.

He landed on the Bass, admired some nesting gannets which 'refuse to move until kicked off the nest ... their threatening antics were ludicrously pompous', obtained some 'specimens', and described an 'inviting snip of rock' on which 'gentlemen often practise rifle-shooting at the geese'. This was probably around 1860.[29]

Shooting nesting seabirds, a royal sport in the past, became a royal sport again. The Prince of Wales, future Edward VII, accepted an invitation in 1859 from Sir Hew Dalrymple, took a special train to North Berwick, embarked with his entourage on his host's yacht at Canty Bay, and, escorted by a gunboat, HMS *Louisa*, made for the Bass: 'there the prince enjoyed some shooting and brought down a number of gannets'.[30] If one is to believe contemporary comments, however, the real damage that was done to the colony in these years was done not by princes and gentlemen but by less elevated folk. Cosmo Innes spoke of 'mischievous idlers shooting at them from boats', and E.T. Booth, a gentleman sportsman himself, spoke of eggs taken by 'irresponsible tourists' and of 'boat loads of strangers who expended a quantity of ammunition in blazing away at the busy swarms engaged with their nests'. One year, according to Colquhoun, 'the whole west side of the rock was depopulated from fishermen and others'. The tenant in 1860 charged 2s 6d for each adult gannet shot (considerably more than he could then get for sale of the young), but in 1876 a different tenant deplored the numbers shot after 1 August while still feeding young, 'adding with emphasis that he had seen the sea strewn with dead gannets which the shooters did not take the trouble to gather up'.[31]

In England, as mentioned above, there developed a movement to put an end to this slaughter of seabirds at their nesting cliffs, pioneered in Yorkshire in order to defend the colonies around Scarborough, and championed by Professor Alfred Newton of Cambridge University. This led to the Sea Birds Preservation Act of 1869 and a close season for gannets up until 1 August. This, however, did not get much sympathy in Scotland, where in 1877 a government report on the herring fisheries recommended the repeal of the Act for 'gannets and other predacious birds' so far as it applied north of the Border. This was only prevented by a strong campaign by Professor Newton and his allies in the press and in the British Association. Unfortunately, the close season for gannets was only from 1 April to 1 August, which proved too soon to put a stop to the killing.[32]

Apparently the only person in Scotland to write in defence of the birds on the Bass was Emily Bowles, who dedicated her pamphlet *A Trip to the Bass Rock*, 1870, to Sir Hew Dalrymple, 'by permission'. Basically it was a topographical and historical account strongly flavoured with religious sentiment, but it pulled no punches about the treatment of the birds:

> As any boat nears the rock, the air suddenly fills with flitting and soaring snow, and the singular creaking and laughing cry of thousands of solan-geese, guillemots, razor-bills and gulls grates upon the ears and nerves. The absurd habit of popping off guns from the pleasure-steamers and boats, and the landing of still more absurdly-ignorant and heartless tourists, who shoot the birds and capture the eggs and young, has greatly lessened the numbers of the fowl, and threatens in a few years to drive them

entirely away, as they have been driven away from other parts of Scotland. It is much to be regretted that the owner of the Bass [here in a footnote she gives his name and address] does not take some interest in preventing the mischief done both by the tourists and those who have charge of the boats, as it is seldom that so fine a haunt of sea-fowl can be studied with safety and ease.[33]

The gannet, she added, deserved 'more than common care … to preserve, or at least to defend, the breed'.

The situation, however, was about to change, and Emily Bowles' view to prevail over that of the self-indulgent sportsmen. Over the next century recreational observation completely displaced recreational killing as a way of enjoying the gannets of the Bass. By 1905 the colony had legal protection throughout the breeding season, and though disturbance did not immediately cease, the tourists who came to sail round the island, and sometimes to land, now did so peacefully to admire the birds. In the second half of the century, bird-watching grew as a popular pastime, and the reputation of the Bass was enhanced both by the sojourn of Dr Bryan Nelson and his wife to study the breeding biology of the gannets on the rock in the 1960s, and by the visit of a black-browed albatross that spent the summers of 1967 and 1968 on the Bass, under the evident impression that it was a gannet, and drew boatloads of twitchers to add it to their lists.

The gannet colony, however, only materialised into a hard economic asset when Bill Gardner, community councillor and businessman in North Berwick, had the inspired idea that video cameras could be put on the rock, and the close-up sight of the gannets and the other breeding birds enjoyed and interpreted in comfort on shore. The upshot, after ten years of hard work, was the construction of the Scottish Seabird Centre, which was opened by Prince Charles in May 2000. It was appropriate symbolism that he took a trip round the island with his binoculars where nearly a century and a half before a preceding Prince of Wales had taken his gun. The Seabird Centre has been an outstanding success. With a turnover of about £1 million, a staff of 50, 300 volunteers, 6,000 members and (in 2007) 284,000 visits, it has helped to transform the economic fortunes of North Berwick, and proved a major attraction in a part of the country that had been losing out on the modern tourist business. It now has cameras on several other islands, including the Isle of May, which show the breeding grey seals in early winter.

THE BASS AND ITS GANNETS

Opposite: Fig. 9.2 – The Bass around 1903: the seabird colony is limited to one part of the cliffs.

Fig. 9.4 – The Bass from the air in 2009: gannets cover every foot of space except in the former lighthouse garden.

Opposite: Fig. 9.3 – The Bass from the air in 1969, the gannets are increasing.

Never in six centuries have the gannets (in the words of the statute of 1592) been more 'proffitable for the commoun weill of this realme'.

3. The Many Consequences for the Gannets

So far we have concerned ourselves only with the benefit that humans had from the Bass gannets, but what was the consequence of human activity for the birds? We first need to know how many gannets there have been in the past. It is easiest to work this problem out backwards, starting with the relative certainties of the present. Seabirds nesting at high densities are hard to count, and without the assistance of cameras and slow-flying light aircraft, doubly so. The most recent census, an aerial survey of 2009, counted 55,482 apparently occupied breeding sites: and the contrast with pictures of the island taken earlier in the century are manifest (Figs 9.2–9.4 and Plate 4) The figures for the twentieth century are given in Table 9.3: all are based either on photography or on estimates likely to be accurate within a 10 per cent margin. As can be seen, the twentieth-century increase is sensational: starting at a level of around 3,000 pairs, the average annual rate of increase between 1904 and 1977 was around 3 per cent, accelerating

Table 9.3 – Gannet nest counts. Source: R. Murray, M. Holling, H. Dott and P. Vandome, *The Breeding Birds of the South-east Scotland* (Edinburgh, 1998); Nelson, Gannet, p. 70; S. Wanless, S. Murray and M.P. Marris, 'The status of Northern Gannet in Britain and Ireland in 2003/04', *British Birds*, 98 (2005), pp. 280–294; S. Murray, 'An aerial survey of the Bass Rock gannetry in 2009', *Scottish Birds*, 31 (2011), pp. 220–225.

Gannet nest counts (apparently occupied sites unless otherwise stated)

1904	3,000 pairs
1913	3,250 pairs
1929	4,147 nests
1936	c.4,150 pairs
1939	4,374 nests
1949	4,280 nests
1962	c.7,000 (6,690–7,126)
1968	c.9,000 (8,000–10,000)
1974	c.10,500 (9,500–11,500)
1977	13,500
1985	21,589
1994	39,751 (34,397 nests)
2004	48,065
2009	55,482

to around 7 per cent between 1985 and 1994, and currently down to nearly 3 per cent again as the rock apparently nears its maximum capacity.

When the expansion began, the birds were entirely restricted to the cliffs, and the seven acres of the sloping surface was occupied by 'enormous numbers' of rabbits, which were shot by the lighthouse keepers who had first been established there in 1902 (Fig. 9.5).[34] The rabbits were killed off in the myxomatosis outbreaks of the 1950s, and the lighthouse keepers departed in 1988. This left the island entirely to the gannets, which had been legally protected since 1905, and their numbers expanded accordingly all over the deserted slopes.

Fig. 9.5 – Lighthouse keepers and lads out shooting rabbits on the Bass around 1909.

But how many gannets had there been before the idle slaughter of the later nineteenth century reduced the total to some 3,000 pairs? There are several Victorian guesses that are not worth credence, but one that does deserve attention is Professor Fleming's suggestion of 5,000 pairs in 1847, which was an extrapolation from about 1,800 young birds taken annually. It is considered a reasonable calculation by modern biologists, though 'it is unlikely that numbers could have been much lower than this at this level of exploitation'.[35] However, the colony might have been larger if something less than maximum sustainable yield was being sought. Clearly, adult survival was being affected by high mortality – already Fleming was deploring 'unfeeling sportsmen shooting the parent bird throughout the period of incubation', apparently from the sea as it was not yet allowed on land.[36] The naturalist MacGillivray estimated some 20,000 gannets on the Bass in 1831; his observations perhaps deserve more credibility than most. Sarah Wanless remarks 'it is difficult to extrapolate from the number harvested to the total population with much greater precision then that numbers were likely to be somewhere between 5–10,000 pairs'.[37]

Thereafter, the cull of young birds steadily declined: estimates included 1,700 in 1850; 1,500 in 1865; 800 in 1874 and 1876. After about 1885 no young were culled, but 2,000 eggs were taken. By the early twentieth century, eggs were no longer removed. As gannets will often lay again should the first egg be lost, a figure of 2,000 eggs does not equate to the same number of young birds or breeding pairs. In earlier centuries, the take of juveniles had been somewhat below that of the early nineteenth century. In 1674–77, it had been an annual average of 1,078, around 1764, 1,296. If that was maximum sustainable yield, we may imagine a population proportionately smaller.[38]

This presumes that the tenants of the Bass harvested to complete capacity without running down the stock. But what if they operated well below that level, either to maintain the price of young gannets or because they could not catch them? The likely maximum population in these circumstances would have been determined firstly by the area of the rock available for them to nest in, and also by the carrying capacity of the marine ecosystem in which they foraged. We can perhaps disregard the second factor in view of the recorded abundance of herring and other marine life in the North Sea before the onset of modern trawling.

Today, gannets nest on almost the whole of the rock apart from the areas of the cliff splashed by the sea, and part of the sloping areas occupied by former garden ground or by buildings and their immediate appurtenances. In the later nineteenth century and earlier twentieth century they occupied only the cliffs, having deserted, sometime after 1862, some more accessible areas near the summit as they were shot out. When Thomas Pennant sailed round the island in 1769, he saw 'multitudes . . . sitting on their nests near the sloping part of the isle' (he does not say *on* the sloping part), and Jardine in 1816 wrote of there being nests on the summit of the rock. All this suggests that they mainly nested on the cliffs but with some overspill onto the edge of the slopes.[39]

Two earlier observations, however, were cited by Gurney to show that 'the gannets' breeding ground was evidently much more extensive than it is now' and 'must have reached over all the upper portions of the Bass'.[40] The first witness was William Harvey, who, in June 1633, landed on the Bass and described the view from the top of the cliffs. He says 'the surface of this island, in the months of May and June is almost entirely covered over with nests, eggs and chicks, so that for their very great numbers you can scarcely anywhere set your foot on an empty space'. He does not say that the birds he walked among were gannets; they could equally have been gulls or terns. A certain hyperbole is evident when he adds that 'such a mighty flock hovers over the island that, like thick clouds, they darken and obscure the day . . . you can scarcely hear the words of them that stand next to you'.

Gurney was misled by an inadequate translation from the Latin to understand that Harvey also said that the whole island was white with guano, but this is what he actually says:

> As you approach the island, it glistens with a shining whiteness, and the cliffs shine as if they were made of the whitest chalk, though the natural colour of the rock is dark and black. That which makes the island white and shining is a very white crust which sticks to it and is friable and is of the same consistency, colour and nature as egg-shell, and all the sides of the island are covered with this hard coating and daubed with this white and friable crust.[41]

He goes on to say that the parts of the island washed by the tide retain a natural black colour, which 'plainly shows that the whiteness on the summit is artificial'. This surely makes it clear that it is only the cliffs that he saw as covered with guano.

The second observer was the engraver John Slezer, who in 1693 speaks of the 'surface [of the Bass] being almost covered with their nests, eggs and young birds', which may be directly copied from Harvey. His illustration of the rock in *Theatrum Scotiae* does not clearly show this, and a contemporary late seventeenth-century painting (Plate 13), obviously in summer, shows no birds at all on the grassy slopes, which are green and not covered with white guano.[42]

There is in fact strong circumstantial evidence to suggest that throughout the period down to the later twentieth century most of the colony was restricted to the cliffs with only a relatively small number outside. Firstly, the Bass supported a flock of sheep which occupied much of the accessible area. Sheep were first mentioned by the English traveller Sir William Brereton in 1635 (two years after Harvey's visit). There is no reason to think that they were a novelty and there are many subsequent records of grazing as an element in the rent, though usually set at a trifling level compared to the value of the gannets (in 1767, 'sheeps grass' was worth £5 compared to £125 for the take from the birds). The numbers of animals were around 30 according to Ray in 1662, 20–30 according to the Francis Grose in 1789 and also to the *Statistical Account* of 1791–2 and 30 (sometimes only 20) according to Fleming in 1847. The extent of their pasture was put at about seven acres both in the *New Statistical Account* of 1835 and by an anonymous author of 1912, which is almost equal to the total area of the Bass at 7.4 acres.[43]

It is likely that the presence of the sheep kept the gannets off most of the island, but not all. On the other hand, sheep can co-exist with terns, as on Shetland today. MacGillivray in 1835 found 300 pairs of gannets nesting below the castle, an area to which the sheep had no access, and Fleming in 1847 spoke of 'two colonies which, unable probably to find space among the cliffs, confine themselves to the very margin' of the rock, implying that they were quite limited in extent. The sheep were very highly regarded for the quality of their meat, but they were withdrawn from the island at the end of the nineteenth century. This left the pasture to the rabbits, which are recorded in the seventeenth century and had been there probably since medieval times, but apparently became much more numerous in the nineteenth century. By 1912 they were swarming all over the slopes.[44]

Secondly, there is the point that the young were harvested by rope and basket, later by rope alone – a very dangerous practice for all concerned, which would not have been necessary had the cull of 1,000–2,000 birds been possible without it. Use of rope and basket is first mentioned by William Turner in 1544,[45] and use of the rope alone continues right down until the end of culling in the 1880s. William Daniell in 1822 explained how the catcher wore thick worsted stockings to get a firmer hold, using the rope mainly to steady himself. A typical account of what happened is that in the *New Statistical Account* in 1835:

> They are taken from the rock by the keeper, who descends with a rope fastened around his waist, and held above by his assistant; another rope is fastened to the rock above, which he holds around his hand to facilitate his movements. He lays hold of the

bird with a hook, draws it towards him, and kills it with a stroke on the head; then with great force throws it from him over the projections of the rock to the sea below, where the men in the boat are prepared to pick it up. The act of throwing, the keeper tells us, is the most difficult and perilous effort in the process.[46]

Ray in 1662 said 'seldom a year passeth but one or other of the climbers fall down and lose their lives, as did one not long before our being there'.[47] Others spoke of the peril of being in the boat below when the heavy birds were raining down from an unseen source above.[48] No one would have undertaken such a task if it had been possible to harvest enough from the slopes alone.

This suggests that only some hundreds of pairs were nesting outwith the cliffs. If the population was otherwise confined to the cliffs, what is the carrying capacity of the ledges? When Bryan Nelson was working on the Bass in the 1970s, he reckoned that about 6,000 pairs nested on the cliffs.[49] So, calculating by this method, we arrive at a similar conclusion as before – the historic population level of the gannets before the twentieth century was within the bracket of 5,000–10,000 and probably 6,000–7,000 pairs, compared to some 55,000 pairs in the early twenty-first century. So the population was kept well below its apparent potential by a combination of human predation and human competition for use of the breeding area. That does not mean that exploitation was, in modern terms, 'unsustainable'. Until the collapse of gannet prices in the mid nineteenth century it was apparently perfectly sustainable, in so far as the colony would consistently yield 1,000-2,000 birds for the table for several centuries without diminishing or disappearing. But this example shows how a natural resource can be simultaneously sustainably exploited and held well below its potential by that exploitation. The population began to recover after effective protection had been introduced in 1905, and by 1962 had regained or exceeded its historic level. However, it continued to grow at the rate of 3 per cent or more until the present day. This, however, is now unlikely to continue, and not only because the gannets are running out of room on the Bass on which to nest. Evidence from recent years suggests that almost half the pairs attending a nesting site are failing to breed, whereas in the 1960s and 1970s virtually all occupied nests had young. This strongly suggests some new limit on their success, perhaps related to food supply.[50]

The gannets may travel 100 miles or more in search of prey for their young, but the fact that the population on the Bass, up to now, has increased so far and so fast above its historic limits is at first sight surprising, considering the extent of overfishing in the North Sea in modern times. Is it possible that it has even benefited from modern fishing practices? It is common, especially in April after their return from winter quarters, to see flocks of up to 500 gannets following the fishing boats back up the Firth of Forth towards Pittenweem, and feeding on discards and offal thrown overboard. Recently in the North Sea, 54 per cent of cod caught, 36 per cent of haddock, 20 per cent of whiting and 14 per cent of herring (55,000 tons of fish), are thrown away as discards by Scottish fishermen alone. Yet it seems from Robert Furness's research that the gannets feed very little on discards in summer, and that they feed their young on sandeels when they can get them, or on herring and mackerel that they catch themselves. It is in winter and early spring, when the sandeels have buried themselves in the seabed and mackerel migrated to the south-west of Europe, and herring become less available, that the gannets turn to discards and out-compete other birds round the trawling fleets. However, contrary to popular belief, discard volumes were at their highest in the 1960s and 1970s, and have been very considerably reduced in the northern North Sea in recent years. Numbers of gannets wintering in the North Sea and attending the trawlers have declined, so it appears that their dependence on discards is probably decreasing. So we really cannot tell whether or not the colony on the Bass could have reached its present size in the absence of modern industrial fishing, but it seems perhaps unlikely that discards were a critical factor.[51]

Such a huge volume of birds crowded on seven acres of rock also has its downside. The grass and plants of the former pasture have been wiped out, and visiting the colony can seem quite hazardous, so overwhelming is the mass of highly protective and well armed gannets. Then there is the smell. It was evident even in the nineteenth century, when Gurney said the fishermen could sniff it in the air from several miles off, and E.T. Booth in 1887 commented that 'the stench that arises from the steaming nests under the scorching rays of the summer sun after a wet morning is positively overpowering'.[52] Now that there are more than ten times as many birds present, the emissions are astonishing. It has been demonstrated that the colony releases 132 tonnes of ammonia into the atmosphere each year, the equivalent of 10 million broiler chickens or 20 average-sized chicken farms.[53] Truly the Bass is a wonder of nature, but it is lucky that no one lives nearby and that the prevailing wind is offshore.

CHAPTER 10
The Isle of May and the Other Seabird Colonies

1. Introduction

The Firth of Forth is studded with islands, 'emeralds chased in gold', as Sir Walter Scott called them when the hero of *Marmion* surveyed the scene from Blackford Hill in Edinburgh. The biggest and the most exposed is the Isle of May, lying across the entrance of the Firth; seen from Crail, five miles away, it is more like a breaching sperm whale than an emerald. A long volcanic rock of 134 acres, it is the largest island off the east coast of Britain, with cliffs on the western side 164 feet high, sloping gradually to a small beach on the east (Fig. 10.2). The next largest (but only half the size of the May) is Inchkeith, rising to 180 feet, in the middle of the upper

Fig. 10.1 – Islands of the Firth of Forth (not to scale)

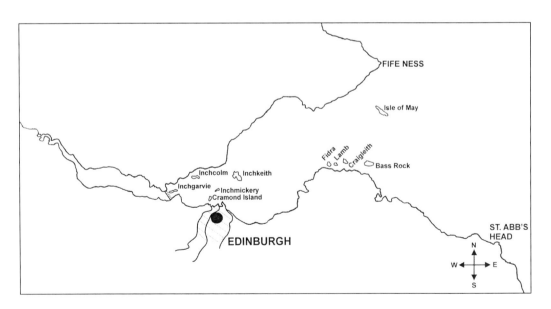

Fig. 10.2 – The Isle of May from the air, 1979.

Firth between Kinghorn and Edinburgh, and the third is Inchcolm, just off Aberdour. Apart from the Bass, three small islands lie off North Berwick, Fidra, Craigleith and the Lamb, and two more off Edinburgh, Inchmickery, hardly more than a long rock, and Cramond Island, which is accessible at low tide. Inchgarvie is the last small island, today attached to the Forth Rail Bridge at Queensferry. There are also half a dozen other less significant rocks and reefs, too small for occupation except by birds and seals. None of the islands are inhabited by people today, apart from the Isle of May, which supports a small colony of non-breeding scientists in summer.

Nevertheless, all except the rocks have had a long history of human habitation, in the case of the May and Inchkeith at least going back to Neolithic times. In the Middle Ages there were monasteries on the May and (surviving for longer) on Inchcolm, and chapels on the May, Fidra and Craigleith, with an established pilgrim traffic to the May and a less important one to Fidra. The May was probably always too exposed to ripen grain, though herbs and vegetables were grown in walled gardens. On Inchkeith, there was excellent grass but no cultivation in the mid seventeenth century 'because of the lack of interest of the owners'.[1] Yet there was cultivation in fertile pockets there in the early nineteenth century, and Thomas Carlyle, visiting in 1817, described 'some patches – little bed quilts as it were – of weak and dishevelled barley trying to grow under difficulties; these, except a square yard or two of potatoes, equally ill off were the only attempt at crops'. There was still some arable early in the twentieth century. On Inchcolm it was said in 1899 that the soil close to the monastic ruins was 'very fertile, producing excellent crops of potatoes and turnips; oats used to be grown'.[2] There are few trees today on any of the islands except Inchcolm and Cramond Island, both relatively sheltered and inshore, but bushes have been planted by the bird observatory on the May to shelter the falls of small migrant birds, which strike the island in south-easterly weather.

Most of the islands are capable of growing grass as well as thrift and sea campion, and were grazed by rabbits, sheep and sometimes cattle until the start of the last century, even when there were no resident farmers. The rabbits are still present on some of them. Before the middle of the eighteenth century, the May had an active fishing community, said to have consisted of up to 15 families. It sent boats to the herring drave, and sometimes one of the fishermen served as chairman of the jury in the local Admiral's Court. In 1724 eight fishermen and the lighthouse keeper were imprisoned by the Admiral's representative in Anstruther tollbooth for not paying him for a whale they had caught, 'with great pains, trouble and damnage' they said, 'in the full sea' – they denied that it was a whale anyway.[3] The villagers were also active smugglers, particularly in the early eighteenth century when high duties after the Treaty of Union made smuggling both profitable and patriotic. Lighthouses were built in the nineteenth century on many of the islands and rocks, but the first one on the May was a coal-fired beacon dating from 1636, the earliest in Scotland. It was replaced in 1819 by an oil-fired light with reflectors, leaving nearly two centuries of cinders as a covering to the crown of the island. The lighthouses were manned until the 1980s by up to seven keepers, sometimes also with wives and families, who cultivated small gardens and kept animals: in the interwar years on the May these included goats, cats, dogs, ferrets and a horse, and they had guns to shoot the rabbits.[4]

Warfare and piracy were another constant theme in the history of the islands for more than 1,000 years, at least from the traditional date of a Viking raid on the May around 875, when they supposedly sacked the church and slaughtered the monks, until the Second World War, when the May, Inchkeith, Inchmickery and Inchgarvie were manned and fortified to protect the approaches to the naval base at Rosyth. Finally, several islands were repeatedly used in the sixteenth and seventeenth centuries to quarantine ships suspected of bearing plague: the captains were ordered to lie at Inchkeith, Inchcolm or elsewhere until their crews had either developed the infection or were proven to be clean. Inchcolm today contains a small colony of black rats, the species that once carried the plague-transmitting fleas: it is one of very few such colonies left in Britain.[5] Inchkeith has only brown rats, which replaced the black rats in the eighteenth century and never carried plague fleas, so they must have arrived later. It also was for a time in the 1980s home to feral cats that escaped from an animal 'sanctuary' established by an eccentric tenant, and which decimated the fulmar chicks.[6] The May, fortunately for the seabirds, has no rats (or cats) at all now, though rat bones have been found in the old middens. Cats were at one time kept by the lighthouse keepers there and one keeper introduced hedgehogs in the 1970s to eat the snails, but they were quietly recaptured by the nature reserve warden and returned to the mainland.[7]

There is another habitat that can be linked to the islands, in the sense that it provides a similar niche for marine birds and mammals, relatively free from the depredations of land-based predators like people, foxes, hedgehogs or cats and dogs, because of its vertiginous character – the great Berwickshire sea-cliffs and caves around St Abb's Head and Fast Castle, which rise sheer to almost 300 feet. As we shall see, these also hold large colonies of seabirds, and also seals.

So, until the second half of the twentieth century, mankind was a very active resident of the islands and to some extent a trespasser on the cliffs as well. Since then, the main human visitors have been summer tourists on carefully managed boat trips to the larger islands, especially the May, and the wardens, bird-watchers and ecologists who look after and study the National Nature Reserve which was declared there in 1956. St Abb's Head is now also a National Nature Reserve, wardened and protected by the National Trust for Scotland. The human impact has not been completely withdrawn from the islands themselves, but it is very much less than it was historically. On the other hand, as we have already seen, the human impact on the waters on which the birds and seals depend has never been greater.

2. The May in the Middle Ages and Later

Of all the islands, the Isle of May has been the best studied. Between 1992 and 1997, a very expert and thorough archaeological excavation was carried out.[8] Firstly, it produced quantities of high quality flint imported from England during the late Neolithic or early Bronze Age, suggesting that the island was then already an important locality on a trade route between Lothian and Fife. The dig was focused on the early ecclesiastical site, and revealed traces of an early Christian cemetery with origins in the fifth or sixth century, along with a stone-built church begun in the tenth century round the shrine of the seventh-century St Ethernan, later called (or conflated with) St Adrian. This was replaced around 1145 by the Cluniac priory founded here by monks from Reading, invited over by David I. At the end of the fourteenth century, English attack and general piratical disorder persuaded their Scottish Augustinian successors in ownership, to concentrate on a safe haven ashore at Pittenweem, leaving the monastic buildings to decay. After that, there was a single canon or resident 'hermit' from Pittenweem priory picking up the profits of the many pilgrims who came to worship at the reliquary chapel. In the later sixteenth century, the remains of this church were converted into a fortified manor house. The inhabitants of the May from then until the twentieth century were entirely secular, at first farmers, fishermen, smugglers and lighthouse keepers, but, after the death of the last villager before the middle of the eighteenth century, only the lighthouse keepers were left.

The excavation also revealed a great deal about the environmental history of the Isle of May in medieval and early modern times. Some of it was unexpected. Although the landscape was basically open, dominated by marshy grassland and heath, fossilised tree pollen from early times indicated that there had been, in places, sparse alder and hazel scrub, with some willow and birch, possibly even oak. Peat began to form around the tenth century as this woodland declined, but only from the mid fifteenth century onwards did trees completely disappear. There are no unequivocal signs of cultivation of any grain crop at any time, but the bones of pastured domestic animals were common in the medieval middens at all periods. Most numerous were sheep or goats, which are difficult to separate, but almost all were considered to be sheep. Cattle were the next most numerous, and the bone evidence suggests that these were from a small milking herd. Pigs were not plentiful, though found at most levels, as were horses in very low numbers, and some chickens occurred throughout.

The remains of wild creatures eaten include a good many seals, assumed to be grey seals, including pups, and the butchery marks indicate that they were skinned for their pelts as well as used for meat. Seal meat would have been important to the monks, as it was allowed in Lent, on Fridays and other fast days, because it was canonically classified as fish. In 1508 the hermit brought a seal to James IV when the king was on pilgrimage to the May, and was paid 13s. Real fish were also extremely important, especially cod and haddock caught by line either from the shore or (the larger specimens) from boats. Herring bones were present but less numerous, and at least another 15 species were eaten. Among shellfish, the most frequently found in the middens were limpets and periwinkles.

The bird bones contain some surprises. Small numbers of great auk bones were found in all periods between the early Middle Ages and the fifteenth or sixteenth century. There were 14 bones in all, from at least four individuals, compared to guillemot bones, which totalled 694, from at least 76 individuals, so the great auks were clearly much rarer. These are the first indication that this species (globally extinct since the nineteenth century) lived in the east of Scotland later than prehistoric times, though remains have also been found in Iron Age middens near Dunbar. They could have been from wintering or moulting birds rather than indications of a breeding colony on the island. Puffins, on the other hand, for which the May is now so famous, left fewer remains in the middens than great auks, and none at all for the period before 1150. It is not because they are inedible, since they were eaten on St Kilda up to the time of its evacuation in 1930 and even today they are considered as good food in the Faeroes. By contrast, guillemot bones were among the most numerous found throughout the Middle Ages, on average outnumbering its congener the razorbill by more than ten to one, as one would expect from the composition of most modern colonies. Shags and cormorant bones were relatively plentiful, the latter being the more numerous before 1150 though they do not now breed on the island: possibly, like the gannet remains, the cormorants were brought from elsewhere in the Firth. The gulls show a variable pattern, with herring or lesser black-backed gulls being the most numerous bird bones in the deposits before 1400, with a sharp drop in relative numbers thereafter. There are only a few bones from great black-backed gulls and kittiwakes in each period.

The biggest ecological change on the May during the Middle Ages was the loss of the woodland scrub and the associated arrival of grazing animals, rabbits and sheep. Rabbits were introduced to the British Isles by the Normans, and the earliest reference in Scotland is to the protection of royal warrens at the start of the thirteenth century. The first mention of rabbits on the May was in 1329, but they had probably been introduced earlier by the

Cluniac monks. When the prior of Pittenweem feued the island to Patrick Learmonth in 1549, among the reasons he gave for disposal was the damage to the warrens, 'irreparably destroyed' by English incursions. Without the rabbits, the principal value of the island was said to have gone. It is hard to see what even the most malicious Englishmen could do to a colony of rabbits, and this was probably an exaggeration made to justify selling a monastic asset. Certainly the late medieval and early modern midden deposits had very numerous rabbit bones, some with evidence of skinning. There was a flourishing trade from east Fife ports in the late sixteenth and early seventeenth centuries in selling many thousands of rabbit skins of various colours (black, brown and grey) every year to Königsberg in the Baltic, where they were manufactured into fur goods. Some of these would undoubtedly have come from the May, though there had long also been warrens on the mainland, as on the Kingmuir of Crail. In 1552 the town council of St Andrews granted the archbishop a licence to introduce and farm rabbits on the northern part of the town links, now part of the Old Course of the Royal and Ancient Golf Club. The links were common land, and the archbishop was not to build dykes, interfere with the citizens' rights to graze cattle and take turfs, or obstruct the townsfolk playing golf and football, indulging in shooting games or 'all uther maner of pastyme as ever they plais'. In return, no-one was to catch his rabbits. [9]

It is possible that there was a relationship between rabbit farming on the May and the decline of the herring gulls, as, though the two species can co-exist, the managers may have tried to clear the gulls to give more room for their animals; equally, there could have been some intensification of sheep farming that had the same effect on the gulls. The rabbits on the May are the oldest single continuous colony of which we have records in Scotland, and for this alone have historical and biological value.

Between 1600 and 1800, we know less about the May than about the Bass, partly because it did not have the same attraction to visiting naturalists, and partly because it lacked a single sought-after product, like the gannet. But the evidence of the seventeenth and eighteenth centuries is of strong continuity from the past. The island certainly continued to have a reputation for high quality sheep pasture. 'Maia ubi-sheepifeda', William Drummond of Hawthornden called it in his macaronic dog-Latin poem of around 1630, and Robert Gordon in the 1640s said the island grew no crops but had good grass for 100 sheep and 20 cows. The *Statistical Account* of 1792 said that it had a reputation for delicious meat and for transforming the coarsest fleeces into wool 'as fine as sattin' by one season's grazing, and in 1803 it held 80 sheep of a larger size than before. By 1868 the stock was down to 60, due to deterioration of the pasture attributed to a plethora of anthills ('the whole island is covered with their hillocks').[10] The ants have now disappeared. The sheep were taken off around 1920, but the lighthouse keepers kept a flock again between 1955 and about 1960, after which they were removed once more. Grazing at this intensity must have had an effect on the amount of ground available to nesting birds. When the ground was highly valued for sheep, there may have been an attempt to control the rabbits, which were described in 1803 as confined to the rocks and cliff edge. But when they were removed the numbers exploded: 'the island literally heaved with rabbits', wrote Ruth Dickson in her autobiographical memoir of a lighthouse keeper's daughter in the 1920s, describing how the men waged a constant war against them with ferrets and snares: the rabbits 'came in black, white, grey, brown and with patches of all'.[11] The way they were caught would probably also have discouraged any establishment of a puffin colony in their burrows.

Sir Robert Sibbald is recognised as Scotland's first naturalist, but in his *History of Fife and Kinross* (1710) in his accounts of the May and Inchkeith he mainly repeated in English (without acknowledgement) what Robert Gordon had published in Latin in the *Blaeu Atlas* of 1654, that account probably dating from the early 1640s. He and Gordon both said of the May that 'many seals are slain upon the east side'.[12] This must imply breeding, as they would have been difficult to catch otherwise. There were also 'many fowls [that] frequent the rocks of it, the names people give to them are skarts, dunturs, gulls, scouts and kittiwakes', the same list as given by Robert Gordon with the addition of gulls. The 'skart' was the cormorant or shag, perhaps more likely to be the latter. The 'duntur' was the eider and the 'scout' the guillemot. There are no surprises here. He makes no specific mention of the puffin on the May, but later in the *History* he describes 'culternebs' which are clearly puffins, and which he knows to be the same species as that on the Farne Islands (see Fig. 10.3). He says they come in summer to breed, and 'haunt much this firth', without being specific as to where the colonies are. As he refers to the 'marrot or auk or razorbill' in much the same terms, and as that would certainly have been on the May, we might assume that the puffin could have been there as well. However, if it had been present in numbers comparable to today, he or Robert Gordon would surely have given it a more specific mention in their accounts of the island. In 1791

Fig. 10.3 – Creatures of the Firth of Forth, from Sibbald's *Scotia Illustrata*, plates by John Reid, 1684. The puffin is unmistakable, but it is pure white below (not barred), so it was probably unfamiliar to the artist. The middle plate is a kittiwake, clearly taken from the nest before fledging – the adults are white. At the foot is the sea-wolf or wolf fish, a fish that lurks in holes in the rocks and is still common off the May and St Abb's Head.

the minister of Anstruther Wester in the *Statistical Account* commented on the 'great variety of sea fowl, such as kittiewakes, scarts, dunsters, gulls, sea-pyots [oystercatchers], marrots etc'. It was a similar list to Sibbald's and he did not mention puffins either. Nor did he mention seals, which he surely would if they had still been breeding.

No other accounts were made of the birds and mammals of the Isle of May in the eighteenth century, but in the course of the nineteenth and early twentieth centuries it became steadily better known.[13] Distinguished Victorian ornithologists like William MacGillivray and William Jardine paid visits, and highly competent authorities like William Evans and J.A. Harvie-Brown gave detailed accounts of the insect life and flora as well as of the mammals and birds. From the 1880s the British Association published investigations into bird migration along the British coasts, and, in the words of W. Eagle Clarke, who was a moving force behind the project, 'the Isle of May furnished some of the best returns'. He mentioned bluethroat, black redstart, ortolan, red-backed shrike, pied flycatcher, wryneck, honey buzzard and turtle dove, still the characteristic kinds of rarities that turn up on the May on passage.[14] This work inspired Evelyn Baxter and Leonora Rintoul to undertake systematic spring and autumn visits from 1907 to 1933, interrupted only by the war, often staying with the lighthouse keepers, and they began to question the prevailing idea that bird migration was channelled along defined routes irrespective of wind direction. Their studies led to the modern idea that much of what is observed on the ground in Britain is 'drift migration' of birds displaced from an intended course further south by wind and rain, particularly, in the case of the May, from the south-east.[15] This in turn led to the establishment of a bird observatory in 1934, still normally manned from April to the end of October (Fig. 10.4) It also began to lead to much closer observation of the breeding birds and mammals.

What was found in respect to breeding fauna was not encouraging, and

Fig. 10.4 – The Low Light on the Isle of May, built in 1843–44, now the bird observatory.

suggests a low point in the middle and end of the nineteenth century.[16] Seals did not breed at all, as far as is known, between the eighteenth century and 1955, though they still visited. Shags and cormorants were both reported as breeding in the early nineteenth century, but by 1907 there were only two or three pairs of shags and no cormorants. A few gannets had bred briefly in the early nineteenth century, probably when their persecution on the Bass was at its height; they did not persist. Of the auks, estimates of guillemots and razorbills were of hundreds of pairs: as both nest on the cliffs, they were hard to count and possibly harder than some other species to kill, though it was possible to rob their nests by going over the cliffs with ropes. As for puffins, there were no other records before the 1880s, when estimates varied from 20 to 40 breeding pairs; in 1892 A.H. Meiklejohn found them 'very plentiful, but their holes were mostly on the isolated stacks', which suggests that they found life on the island incompatible with the sheep and rabbits. By 1921 they were down to six pairs. A pair or two of black guillemots nested frequently on the island in the first half of the nineteenth century but have not done so since.

The fortunes of gulls were no better. The cliff-nesting kittiwakes survived despite persecution and 'some hundreds' were reported nesting by the lighthouse keeper in 1883. On the other hand there were no ground-nesting gulls recorded at all in the nineteenth century before Baxter and Rintoul found a single herring gull's nest in 1907, a situation which was obviously different from that in previous centuries. Perhaps because of the absence of gulls, William Jardine found four species of terns (sandwich, common, Arctic and the rare roseate tern) breeding in some numbers in the 1820s, but they had all gone by the mid nineteenth century. Eiders were not counted, but Harvie-Brown in the 1880s considered that their numbers were declining due to persecution.

The reasons for this dire state of affairs were probably twofold. Firstly, the use of the island for sheep and rabbits, combined with a relatively large population of lighthouse keepers and their families, especially after 1886 when electricity was installed (population grew from 12 in 1838 to 27 in 1891), meant increased pressure on ground-nesting birds. Secondly, tourists increased over the century, often unscrupulous and increasingly armed with effective cartridge-loading shotguns. Kittiwakes, as elsewhere, were a favourite target. An account of 1803 said that young kittiwakes were a favourite dish with many people and that the 'the shooting of them when they come new-fledged from the nests on the cliffs is deemed excellent sport'. William MacGillivray in 1840 described how the fun of killing was itself sufficient for some people:

> I have seen a person station himself on the top of the kittiwake cliff on the Isle of May, and shoot incessantly for several hours, without so much as afterwards picking up a single individual of the many killed and maimed birds with which the smooth water was strewn beneath.[17]

Ground-nesting terns and eiders would have been at still greater risk from egg-collecting, dog-owning and gun-toting tourists. A Victorian mania for collecting ferns also led to a destructive quest for the local sea spleenwort, which, however, survived on cliffs and in the caves.

The sort of wanton destruction witnessed by MacGillivray was common throughout British coasts, and led ultimately to the Sea Birds Preservation Act of 1869, after which such killing in the breeding season became illegal. It took some time for this to be enforced, and often local by-laws to make it effective, but by the early twentieth century its effects were just beginning to be felt. Several species of seabird increased in numbers on the May between about 1907 and 1945, most notably terns and gulls. The common terns returned around 1914–18, and after fluctuating, built up to a dramatic colony of 5,000–6,000 pairs in 1946–47, with some hundreds of pairs of Arctic terns among them. Sandwich terns reached 1,400–1,500 pairs in

1946. None of the terns were ever to be so prolific afterwards, probably because of another success story from two other species which killed the tern chicks and replaced them on their nesting ground. The herring gull population built up from a single pair in 1907 to 58 pairs in 1924 and 760 by 1947, and the lesser black-backed gull, which began to nest for the first time in 1931, had achieved 100 pairs by 1948.[18] Both were about to hit the thousands, and to displace the terns. But in the second half of the twentieth century many other astonishing transformations were to occur on the May and elsewhere on the islands of the Firth of Forth, apparently without parallel in their environmental history.

3. The Seabird Explosion of the Later Twentieth Century

In the second half of the twentieth century, there were (as we saw in the last chapter) unprecedented increases in the numbers of the gannets on the Bass. Exactly the same was true of a dozen other species of seabirds associated with the Forth islands and the Berwickshire cliffs. The reasons are complicated and occasionally unclear or disputed, and the timing of the population explosion differed in each species.[19] Behind them all, however, lies the fundamental fact that, for the first time in history, man had suspended his exploitation of seabirds, no longer seeking to kill them either for food or for fun. Gradually, as the colonies recovered from centuries, even millennia, of human predation, their potential began to be realised. There is also a seeming paradox that this demographic explosion took place against a background of overfishing and a consequent transformation of the marine ecosystem: how then could there be enough food to support a population of seabirds at the end of the twentieth century possibly 40 times larger than it was at the beginning? The answer is partly that the fish removed, like cod and herring, were mainly the large ones that fed on the smaller fish like lesser sandeels, sprats and saithe, which multiplied as a consequence. Apart from that, fishing has long entailed a great deal of waste, discards and offal, thrown back into the sea where it was immediately available to those species that followed the boats. In the case of some gulls, open landfill sites on shore presented additional opportunities to exploit a throw-away society. These strong underlying common features are behind the success of most of the seabirds in the second half of the century. But to enjoy the complexities of the story, it is best to discuss the species separately.

Table 10.1 – Breeding seabirds of the Firth of Forth.

Breeding Seabirds of the Firth of Forth

	c.1935	1969–1970	1985–1988	1998–2002
Gannet	4,147	8,977	21,591	48,065
Fulmar	70+	1,646	3,495	3,788
Cormorant	0	153	409	411
Shag	<100	1,467	3,617	1,442
Lesser Black-backed Gull	6+*	2,654	4,208	7,654
Herring Gull	[100s]	24,760	12,887	11,132
Great Blacked-backed Gull	0	5	9	42
Kittiwake	[1000+]	16,053	33,638	26,192
Guillemot	[1000s]	11,689	33,246	54,306
Razorbill	?	553	2,782	5,594
Puffin	<100	3,084	14,377	72,136

* plus 'numbers on Craigleith'

Notes: Figures refer to apparently occupied nests or burrows, in the case of guillemots and razorbills arrived at by multiplying the number of individuals by 0.67: see C. Lloyd, M.L. Tasker and K. Partridge, *The Status of Seabirds in Britain and Ireland* (London, 1991).

The figures for 1935 are estimates from the accounts given in L.J. Rintoul and E.V. Baxter, *A Vertebrate Fauna of Forth* (Edinburgh, 1935).

The remaining figures are calculated from R. Forrester, I. Andrews *et al. The Birds of Scotland* (Aberlady, 2007), which reproduce the counts of three national surveys of seabird populations spanning the years cited. The gannet figures, however, relate to 2004.

The comments that follow should be considered along with Table 10.1 and Figure 10.5, which give an overview of what happened.

Since 1973, the Isle of May has been the subject of annual study by scientists working for the Centre for Ecology and Hydrology (formerly the Institute for Terrestrial Ecology), gathering the most detailed long-term data set on any seabird colony in Europe. National counts of breeding seabirds also took place in 1969–70, 1985–88 and 1998–2002. Thus the quality of the information available over the last 40 years has been much better than it was earlier, but the size of the changes since the start of the twentieth century was such that nothing can obscure the scale of what happened.

(a) Eider

The eider duck is the only species considered here that is not scientifically classified as a 'seabird', though it is entirely marine, feeding on shellfish such as mussels and small crabs. It nests mainly (but not only) on islands where it is safe from dogs, foxes and mink, the duck bringing her train of ducklings to the nearest shore as soon as they are hatched. The eider has long been

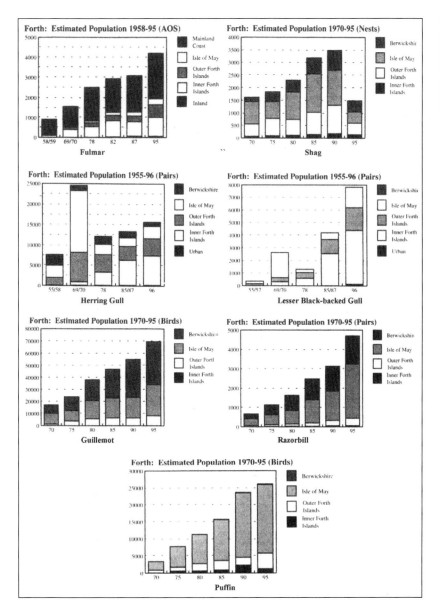

Fig. 10.5 – Breeding populations of certain seabirds in the Firth of Forth, from R. Murray *et al.*, *Breeding Birds of Southeast Scotland* (Edinburgh, 1988).

known on the Isle of May both from archaeological and written records, but by the 1880s it was considered to be declining. This was reversed in the early twentieth century, and by the 1950s there were reckoned to be about 70 pairs breeding there. By the 1960s there were 150 pairs on the May and as many more shared between Inchmickery and Inchkeith. The numbers grew very rapidly until the mid 1990s, when they stabilised at around 1,200 pairs on the May, with another 1,000 pairs or more around the Firth.

Protection and management on the May is responsible for most of this: culling the large gulls from 1972 and cutting back nettles to allow the ducklings to reach the sea without being stung to death both helped. The recovery of mussel stocks from a low point brought about by overfishing for bait in the late nineteenth century must also have played an important part.

(b) Fulmar

The fulmar is the only seabird breeding in the Firth that was completely unknown there before the twentieth century. It is the largest of four species of tubenose (petrels and shearwaters) that breed in Britain and sweep across northern seas and oceans to feed from organisms at the surface of the water. They are all excellent long-distance fliers and visit the Firth, but the fulmar alone regularly nests here. In the eighteenth century the only known colonies in the Atlantic were on St Kilda and on Grimsay in Iceland, but from the latter they began to expand all round the Icelandic coast. From there, in 1839, they spread to the Faeroes, and from 1878 to Foula in Shetland. They were first seen around the May in 1914 – there is no trace of them in the archaeological record either here or elsewhere in Scotland outside St Kilda. The first breeding record in the Firth of Forth area was at St Abb's Head in 1921, and there were several hundred pairs in the Firth area by 1950: by the end of the twentieth century some 485,000 pairs bred in Scotland, of which nearly 4,000 were on cliffs and islands in Fife, Lothian and the Borders. This was a little below the previous count in 1995, suggesting that the expansion had stopped. Fulmars were never primarily a bird of the islands, the biggest colonies always being on the Berwickshire and East Lothian cliffs, yet colonies on the May were established by 1930 and on most of the other Forth islands by the late 1960s, Inchkeith being the largest of these.

The explanation for the extraordinary expansion of the fulmar in the North Atlantic remains a matter for speculation and debate. James Fisher, in a classic study of 1952, related the increase to the new supplies of food, at first of offal from whalers and then of fish and offal discarded from trawlers.[20] Others have supposed that changes in sea temperatures favoured the streams of plankton on which they partly feed, or that a new genotype adapted to colonisation accidentally arose in the population in Iceland.[21] These explanations could be jointly true. The increase of the fulmar is the

Fig. 10.6 – Shags nesting on the Isle of May.

most dramatic in global seabird population numbers ever recorded, but the contribution of the Firth of Forth to the overall growth even within Britain is modest.

(c) Cormorant and Shag

Cormorants and shags, collectively known in old Scots as 'skarts', are two closely allied species. They feed inshore, diving deeply, but seldom fishing more than 25 miles from land, using the waters west of the May more freely than most of the other seabirds which predominantly feed further out. The cormorant in salt water feeds particularly on flatfish, wrasse, saithe and whiting, but also feeds inland on a variety of freshwater fish. It is much the commoner of the two in the estuary above Queensferry. The shag takes smaller prey, such as sandeels, small herring and saithe. On the Isle of May they are fond of dragonet and (particularly in recent years) they also take a lot of butterfish.

Both cormorants and shags were ancient inhabitants of the Firth, their remains on the May going back to prehistory, but in the nineteenth century and as late as the 1930s they were not common. The cormorant, having been restricted at the start of the twentieth century to a few pairs on the Berwickshire cliffs and considered to be declining, was actually extinct as a breeding bird in the region by 1935. The shag (Fig. 10.6) first bred in the Firth in the twentieth century on the May in 1918, and by 1935 there were six pairs on the Isle of May, with about two dozen pairs at St Abb's Head, 30 or 40 on the Bass and a pair or two on Craigleith – below 100 pairs in the entire Firth area.

From this base, both species began rapidly to expand. The cormorant, still not mentioned in the record as a breeding species in the Firth of Forth in the 1950s, increased to a peak of over 500 pairs by 1991, and has been nesting on the Berwickshire cliffs, on Craigleith, the Lamb and Eyebroughty off North Berwick, on Carr Craig and the Haystack off Fife, and on Inchkeith. 'The totals for the whole Forth area fluctuate wildly but continue to increase', it was noted in 1998, and birds would readily move from one island colony to another.[22] By then the area contained about 9 per cent of the Scottish population.

The shag population grew even faster, peaking at over 4,000 pairs in 1992. It was then struck by disaster in the bad winter of 1993–94, which led to starvation and a wreck of large numbers washed up on North Sea coasts. In the breeding season of 1994, the population was found to have dropped to 800 pairs, more than an 80 per cent decline from which there was only limited recovery by the end of the century. At its prime, the May colony of nearly 2,000 pairs was among the largest in Scotland, and shags also bred on the Berwickshire cliffs, the Bass, and several other islands as far up the Firth as Inchkeith and Inchcolm. They accounted for about 11 per cent of the total Scottish population, and tended (unlike cormorants) to be very faithful to their chosen nesting colony.

The increase of both species in the Firth of Forth was probably related both to the decline in human persecution and to a growth of accessible food supplies. In the past, they were easily shot on the nest or at sea, and regarded as pests – as the cormorant still is when it fishes inland. An increase in small fish following the extirpation of the large predatory ones would particularly favour the shag.

(d) The Larger Gulls

The lesser black-backed gull and the herring gull are two large gulls which, in the course of the twentieth century, developed a commensal relationship with man, frequenting towns, harbours, fields and especially rubbish dumps, in a quite new way. In 1900 they were not common. Fishermen in Pittenweem and elsewhere could lay out their catches on the quayside for sale, no doubt watched but not overwhelmed by gulls, and nineteenth-century naturalists noted as remarkable any gathering above a score. Thus Robert Gray in 1871 observed that herring gulls were nowhere more common in early spring than in Fife; 'on Leven sands they assemble in companies numbering thirty or forty birds'.[23] By 1935, Rintoul and Baxter said that flocks there of three or four hundred were 'nothing surprising', and they had once seen 'thousands' in winter on the Forth below Leith, no doubt where the sewers ran out. At the end of the twentieth century, numbers in winter had grown by yet another order of magnitude. A count in late January 1994 recorded 27,400 herring gulls in coastal Fife, of which 4,150 were east of Leven and 19,700 were at Longannet in the estuary; inland, gravel pits near Cupar regularly attracted 1,500 herring gulls, and in 1994 they peaked at 5,000 there.[24] Everyone knew the 'seagull', even if a notice in Anstruther that urged the public not to feed them actually portrayed an albatross.

Their story as a breeding bird in the Firth of Forth, however, is complex. It seems from ringing returns that many herring gulls that nest here and elsewhere in the south of Scotland move away in winter, to be replaced by a different population from further north in Scotland and possibly from Scandinavia.[25] Almost all lesser black-backs move south, and it is an uncommon bird in winter. The two species often breed together on flat, turfy, surfaces, but need to be secure from foxes and other land predators. Islands are ideal, but around 1980 there was an inland colony of some 8,000 pairs of lesser black-backs on Flanders Moss west of Stirling, which had been founded by a few pairs around 1930. It was later persecuted out of existence. Both lesser black-backs and herring gulls have since taken to breeding in relatively small numbers on flat roofs in coastal towns: one of the first of these colonies was discovered at the naval base in Rosyth in 1988,[26] but the combined total of all such sites by the 1990s was only a few hundred pairs.

So the islands remained the gulls' stronghold. Around 1950, there were in the region of 100 pairs of lesser black-backs on the May and probably a little over 1,000 pairs of herring gulls. Neither had been nesting on the island in 1900. The colony then increased rapidly by 1972 to about 2,500 pairs of lesser black-backs and 15,000 pairs of herring gulls, as well as some 8,000 non-breeding herring gulls. At this point the Nature Conservancy began a policy of culling the gulls, partly in the hope of tempting return by the more highly valued terns, which had again ceased to breed in 1957 as gull numbers multiplied, but also because gulls were causing soil erosion and a change in vegetation: large areas of turf formerly dominated by thrift and sea campion had been replaced by coarse grasses and nettles. It was not a pretty task: there was public disquiet when visitors saw the gulls flapping round in their death throes, or piles of bodies awaiting disposal on the shore at Crail. Almost 44,000 gulls were killed in five years, and it was estimated that a further 27,000 birds were deterred from joining the colony. Culling continued until 1986: it eventually brought the size of the colony down to under 3,000 pairs, a level which it was intended to maintain, though this proved to be difficult (on the May in 2010 there were 3,200 pairs of herring gulls and 2,600 pairs of lesser black-backs).[27] The terns indeed began to return from the 1980s, though not to their former numbers: by 1999, there were on the island some 110 pairs of Sandwich terns, which only bred for two or three years, some 400 pairs of common terns and over 700 pairs of Arctic terns, which have formed a more lasting colony of varying size.

As the colony of gulls on the May declined, and others were displaced from the Bass by the spread of the gannets on to the former turf surface, some of the intimidated gulls moved elsewhere, especially to the inner islands. Inchkeith by 1996 was the largest colony, with nearly 5,000 pairs of herring gulls and 2,600 pairs of lesser black-backs: 'they have imposed themselves over the whole of the island's surface during the breeding season', commented Ron Morris in 2003.[28] By the end of the century, herring gulls in the Firth were still fewer than half their numbers of 30 years before, but lesser black-backs had not been set back in the same way, except temporarily. The ratio between the two species as breeding birds changed in the Firth of Forth from 10:1 in 1970 to 3:2 in 2000. The lesser black-back, in fact, had almost tripled its numbers at the same time as the herring gull declined.

What were the reasons for the great increase in large gulls and the subsequent check, especially to the herring gull? On the one hand, there was the decline of human persecution before the Nature Conservancy's cull, and this obviously permitted expansion, until the critical point was reached on the May in 1972. However, a more important factor is surely the changing dietary habits of both gulls, but especially the herring gull, as they became more and more dependent on food directly provided by man. Many of those breeding on the islands fly to the mainland or the coast to forage, even in summer, though the lesser black-back seems still to be more marine than the herring gull, and when it flies inland it appears to be relatively more attracted to agricultural fields. However, the herring gull is particularly strongly attracted to the detritus of sewage outfalls and rubbish dumps, which became more and more available in the first three-quarters of the twentieth century. The check to the expansion of herring gulls in the long term probably had much less to do with the cull than with the cleaning up of sewage disposal systems from the late 1970s, and the move to cover the landfill sites as soon as possible after rubbish had been deposited. Regulation should have improved in 1974 under the Control of Pollution Act which called on local authorities to cover the tips every day, but it was not until SEPA took over the regulatory role in 1996 that slack management was thoroughly eliminated.[29] Since both gulls are still also fish eaters, but only take food that they can reach from the surface of the sea, the opportunities afforded by fisheries by-catch were also a help. But the gulls were limited by their inability to compete with the bigger and even more aggressive gannets, when both were following the same trawler. Life became harder for gulls towards the end of the twentieth century.

A third species of large gull, the greater black-backed gull, is a common winter visitor but not yet a common breeder. A few pairs were said to have bred on the Bass in the early nineteenth century, but no further breeding was recorded until a pair on the May in 1962: even by the end of the century, only 25 pairs nested in the Firth of Forth.

(e) Kittiwake

Kittiwakes are much smaller gulls than those discussed above, entirely maritime in their distribution and cliff-nesting in the breeding season. Their food in the breeding season on the Wee Bankie and elsewhere in the approaches to the Firth of Forth is almost exclusively lesser sandeels. In April and May the adults feed on sandeels above a year old, but in June and July they feed both themselves and their young on juvenile sandeels less than a year old. These are picked from the surface of the sea: kittiwakes take their prey without diving or plunging, so are likely to be more vulnerable than auks to changes in the availability of sandeels. They are mainly summer visitors and passage migrants in the Firth of Forth, though in stormy winter weather flocks may be driven in from the wider North Sea.

The remains of kittiwakes are found, albeit in small numbers, on the May in prehistoric and medieval times, and they are mentioned by Robert Gordon in his account of the island in the 1640s as a delicacy 'very eagerly eaten in the month of July'. There also are eighteenth- and nineteenth-century accounts of colonies on the Bass and on St Abb's Head, where it was said in the *Statistical Account* in the 1790s that they raised 'incredible numbers of young'. As in Yorkshire, in mid Victorian times they were shot on the cliffs from passing steamers, and, as we have seen, they were also the object of killing excursions to the Isle of May.

At the start of the twentieth century, numbers in the Firth of Forth were probably under 1,000 pairs. On the May, they were recovering from a low level of 'some hundreds' reported in 1883: Rintoul and Baxter reported increases there between 1909 and 1921 to 1,900 pairs. Evans in 1911 estimated 200 pairs from St Abb's Head, and there were small numbers on the Bass. In the 1930s, there was modest expansion: Craigleith, ledges on the harbour buildings at Dunbar and the cliffs at Fast Castle were all occupied. Then, in the 1950s, the real take-off began, with the Lamb, Inchkeith and Fidra all colonised before 1964. The Berwickshire cliffs quickly came to eclipse the May as the main colony, holding two-thirds or more of the entire population: the stretch near St Abb's Head alone held 4,600 pairs by 1957, with numbers doubling by 1976 and doubling again by 1988. By 1990, there were reckoned to be some 42,000 pairs in the Firth of Forth, of which some 30,000 were on the Berwickshire cliffs and over 8,000 on the May. That was around 12 per cent of the Scottish population.

The main cause of this enormous long-term expansion was probably a delayed but sustained recovery from nineteenth-century persecution, but it could only have reached these heights if food in the breeding season continued to be plentiful. Particularly, kittiwakes and puffins depend on sandeels to feed to their chicks, and the sandeels perhaps themselves increased due to the removal by overfishing of larger fish. However, in 1990 the Danes pioneered a new industrial fishing for sandeels, concentrated on the fishing banks east of the May that were also used by most of the seabirds (the Wee Bankie, the Marr Bank and the Berwick Bank). This reached its peak in 1993, with a catch of over 100,000 tons, but lasted until 1999, when the catch reached 40,000 tons, after which the fishery was closed by European authorities. This followed pressure from conservationists concerned about the birds, and from Scottish fishermen who feared that the Danes would imperil any recovery of stocks of larger fish like haddock which also depend on sandeels for food. It was striking in those years to see local fishing boats flying the Greenpeace flag.

The breeding success of the kittiwakes was seriously affected, evidently by the over-exploitation of the sandeels, and their numbers began to decline. Following the closure, their breeding success began a limited recovery, though overall numbers did not at once increase. No other species of seabird seemed to be hit so badly by the fishery.[30]

However, there turned out to be more to the story than simply overfishing, as the sandeels were also vulnerable to changes in water temperature, failing to breed in large numbers in warm winters. The 1990s were a decade when climate change began to make an appreciable impact on the North Sea, and it was the combination of the sandeel fishery with warmer waters that had such a fatal effect, switching the population growth rate in the Isle of May population from an 8 per cent annual increase in the late 1980s to an 11 per cent annual decline in the late 1990s. Results appeared to be much the same elsewhere in the North Sea.[31] Stopping the fishery removed one cause of the decline but could not influence the other, a point we shall consider again later.

(f) Guillemots and Razorbills

Two related species of auk, common guillemot and razorbill, often nest on cliff ledges alongside the kittiwakes, but depend on a slightly more varied diet of small fish, including sprats as well as sandeels as the most important. They both dive for their prey, yet differ somewhat in their foraging preferences, the razorbill being found more often close to the shore. Neither species typically flies great distances in search of food. In most British colonies, numbers of guillemots have always vastly exceeded those of razorbills. They were both common in the archaeological deposits on the May and perhaps often got lumped together in early seabird lists. The naturalists of the nineteenth-century Firth knew them well, and although no one estimated the numbers of either species on the May as more than some 'hundreds' of pairs, Joseph Agnew, the lighthouse keeper in 1883, reckoned the guillemot to be the most numerous nesting species on the island. There could have been 2,000 pairs or more by the first decade of the twentieth century. Elsewhere, there were colonies of hundreds of pairs on the Bass and on St Abb's Head, numbers being depressed by egg-collecting from the cliffs until this became illegal early in the twentieth century.

By the mid 1950s, there were still about 2,000 pairs of guillemots and some 375 pairs of razorbills on the May, but westward expansion began with breeding on Craigleith in 1934, on the Lamb in 1963, on Fidra in 1964 and on Inchkeith in 1976, guillemots leading the way in most cases. The Berwickshire cliffs, with more room, steadily took a bigger proportion of the guillemots: by 1996, the total was estimated at 46,600 pairs, of which half were on the Berwickshire cliffs and about a third on the Isle of May. Razorbills had not been so successful, with some 4,250 pairs in total by 1996, of which two-thirds were on the May and little more than a fifth on the Berwickshire cliffs.

It is something of a mystery why the guillemot is so much more successful than the razorbill, given that they both depend primarily on sandeels and sprats, and both benefited in the same way from complete protection of their colonies from the early twentieth century. Some 6 or 7 per cent of the Scottish population of both species are found in the Firth of Forth.

Another relative, the black guillemot, did not increase at all in the Firth in the twentieth century. Its bones have been found in archaeological deposits on the May, it was described from the Bass in the seventeenth century, and a pair or two could be found breeding on the May in the nineteenth century. It does not breed in the area now, though isolated individuals are occasionally seen offshore in summer.

(g) Puffin

Puffins, most iconic and popular of all seabirds, are the smallest of the three auks breeding in the Firth of Forth, but they nest in burrows. In the breeding season, like the kittiwakes, they feed on juvenile lesser sandeels caught to the east of the May, so are less often seen from the land than other auks that have a more variable diet. They dive deeply on the sandeel shoals and thus can take fish not available to kittiwakes, which can only feed close to the surface. They are the auk least likely to be encountered in winter, as they arrive on the breeding colonies in late March and are off to the wider seas again in August.

Puffin bones were rare in the middens of the May in prehistory and the Middle Ages, and the birds are not much mentioned locally before the nineteenth century. In the 1880s, there were up to 40 pairs on the May, a similar sized colony at St Abb's Head and smaller ones on the Bass and on Craigleith. The first half of the twentieth century even saw a decline in the puffin in Britain, and it did not then do well in the Firth of Forth either. As late as 1959, Eggeling reported only five pairs on the May. They nested there not on the sward but in fissures in the rocks on the cliff edge, but he had witnessed an attempt by at least 50 pairs to colonise the turf in 1957 and 1958, frustrated, as he believed, by the gulls.

Shortly after that, there was a dramatic increase. The May reached an estimated 3,500 pairs in 1972, 10,000 pairs by 1981, 20,000 pairs by 1992 and a peak of 69,000 pairs (strictly, occupied burrows) by 2003, which made it the largest puffin colony in Britain after St Kilda, and the largest colony of any seabird within the Firth of Forth. The rate of increase on the May of 22 per cent per year in the 1970s was beyond what the birds could have achieved without immigration, and the source of the new recruits turned out to be the rapidly growing Farne Islands colony off the Northumbrian coast, puffins in the east of Scotland and the north-east of England being 'a fairly closed unit'.[32] The population of this entire unit was growing at the rate of 8.4 per cent a year for three decades from 1970, and it is interesting that they did not experience the same setback in the 1990s from the Danish sandeel fishery as the kittiwakes. Perhaps their ability to dive for food accounts for the difference. After 2003, however, as we shall see, there was also a decline in puffins.

Since the 1960s, the May has always been the main colony, often holding three-quarters or more of all the birds. Craigleith was second in importance and Inchkeith, colonised in 1965, third, neither considered in 1994 to hold more than a few thousand pairs. Fidra, founded in 1966, had grown by then only to a few hundred. The Bass and St Abb's Head, though old colonies, failed to grow beyond a few score pairs, limited perhaps in the first case by competition for space from the gannets and in the second by the danger of rats and foxes on a mainland site. New colonies elsewhere on the Berwickshire coast, on the Lamb (1984), Inchmickery (1991) and Inchcolm (1992) were also still quite small.

After 1994, however, the puffin colony on Craigleith underwent a sudden expansion and became the cause of an ecological conundrum. Having been estimated at some 4,000 pairs in 1995, the population suddenly expanded to 28,000 pairs in 1999 (presumably by immigration), whereupon it came into conflict with the tree mallow on the island.[33] The tree mallow is a handsome plant, growing to well above six feet in favourable conditions, native to the west and south-west of England and Wales but apparently introduced into the Firth of Forth centuries before.[34] John Ray noticed it growing plentifully when he visited the Bass in 1662, and Victorian botanists admired its 'great profusion' and 'gorgeous appearance, with its rose-coloured flowers streaked with darker veins'. It did little harm on the Bass, and when it was first noticed on the nearby island in 1966, its 'abundance and luxurious growth' was called the 'glory of Craigleith'. In both places, the tree mallow had been kept in check by rabbits, and before they were taken away, by sheep. However, an outbreak of myxomatosis on Craigleith in the 1950s also removed the rabbits, a series of mild winters in the 1980s and 1990s prevented any frost check to the plants, and an increase in the gull colony enriched the ground. The puffins themselves added to the guano, and by digging holes in the turf, conveniently provided an ideal seeding ground. The tree mallows suddenly became extremely invasive, and by 2006, three-quarters of Craigleith (described in 1899 as having 'little vegetation'), was completely covered by tall, dense and bushy plants. The puffins declined from 28,000 burrows in 1999 to 12,100 in 2003, eiders fell by 90 per cent and herring and lesser black-backed gulls by more than half. Following a public consultation in the North Berwick area, it was agreed to try to control the invasive 'alien' tree mallow to favour the no less invasive but native puffin, but in a gradual way that will keep conservation volunteers busy on the island in summer for many years to come. The option of reintroducing the rabbits was considered and officially rejected. Had the rabbits themselves been a native species and not also an 'alien' (albeit a medieval introduction in Scotland), there is little doubt that they would have been used, and they might have kept the seed bank from regenerating. In the event, rabbits were illicitly reintroduced, and what will happen now remains to be seen.

The general and sudden increase of puffins in the Firth of Forth in the second part of the twentieth century is a puzzle, even allowing for the immigration from the Farnes. It is hard to see it as primarily due to decline of persecution, as puffins were not widely eaten in eastern Britain in historic times, and do not feature much in stories of seabird killing for sport either, though probably only because they were not common enough to present frequent targets. Competition for the breeding turf may be a factor: they can thrive alongside sheep, rabbits and gulls once they are established, but the manner in which for so long they had to keep to crevices on the cliff edge on the May suggests that they might have found it hard to get established in a new place in the face of bigger users. Generally sheep were withdrawn from small islands in the twentieth century, and the cull of gulls on the May coincided with the start of the puffins' success there. Though rabbits were severely reduced by myxomatosis at much the same time, puffins can normally out-compete them in competition for holes and so this last was probably not a factor.[35] It has been suggested, as with other species of seabirds, that the population could grow because sandeels became more available with the decline of larger fish, and because the seas became colder for a time from the 1950s, which arguably favoured the supply of the plankton on which the sandeels depended. Against this latter point, puffins multiplied rapidly from the 1970s until after 2003, even though for most of that time sea temperature was clearly rising again. As we have noted, the advent of the Danish sandeel fishery between 1990 and 1999 did not hold up the increase in the population. Whatever the reasons, the puffin catapulted itself from being a rarity to being the most abundant breeding seabird of all in the Firth of Forth. It had also stormed the affections of the public, and a notice in Anstruther encouraging tourists to visit the Isle of May in the 1990s read, simply: 'fifty thousand puffins can't be wrong'.

4. Problems of the New Century

In the closing years of the twentieth century, but especially in the first ten years of the new millennium, it became clear that the population explosion of seabirds that had marked most of the previous half century was not going to continue, despite the decision in 2000 to close the sandeel fishery. In this, the story of the Firth of Forth was no different from that of the remainder of the British North Sea coasts. Most species were found to be declining; especially, the colony of puffins on the Isle of May dropped by about 40 per cent between 2003 and 2008. Similar, though less extreme, downward trends were found for shags, kittiwakes, guillemots and razorbills on the May. Conservationists are alarmed, though the declines do not yet come near to wiping out the gains of previous decades.

Evidently the problem revolved around the plight of the lesser sandeels at a time of climate change. It is the class of sandeels less than one year old which are so significant to seabirds, and the average size of individuals fell between 1973 and 2002 from an average of about 75 mm to about 65 mm, a 15 per cent drop in length equating to a 40 per cent drop in energy content. Diminished size not only provides less nutrition, it is also likely to make surviving fish mature later and to reproduce less, ultimately leading to a decline in the total stock.[36] This could be related to a change in the quality and species composition of the plankton on which the fish feed, which is likely in turn to have been affected by the temperature of the water and the character of the currents in the North Sea. Possibly this is a direct consequence of global warming, possibly a related effect of variations in the North Atlantic Oscillation also partly linked to climate change. The latter is a meteorological phenomenon that results from fluctuations in barometric pressure between the permanent low over Iceland and the permanent high over the Azores, and by controlling the strength and latitude of storm tracks across the Atlantic, it determines the inflows and upwellings of water around us. In years when the oscillation is positive, we get a wet summer and a lot of warm water from the north-eastern Atlantic pushed into the North Sea.

At times in the last ten years, the desperation of the seabirds looking for food in the breeding season has been palpable. In 2006 in particular, puffins and kittiwakes, faced with a shortage of properly-sized sandeels, turned to a warm-water fish which had for the first time appeared in numbers in the North Sea, the snake pipefish (Plate 16). But the young found their spiny and bony bodies impossible to eat, or of very poor nutritional value, so they choked or starved surrounded by piles of uneaten pipefish. Both bird species had a very bad season.[37]

Shortage of food can also disrupt the social behaviour of a seabird colony in unexpected ways. Among guillemots, which hatch a single chick each year, normally one parent goes off to catch food while the other stays behind to brood and defend it. In 2007, the guillemots on the Isle of May found food so short in early July, at the time of fledging when nutritional demands were at their peak, that in 60 per cent of pairs both parents went to forage. The unattended chicks then frequently suffered attack from neighbouring adults that had their own chicks. Two-thirds of all the chicks that hatched died that year: in 69 per cent of the cases where the cause of death was known, it was from assault by an adult guillemot, and in the remainder of cases it was apparently due to exposure or starvation.[38]

The three years since 2009 have been rather better than earlier in the decade, though the kittiwakes failed again in 2010 and razorbills in 2011. Some birds have tried partly to replace sandeels with sprats, the guillemots since 1997, and kittiwakes, razorbills and puffins in 2010, though with mixed success. More importantly, there may have been a partial recovery in the numbers of lesser sandeels, though how lasting this will be no one can tell.

The fact is that the sea remains a mystery: we know too little about what drives its ecosystems even in an area as intensively studied over many years as the Firth of Forth and its offshore banks. The seabird colonies are age-old but far from unvarying. In the whole of history, they probably never did so badly as in the nineteenth century or so well as in the second half of the twentieth century. Though they face an uncertain future now, their problems today appear at first sight to be relatively small compared to the times when kittiwakes were shot for food and pleasure, and puffins had to compete with sheep for a place to breed. If, however, global warming continues at its present rate or accelerates, the consequences for the seabirds of the Firth of Forth could be totally catastrophic.

CHAPTER 11
SEALS: THE BONE OF CONTENTION

1. The Grey Seal in the Firth of Forth

The grey seal is a predator of enormous size compared to the fish-eating birds we have been considering. A puffin averages about 400 grams: the average grey seal weighs about 120 kilograms, the same as 300 puffins, though mature bulls average about 230 kilograms and breeding females about 180 kilograms – a great many seals in a population are half-grown immatures, as they mature slowly (Plate 5). However, because there are biological economies of scale in a big animal, this average seal only needs to consume as much fish for its energy requirements as 72 puffins.[1] Now, 4,000 pups were born in the Firth of Forth in 2009, and there are probably over four seals in the area for each pup born. So the local grey seals probably eat as much fish as would well over a million puffins. On the other hand there are many other seabirds, some of them, like gannets, very much bigger than puffins, and the total catch of fish by seals in the North Sea does not equal the catch of all the seabirds.

What the seals eat, and in what quantities, are long disputed matters, but authoritative research published in 2005 suggested that in the North Sea more than half their diet at all times of the year consisted of sandeels, though they are generalist predators with more than 25 other species known to be part of their prey. Their second choice after sandeels was cod and other fish of the gadoid family, like haddock and whiting, amounting to more than a third of their diet in winter, which brings them into potential conflict with fishermen. The study nevertheless concluded that seals were unlikely to have caught more than 3.7 per cent of the North Sea cod total biomass and 2.7 per cent of the sandeel biomass, much smaller proportions of both than the commercial fisheries caught. By far the biggest predators of fish are other, bigger fish (often of the same species, as fish are inveterate cannibals), and some of these will of course be eaten by the seals. One critical question, whether the seals were limiting the ability of cod to recover from earlier overfishing of their stocks by man, the so-called 'predator pit', could, however, not be confidently answered.[2] A lot may depend on how the seals respond when a prey species becomes scarce: if there are abundant alternatives, the effect may not be very great. The proportions of fish consumed do not necessarily apply to other areas where the grey seal is common, as in the Outer Hebrides where gadoid fish make up 40 per cent of the diet. A complicating factor is that the grey seal is a carrier of the cod-worm parasite. Although infestation of cod has not increased in proportion to the rise in the number of seals, the presence of so many seals and so few remaining cod could provide a difficult environment for recovery.

Grey seals are not usually migratory in any regular or seasonal way, though they can be great travellers. The breeding females are ashore for less than a month. In the Firth of Forth, they come to their nursery sites in early November and pup within a day or two of landing: they feed their young on a very rich milk for between 18 and 21 days, remain on land to mate with a bull for the next day or two, and return to the sea. The males stay on shore rather longer, competing for females – for 36 days on average. Neither sex feeds during their spell on the breeding grounds. Otherwise, the seals spend two-thirds of their time at sea, and the remainder hauled out on shore, on stretches of rock and sand where they cannot readily be disturbed: the sandbanks off Tentsmuir at the mouth of the Tay are a favourite spot, and the Isle of May another, though less important one. The animals on the Tentsmuir sands do not necessarily breed locally – some have been shown to come from Orkney, for example, and to return there in winter.[3]

From the haul-out sites the seals go on expeditions. Some of these entail long-distance visits of several hundreds of miles to other haul-out sites and distant fishing grounds, as in Orkney or the Farne Islands, so there is no ecological divide between the Firth of Forth seals and those of other parts of the North Sea. But the great majority of trips (calculated at 88 per cent) last only a couple of days and go to sites on average only about 20 miles away from the haul-out site, usually to sandy and gravelly bottoms where the sandeels congregate.[4] Recent work has refined this picture and shown how the population that hauls out on sites in or close to the Firth of Forth tends to feed within quite a discrete area, most trips being focused close to the haul-out sites themselves, but with a penumbra of foraging 60 or 70 miles to the east.[5]

The latest estimates suggest that there may be around 120,000 grey seals around the British Isles,[6] the ones round British coasts amounting to most of the eastern Atlantic population, of which the North Sea is reckoned a part, and to some 45 per cent of the world population. Other discrete populations are in the western Atlantic, particularly centred off Canada, where it is also expanding rapidly, and in the Baltic, where it is now also expanding following a long period in the late twentieth century when it was not thriving.[7] Overall, the grey seal is not now at risk, though the British population was considered endangered a century ago. It can even be described as locally abundant, but it is certainly a species for which Britain has a particular international responsibility. Most of the British seals breed in Scotland, and around 8 per cent of the UK population breeds in the Firth of Forth.

The grey seal has had a chequered history here. Prehistoric remains were frequent in the middens of the May, and also occurred on Inchkeith and in a cave on Fife Ness. Before the eighteenth century, grey seals remained familiar at least on the May, where the colony co-existed after a fashion with the monks and then with the resident fishermen. Then came an unrecorded disaster, and the colony was eliminated. William Evans, writing his *Mammalian Fauna of the Edinburgh District* in 1892, knew of no record at all of a grey seal from the Firth of Forth. Rintoul and Baxter as late as 1935 knew of only two that had been shot in the Firth, and though they claimed to have seen them in the estuary where harbour seals (then called common seals) are more likely, they made no mention at all of any but harbour seals on the Isle of May, though these they had 'often seen'.[8] The two species were probably often confused, but nothing suggests that grey seals were at this time anything but unfamiliar and rare. At that time the only colony known to be breeding on the east coast of Britain was on the Farne Islands.

The story of the recolonisation of the Firth of Forth from the Farnes, in the later twentieth century, is every bit as dramatic as that of the puffins or any of the other seabirds. When the bird observatory on the May was reopened after the war, in 1946, grey seals were occasionally seen, and as late as 1950 a party of three in March and three later sightings of singles, were the only ones reported in a year. Increasing numbers began to be seen hauled out in summer, including 36 in June 1956 and then, the following December, 'the principal lighthouse keeper found a white-furred pup . . . picked it up and showed it to his wife . . . it cannot have been more than two or three days old'.⁹ That was the first indication of an impending influx, but as far as could be judged in the absence of wardens on the island in winter, pup production was limited to one or two a year before the late 1970s. It became clear from tagging, and later from blood tests, that these animals were coming from the Farnes. Here the National Trust had begun a programme of culling in 1963, in response to concerns about seal attacks on salmon heading for the Tweed, followed by two large culls of adult females and a few males in 1972 and 1975 which were primarily an attempt to stop erosion of the area where the seabirds bred. The Trust also attempted (in vain) to fence off the breeding areas.¹⁰ When none of these things worked, they placed wardens on the islands at the start of the breeding season, and any seals that ventured ashore were shot. These measures eventually led to the colonisation of the May.

In 1977, 30 pups were born on the May and in 1979 about 300, according to the lighthouse keepers (Fig. 11.1). In 1980 the Sea Mammal Research Unit counted about 500; by 1993, this had grown to 1,500, and the area occupied by the colony increased as the choicest spots, low-lying and closest to the sea, filled up, though, unlike on the Farnes, the seals did not trespass on the area occupied in summer by puffins.¹¹ In the late 1990s, numbers peaked on the May at around 2,000 and have held relatively level at this figure or a little below.

Scottish Natural Heritage are proud of the colony on a National Nature Reserve now, but it was not always so: when the animals first began to come ashore, their predecessors, the Nature Conservancy Council, feared that they would cause the same sort of erosion that had occurred on the Farne Islands, and the NCC land agents suggested keeping them off with an electric fence, or with concrete posts, or with a wall 300 yards long and five feet high – there was, they said, 'plenty of stone on redundant walls round St Adrian's chapel'. Morton Boyd, the Scottish Director, suggested a full-time warden with a 'good wild dog who would disturb the site until the seals were forced to go elsewhere'. None of these ideas were pursued, and the colony developed undisturbed. Any erosion was limited to the small area of the island called Rona where most of the animals breed.¹²

In the late 1990s females from the May also began to colonise the caves and coves of the Berwickshire coast at Fast Castle, where breeding was first reported in 1995, but which by 2009 was producing over 1,700 pups, so beginning to rival the May. Small numbers also began to breed on Inchkeith in 2001, reaching over 200 pups by 2009: grey seals had not been

Fig. 11.1 – Grey seals and gulls, Isle of May, October 1979: there was press interest as the seals had just begun to recolonise the Firth of Forth after disturbance on the Farne Islands, Northumberland.

known there since the Neolithic.[13] Pup production in the Firth of Forth colonies had still been growing at over 5 per cent a year between 2003 and 2008, while in the other main colonies in the Inner and Outer Hebrides it was virtually stagnant, and in Orkney increasing at a much lower rate. It is not known why the Forth population should continue to grow so fast, though at the start of the twenty-first century the new colony at Donna Nook in Lincolnshire was growing even faster.

2. The Seal Controversy in Scotland

Grey seals have been controversial for decades, not just in the Firth of Forth, but around the British Isles. A demographic explosion occurred in the twentieth century at colonies all round the coasts, and was at least partly rooted in a switch from persecution to protection. The story does not relate just to the Firth of Forth, and indeed because by the time the May colony began to grow to a substantial size, the critical conservation decisions had already been taken because of developments elsewhere, it is best told in a wider context.

Seals had from time immemorial been killed on Scotland's coasts, by local people, for their meat, oil and skins, and latterly because of damage to salmon fisheries. A seal hunt in winter was no small undertaking, as seventeenth-century accounts make clear. On Haskeir off North Uist the hunters attacked the seals at the breeding colonies with clubs:

> and so the men and the selchis do fight strongly and there will be innumerable selchis slain wherewith they loaden their boats, which causes many of them oftyme perish and droune in respect they loaden their boats with so manie selchis.

There were similar accounts from Orkney, where the 'monsters, eyeing them with dread and gnashing their teeth with rage strive to get out of the way with wide open mouth, then they attack with all their strength'. At Arbroath, seals were described as of 'a hudge bignes nigh to an ordinaire ox but longer', and the hunters used to enter the caves in the dark with boat and lighted candles, to kill young and old.[14] Perhaps they did the same in Berwickshire. The method mentioned by Robert Gordon, on the May, in the 1640s, is simply shooting. Sometimes the methods were horribly cruel: there are accounts from 1828, and again from the 1840s, of seals in the Moray Firth being stampeded over barbed iron spikes on their haul-out sites, and disembowelled or shot as they struggled to regain the sea.[15]

In the course of the later nineteenth and early twentieth centuries, although small-scale killing seals for local use continued in places like Fair Isle (Fig 11.2), exploitation for distant use apparently became more systematic as a market for grey seal skins opened up in Glasgow and Edinburgh. An article by the charismatic Edwardian explorer and adventurer, Hesketh Pritchard, in *Cornhill Magazine,* in 1913, drew attention to the continued cruelty of the hunt on Haskeir, and to the declining numbers of grey seals in Scotland. It was thought at the time that only about 500 individuals survived, though in reality there were probably considerably more, and it was feared that a new craze for wearing sealskin jackets in the motor car might finally send the animals into extinction. Pritchard's friend Charles Lyell, and a small group of sympathetic MPs and peers, secured statutory protection for the animal in the Grey Seals (Protection) Act of 1914, setting a close season for killing them between 1 October and 15 December. Though they could legally be hunted at other times of the year, it was mainly on their breeding grounds that grey seals were vulnerable. This was

SEALS: THE BONE OF CONTENTION

Fig. 11.2 – Grey seals shot on the Skerry, Fair Isle, c.1920, and towed ashore: a large seal gave about five gallons of oil, and the skins were traded or used in fishing gear.

the first instance of statutory protection of any mammalian predator in the United Kingdom.[16]

The Act was intended to be reviewed annually, though it did not silence criticism of damage done by seals. For example, in February 1918, Colonel Stewart, Equerry to Princess Beatrice (Queen Victoria's daughter), a man 'who knows the coast of Scotland thoroughly', sent a letter to his MP, forwarded to the Scottish Secretary, saying that he had seen as many as two grey seals at one time on the coast of Colonsay. Some were 15 feet long:

> Picture to yourself the enormous quantities of salmon, sea-trout and all sizeable fish (all food provided by a kind Providence) put into the stomachs of such deadly and unerring fishers. They eat and sleep. They do the like all day long.

Such bad animals should of course be destroyed, along with the countless seabirds that also ate too much. Officials decided on this occasion that they need not trouble to respond.[17]

In the event, the statute survived unamended until 1932, though according to a *Times* correspondent on 8 January 1929, in the west of

Scotland it was disregarded and 'almost unknown' to population and police alike. In 1926, the Scottish Office, considering 'the view that the grey seal does considerable injury to fisheries' asked their experienced salmon scientist and inspector, William Calderwood, and Professor James Ritchie of Edinburgh University, to visit the breeding colonies and, for the first time, count the young. Their survey, 'carried out I understand at some risk to life and limb owing to the nature of the animals' habitat and to the rough weather', took two winters. They counted 1,003 young, concluded that this implied a population of 4,300 animals, and that the grey seal was no longer in danger of extinction. However, it was feared that Norwegian sealers, who were alleged to have already wiped out their own grey seals, might raid the Scottish colonies if killing in the breeding season were to be permitted. Although a revised population estimate in 1932 reckoned 8,000 seals, a new statute that year, piloted through Parliament by Lord Strathcona who himself owned a grey seal colony off Colonsay, continued the protection, even extending the close season from 1 September to 31 December. However, as a concession to fishermen, it also gave the Secretary of State power to issue licences to kill seals within this time, except at Haskeir, which gained total protection. These provisions were continued in the 1970 Conservation of Seals Act (minus the detail about Haskeir), when a three-month close season was also introduced for harbour seals.[18] This remained the legislative framework until the Marine (Scotland) Act of 2010, which conferred year-round protection on all seals while giving the Scottish Government power to issue licences to shoot any judged to be damaging fisheries.

After the Second World War, it quickly became evident that the seals were still increasing at an accelerating rate, and conflicts grew, especially with salmon fishermen. In 1959, the Nature Conservancy appointed a committee on Grey Seals and Fisheries, reporting in 1963 that the population round the UK and Ireland had grown to 36,000, of which over 80 per cent were in Scotland, and recommending that an experimental cull be introduced to reduce the numbers in the Farnes and in Orkney. A paper in 1960 by Bennet Rae from the Scottish Marine Laboratory concluded that the seal catch was equivalent to a quarter of all the fish taken by Scottish fishermen in home waters, that it actually reduced the catch of the fisheries by 15 per cent, and concluded: 'statements that seals do no harm to fisheries can only be regarded as irresponsible'. At the time, that was what most scientists thought, though it was agreed that far too little was really known about seal diet and that Rae's paper, being based on seals shot at salmon nets, was likely to rest on a biased sample.[19]

Culls were indeed attempted from 1960 in Orkney under licence from the Scottish Office, and continued for two decades. The Orkney operation differed from that on the Farnes, the aim being to set up a regular harvest of grey seal pups which it was hoped (wrongly) would ultimately reduce the size of the population. Both met with immediate protest, from local people in Orkney and Northumberland, and from others who considered the culls unethical or cruel, but after some hesitation, and continuing pressure from the fisheries lobby, they were nevertheless continued. There was also a pup cull in the Outer Hebrides between 1972 and 1980.[20] None of these culls were successful in limiting numbers, and by the later 1970s it was clear that the grey seal population was still rapidly rising around the British coasts, including in Orkney where the intention of the plan had been to reduce the population by a quarter.

Consequently, in 1977, the Scottish Office, on the advice of the Seals Advisory Committee of the Natural Environment Research Council (NERC) and with the agreement of the Nature Conservancy Council, proposed a new and much more drastic cull, to reduce the population (then estimated at about 50,000) down by nearly one-third over six years, to the mid 1960s' level, which would entail killing the very considerable number of 900 breeding females and 4,000 young each year. The animals were to be taken in Orkney and the Outer Hebrides. The Norwegian firm contracted to do the killing in 1977 concentrated in the Outer Hebrides and was hampered by bad weather. Only 600 animals were killed altogether, but in any case the remote location ensured relatively little comment in the press or other public reaction. The authorities anticipated opposition in the future, but were utterly taken aback by what happened the following year.[21]

Events in Canada shaped the reaction in Scotland. During the 1970s, the new colour televisions had shown scenes, still vividly recalled today by older viewers, of the blood of thousands of baby harp seals turning the Arctic ice red, as they were clubbed to death for their white skins. The Newfoundland hunters had done this since the mid nineteenth century and the largest catches had been at the end of that century and in the 1950s, but only at this point did the scale of the harvest and the profits made from it begin to attract international attention. It was called 'the greatest kill of concentrated wild mammals in the world'. A limit was set to the cull at 160,000 in 1977 and 180,000 in 1978.

The harp seal was not actually endangered: the objection was to the apparent cruelty and the ruthless manner of killing. Millions of viewers were both riveted and appalled by pictures of helpless pups, with dog-like faces, wet noses and big brown eyes, staring up at their killers. The International Fund for Animal Welfare, later joined by Greenpeace, went onto the ice-floes in Labrador in 1976 to interrupt the hunt, and stepped up the disruption in 1978. To the delight of the film crews, Brigitte Bardot, the most famous and beautiful film-star in France, joined the protests. The Newfoundlanders became reviled and the Canadian government in 1977 passed legislation forbidding any aircraft from flying at an altitude of less than 2,000 feet over any seal on the ice, or landing an aircraft within half a mile. They called this the Seal Protection Act, but its intention was to prevent anyone from filming or disrupting the kill.[22] The effect, of course, was to outrage international opinion even more. In 1983, the European Economic Community imposed a ban on the sale of the skins of 'baby' seals. In Britain and North America demand collapsed, effectively reducing the market to between 18,000 and 70,000 skins a year in the 1980s and 1990s. Since then it has not only revived but become larger than ever, currently running at about 300,000 Canadian skins a year.[23] The grey seals taken in Orkney since 1960 had also been used for their skins (sometimes made into sporrans), taken when the pups were three to four weeks old. They were of inferior quality, but all seal skins became pariah overnight, and in the case of Scottish skins no more were ever taken after 1983.

Inevitably parallels were drawn between what was planned for Scotland and what was happening in Canada, though the scale of the intended cull was of a different order of magnitude, the controls on the methods of killing were much stricter, and the intention was not primarily profit but conservation of fish stocks. The Labour Secretary of State, Bruce Millan, announced the culling plan for 1978 in July. Greenpeace sent the *Rainbow Warrior* to Kirkwall in October, to confront the Norwegian boat that had again been contracted to carry out the main part of the cull. Meanwhile, the government had received a petition against the cull from a total of 42,000 people, an extraordinary number before the days of the Internet, and it included 1,200 signatures from Orkney where on this occasion much of the killing was to take place. There were debates in the European Parliament, and voluntary conservation bodies in Britain and abroad for the first time expressed doubts about the scientific justification for the cull.[24]

As tension mounted, there were clashes on the quayside between officials and protesters, with media helicopters hovering overhead. Bruce Millan lost his nerve. On 16 October, he announced to the press that in view of widespread public concern 'I have decided . . . to reduce the size of the cull this year so that everyone will have the opportunity to study the scientific evidence . . . I have decided to withdraw the Norwegian firm.' Instead of nearly 5,000 animals young and old being killed, the number was reduced to 2,000 pups to be taken by local people.[25] In the event, even fewer were actually taken. To the opponents it seemed a tremendous victory, to the fishermen and officials a retreat before public hysteria. To John Lister-Kaye, a conservationist who was nevertheless persuaded by the science used by the Department of Agriculture and Fisheries for Scotland, the public was displaying 'misinformed opinion induced by propaganda which played on the emotions of millions and presented little fact for intelligent consideration'.[26]

Reality was perhaps more complex. The retreat of Bruce Millan did not itself end the cull or the protests. When Mrs Thatcher came to power in 1979, the Scottish Office and the NCC hoped that a more limited cull would at least enable the grey seal stock to be stabilised at its existing level, though the fishermen called this 'a very poor substitute for the significant reduction in seal numbers which the industry considered necessary'.[27] The Conservative Secretary of State, George Younger, announced that he would seek 'a clear long-term management plan to control the level of the grey seal population'. He continued what he preferred to describe as the 'traditional pup hunt' in Orkney: a letter to the Scottish Office from the International Fund for Animal Welfare suggested that describing something as 'traditional' that had only originated in 1960 was 'a misuse of the Queen's English'. In 1982 he was still looking for what an official called a 'quiet way of killing seals'.[28] He did not find it, and the official in question did not expect that he would. Further protests from conservationists and the public, together with the collapse in the price of grey skins and the death of elderly hunters in Orkney, along with a threat of further direct action by the Sea Shepherd Fund (which had disrupted hunts of harbour seals on the Wash) and changing scientific understanding, finally ended the culls in 1983. In subsequent years the Scottish Office generally broadly supported the calls of fishermen for a cull, but the Home Office in London was too nervous of the idea to allow it to go ahead.[29]

In truth, the public was not as hysterical as supposed, nor any more misinformed than the fisheries scientists. It was demonstrating not so much sentimentality as a widespread (though not universal) sensibility about animals, which the officials and ministers in the Scottish Office did not completely understand. Such a sensibility was based on a feeling that the life of wild sensate creatures should not be taken merely for profit or amusement, unless they were being culled to prevent damage to farm crops or wider environmental harm. When scientific proof of harm was claimed, as it was in the case of the seals, to be acceptable to the public it had to be absolutely unequivocal, and a cull had to be certain to work. Especially, it was not acceptable to kill breeding animals when they were helpless and had no means of escape, as in sealing, or to outpace them with a ruthless industrial technology of killing, as in whaling, or to start culling when you would kill a lot of wild animals for uncertain outcomes, as in a badger cull. For these reasons the public, though often accused of having no more than a sentimental 'Bambi syndrome', would accept rabbit and wood pigeon shooting, or even a deer cull, but not a seal cull. The rabbits and wood pigeons were eating crops and deer were harming the environment; none was killed when helpless, and all were adept at escaping.

Of course the public were easily influenced by compelling images in the media, and there were strict limits on their compassion whenever their own self-interest was involved. Lister-Kaye found that half those questioned who were opposed to the seal cull changed their mind when told that opposing it might put up the price of fish.[30] Most people are horrified when they see what happens in slaughterhouses or on intensive poultry farms, but they continue to buy pork and chicken regardless. Stopping sealing and whaling involved no real perceived costs for the public, so – perhaps with typical human hypocrisy – they felt free to indulge their sensibility.

But this sensibility was nevertheless real, honestly and angrily felt. It was also something comparatively new. It developed among the public in the nineteenth and twentieth centuries, first seen in the campaigns for the seabird preservation legislation of 1869 and its successors. It increased over time and was not at all restricted to ignorant or urban people out of touch with the real world of nature, as often alleged. In 1939, for example, an influential and much read ecologist, Frank Fraser Darling, wrote about animal killing, from a personal viewpoint:

> Respect for life is not squeamishness. It is the regard which says in effect that here is an individual born into the world, moving with a quality of life which once taken cannot be replaced. If

I kill knowingly and cut short this individuality, what lasting good can I expect from the act?

He was not at all opposed to killing for food, as we are naturally a predator species, but he denounced killing for amusement or for frivolous purposes, such as to take furs or unnecessary scientific specimens.[31] But he quite accepted that a population of wild animals might have to be reduced by killing if it was harming the wider environment, even arguing himself in the 1950s and 1960s that a seal cull might become necessary.[32] He always believed that wildlife needed to be managed, not just preserved.

There were innumerable letters to the Scottish Office in the 1970s and 1980s expressing distaste for what the writers considered as mindless slaughter, exactly of the kind deplored by Fraser Darling in 1939. A typical one in 1982 was kept by officials as an example (as they wrote in the margin) of the 'tiresome group letters' that they had to deal with. It was from E.J. Smith writing on behalf of the Isle of Wight Ocean Defence, purporting to be a local grouping concerned with defence of the seas, but perhaps consisting mainly of Smith himself. He is recalled on the island today as an enthusiastic supporter of Greenpeace. The Scottish Office had announced the issue of a licence to kill 1,200 seals in the Orkney 'pup hunt . . . for the exploitation of a population surplus of grey seals as a resource' and explaining that no decision had been taken about a wider cull pending 'the results of commissioned research'. Eric Smith's angry letter said:

> no man has the right to use another living creature as a resource, in the same way that we use coal, gas etc. The grey seal offers no threat to man at all, [you are] condoning the murder of the seals for the greed of mankind, who turn their dead bodies into trinketsets [*sic*] etc for sheer profit.[33]

Tiresome it may indeed have been to officials, but it was not irrational. It certainly helped the seals' cause that, as Tam Dalyell put it in the House of Commons, 'we see pictures of beautiful, wide-eyed, cuddly seal pups peering from the pages of many Sunday newspapers'.[34] Yet the nub of the matter was not that the seal had magically become 'iconic', or that baby seals looked like dogs, or human babies, or cuddly toys, but that people throughout the country had witnessed the culls, seen the conflicts reported and heard the issues debated on their television screens and made up their minds. The public was disgusted by the manner of the killing and the failure of justification. This public included Mrs Harold Wilson, who 'made no secret of the fact that she was constantly asking her husband "what are you going to do about the seals?"'.[35] More importantly for the political community, it included thousands, if not millions, of other voters prepared to express their views in a democracy.

In this debate, science was absolutely central. It was slowly recognised that it was extremely difficult to determine how much commercial fish the seals ate, and what difference a seal cull would make to the fishing industry even if one was instituted. In the 1950s, government had set up a committee to advise it, with representatives drawn from their own departments, the Nature Conservancy, the universities and learned societies. This appointed a full-time worker in 1960, and from this developed a Seals Research Unit administered by NERC, at Lowestoft. This in turn in 1977 amalgamated with the whale research division of the Institute of Oceanographic Science and transmogrified into the Sea Mammal Research Unit (SMRU), moving to Cambridge, later to the University of St Andrews, but still always answerable to NERC.

In the early days, the scientific consensus was that seals took a substantial part of their diet from commercial fish, and that this was an appreciable cost to the fishing industry which might be effectively addressed by a seal cull. This was still the position in 1977–78, when the latest scientific opinion was expressed in papers by two government fisheries scientists, suggesting real damage, and by C.F. Summers, the head of the NERC unit, suggesting a great increase in seals and the need for a cull that would not merely take pups, but, if it were to be effective, would have to take breeding females as well.[36]

At this point, other scientists, particularly J.R. Beddington of York University and Jeremy Greenwood of Dundee University, and others associated with the voluntary bodies of the conservation movement, began to criticise the accepted view, on a number of well-founded grounds. They indicated that data on seal diet was biased because it took half the samples near salmon nets, which might only relate to a small number of seals that could be best dealt with by shooting near the nets. They stressed that there were great uncertainties about how much fish seals ate, and of which species, or how much of the fish eaten by seals would ever be available to Scottish fishermen to catch anyway. They furthermore pointed out that cod-worm infection, though undoubtedly spread by seals, was poorly correlated to the numbers of seals, and they emphasised the margins of uncertainty in the data on the size of the seal population.[37] This criticism was formally presented to the Scottish Office in May 1979 by the voluntary conservation bodies, but the evidence had come out piecemeal in the preceding 18 months and been picked up by the media, sowing further doubts in the public mind as to the correctness of the official case. In any case, the public was inclined to the robust view that fish were scarce in the North Sea because the fishermen themselves had overfished it, not because of the seals, and that, in the words of one woman from Glasgow during a phone-in on Radio Scotland, if the government was really worried about the marine ecosystem it would be more effective to cull the fishermen than the seals.

It took time for officialdom to accept these criticisms, but the Secretary of State was happy to agree to the call for more research, and so to delay a decision on a cull that was bound to offend either the public or the fishermen. NERC changed the way they supervised their seal research. Shortly afterwards there was a change in the leadership at SMRU, and new findings from there began to suggest that in several respects the criticisms of the voluntary bodies represented in the Council for Nature had been well founded. In particular, it became evident that sandeels played a much bigger role in seal diet than anyone had ever imagined, and white fish and salmon a smaller role. It also gradually began to become clear that the formula on which fisheries science had been relying was seriously deficient in explaining all the dynamics of the marine ecosystem. When the Conservatives abandoned the idea of a cull in 1983, they could with relief declare that science was on their side. Had the science found that a cull would be effective and necessary, it is very doubtful if the public dislike could eventually have prevailed.

3. Living with Seals

We have dwelt at length on the events of 1977–83 because this was the turning point in the whole debate. It was recognised at the time that if a cull was delayed any longer the grey seal population would spiral to a level where it would become impossible to bring the numbers down without killing adults and young on an enormous scale, which would be 'seen as abhorrent and would be expensive and difficult to mount'.[38] Officials in the Department for Agriculture and Fisheries for Scotland were exasperated:

> Every year's delay means it is more difficult to mount an effective and humane culling operation. Every year's delay means higher and higher numbers will require to be taken simply to *contain* the population…What is special about seals? – the containment of red deer is not now a public issue.

Reluctant to admit the new scientific criticisms, the officials accused the voluntary conservation bodies of irresponsibility, and the Nature Conservancy Council of going back on its previous commitment to bring the population down – 'possibly on emotional rather than scientific grounds'. In fact, the minutes of the relevant committee in NCC show the organisation convinced by the new scientific case made, at least as far as recommending caution and better evidence before authorising a large cull.[39] Ray Beverton, recently retired head of NERC and one of the most influential fisheries scientists and population biologists of the old school in Britain, made a similar point about the need to act quickly, but he blamed the fishermen for not being measured in their approach:

> Because of the resumed growth of the seal population time is rapidly running out on being able to undertake any feasible control…if the fisheries lobby really begins to make its position known to the wider public in a sensible and not exaggerated way, it might still be possible to do something.[40]

These warnings proved apt, and the grey seal population now is several times the size that it was in 1979. For two decades after 1983, the fishing industry every year continued to call for the resumption of culling, but it is now more widely realised that a cull would be impossible, on economic as well as on humanitarian grounds. It is also likely to be ineffective, and perhaps never could have been effective even in the 1970s. A cull would have to involve adults as well as young, and to be undertaken every year to maintain the population at a depressed level. The disposal of the carcasses would now have to involve helicopters and incinerators on an industrial scale, like disposal of the bodies of cattle in a foot-and-mouth disease outbreak. In addition, the disturbed colonies would probably fragment and form new ones, not restricted in size by density dependence as they appear now to be over most of their range. Nor is there any evidence that it would work: grey seals feed on the predators of young commercial fish as well as on the commercial fish themselves. We now know that there are 300,000 harbour porpoises in the North Sea also feeding (in part) on commercial fish, so at least as great a potential threat as the seals. Even the toughest fisherman blanches before the prospect of calling for a cull of porpoises. Still less would anyone dare to call for a cull of bottle-nosed dolphins, which may take more salmon than seals, as they can take an adult salmon in open water.[41] Dolphin sightings have much increased in the Firth of Forth since the early 1990s.

The argument as to whether the grey seals materially damage salmon fisheries is still not completely resolved. Salmon and sea trout form only a tiny proportion of grey seals' diet taken as a whole, but a relatively small number specialise in ambushing salmonids coming into the rivers to breed, and because wild salmon have become so scarce compared to their historic numbers, even a few seals might now have a serious impact on their population. Possibly 2,000–5,000 seals a year were still being legally shot at the mouths of salmon rivers (and at salmon farms) at the start of the present century, though the numbers are not accurately known. Under the 2010 Marine Act, shooting will have to be licensed by the Scottish Government, and it seems likely that numbers killed will be substantially fewer than before. It clearly makes sense to concentrate on removing a few rogue animals in the interests of the salmon population. Attempts further to restrict seal shooting seem very unlikely to be successful, because the public and scientists of all descriptions have always regarded selective shooting at salmon farms and salmon runs as quite defensible when no other remedy is available. It may be, however, that 'pingers', acoustic deterrents on the nets, will in the future prove more effective than shooting.

Meanwhile, in the Firth of Forth, the numbers of seals continue to increase more rapidly than elsewhere in Scotland, and fishermen of many kinds have not been pleased about this for the last 30 years. Lobster pots are ripped open, the catch of sea-anglers is neatly removed from their lines before their eyes, salmon and sea trout coming up the Firth are devoured by seals that wait in the estuary. Nevertheless, in 2010, when the production of grey seals in the area reached new heights, the number of salmon caught in the River Forth exceeded that of any other beat in Scotland.[42]

Until his death in 1999, Councillor James Braid of St Monans was a constant advocate, in the press and in the council chamber of the Fife local authority, of action to limit seal numbers in the Firth of Forth, his opinions expressing the frustration of the fishermen but unable to force action by the authorities. The question has never been whether seals eat fish that fishermen want to catch, but whether killing seals makes any material difference to the amount they can catch, given the present human overfishing of the seas. On this there is one view by the fishermen, and another by the majority of scientists. Fishermen sometimes tell the visitor that it is the over-abundance of seals, not the reckless actions of themselves, that is responsible for a sea empty of commercial fish. Since there were still only a few seals when the Firth ran out of its stocks of herring, cod and haddock, to a historian this does not seem likely to be true.

As a postscript, there is a second, smaller species of seal that inhabits the Firth of Forth, the harbour seal, which, over the last century, has not at all flourished in the same way as the grey seal. It is currently described as being 'in a depleted state' and in continuing decline along the east coast and within the Firth of Forth.[43] An animal breeding on sandbanks rather than on islands, its main haunts in south-east Scotland have always been the Firth of Tay rather than the Firth of Forth, and until recently it outnumbered the grey seal on the haul-out sands that both use off Tentsmuir. It is still common in the upper reaches around Kinghorn and Burntisland where the small Firth of Forth population is concentrated, and these animals have apparently thrived more than their congeners in the Tay and the Eden. Not distinguished from the larger grey seal until the eighteenth century, the two species were readily confused even as late as the 1930s, and early records are not always reliable. Even today, most fishermen and members of the public find it hard to tell them apart. Nevertheless, harbour seals were frequently identified as visiting most parts of the Firth of Forth, including the Isle of May, in earlier decades. Their main food has been shown also to be sandeels, even though they also take a wide variety of fish including whiting and sometimes salmon. In the southern North Sea in 1988 and again in 2002 they were attacked by a new disease, phocine distemper, which wrecked havoc on the species, but was much less serious in Scotland than further south. For reasons which remain obscure, but could include competition with the grey seal, the population on the east coast of Scotland has nevertheless dropped by more than half since 2001, while the Firth of Tay population has dropped even more, by 85 per cent, since 1992. Consequently, harbour seals are now less often reported in the outer parts of the Firth of Forth. Their contrasting fortune serves to remind us how little we still know about why some species around us flourish, and others decline.

CONCLUSION

Geologists call the period since the end of the last Ice Age, 10,000 years ago, the Holocene, the most recent of a succession of glacial and interglacial ages in the history of the Earth. Paul Crutzen, atmospheric chemist and Nobel prizewinner, has with others recently proposed a new term, the Anthropocene, to express a change in the history of the Earth of more recent origin, but of no less significance than the melting of the ice. After the middle of the eighteenth century, for the first time, absolute domination of the global atmosphere by natural forces began to give way to an age where the Earth became increasingly affected by man's behaviour, its onset illustrated by a sharp increase in carbon dioxide and other greenhouse gasses trapped in bubbles in glacial ice-cores. Humanity at this point started a process of climatic and ecological change that it is now apparently unable to stop and unwilling to believe. By the early twenty-first century, mankind holds the fate of the planet in its hands, but, dithering and cynical, still has no effective plan of how to avert catastrophe.[1]

Others have argued that the onset of human influence goes back long before then, to the invention of farming and the deforestation of large parts of the earth several thousand years ago, a point clearly true ecologically but perhaps harder to trace atmospherically.[2] Whatever the details, it is undeniable that in the last two centuries, and particularly since 1900, accelerating use of natural resources, caused by steeply rising population and the expansion of the global economy, has created what John McNeill terms, in a book title of 2000, *Something New under the Sun*: a situation of runaway growth without parallel in human history. Across the world, between the 1890s and the 1990s, population grew fourfold, the global economy fourteenfold, industrial output fortyfold, carbon dioxide emissions seventeenfold and fish catch in the global seas, thirty-fivefold.[3]

Governments pay lip-service to the principle of sustainability while striving incessantly for rates of economic growth that they must know cannot possibly be sustained.

The history of the Firth of Forth must be seen in this context. As far as we can tell, before about 1830 the pace of change in the marine ecosystem and its animal populations was slow, though the changes were in some respects cumulatively important. Certainly there were large declines in whales and seals, and probably in the seabird colonies of the May and elsewhere, through millennia of exploitation for local demand, beginning in prehistory. As early as the eighteenth century, the first troubles of the oyster beds were driven by a new feature of substantial outside demand, combined with climatic stress that in this case was not anthropogenic. This foreshadowed a coming trend, when in the nineteenth century sea-fishing became increasingly orientated towards more distant markets and more intensive in its methods. Firstly, the seemingly inexhaustible herring disappeared from the Firth and then in the later twentieth century from the entire North Sea (at least for a time as a commercial catch), followed by cod and haddock, leaving behind only little fish and prawns. Of the shellfish, mussels came under pressure from the demands of bait for Victorian line fishing, but were saved by the invention of the trawl, though that did great harm in other directions. Oysters finally succumbed completely to human greed. None of these once great resources, apart from the mussels, show much sign of return in the Firth after a century of grossly unsustainable over-exploitation, even though that ended decades ago.

But this is not the whole story. Pollution first of the rivers and then of the sea was a dramatic theme of the years between about 1850 and 1970, progressively poisoning the water for riverine fish like trout and salmon, and having serious consequences on the benthic community offshore from the sewer outfalls. It was seen to be disgusting, but no one seemed to have enough resolve to do anything about it. Then, quite suddenly and quickly, towards the close of the twentieth century, that was reversed. The salmon have a fair run again up the Forth and other rivers, and the benthic fauna and flora off Leith, Grangemouth and Levenmouth have returned to something like their earlier condition. Similarly, the numbers of seabirds on islands like the Bass and the May, already in earlier centuries kept to modest levels by local exploitation for food, dropped sharply in the nineteenth century when the taste for bird flesh declined and irresponsible sportsmen with cheap guns and cartridges shot them for amusement. The

CONCLUSION

killing was halted by legislation and ultimately the decline was reversed. In the twentieth century, gannets, auks and gulls reached numbers not seen for many centuries, if ever in human history – that some species are now falling back from this should not blind us to the fact that their increase has been extraordinary. Likewise, grey seals were hardly known any more in the Firth of Forth by 1800, but following protection in the early twentieth century there began a remarkable expansion, until now many thousands of pups are born in the islands and coves every year, numbers quite unprecedented in the known past.

Even cetaceans are seen more often than they were. Since 1992, when a school of about 60 bottle-nosed dolphins unexpectedly arrived from the Moray Firth to visit the Firth of Forth in summer, they have returned every year. Whales have always been chance visitors, washed up from time to time, as with the blue whale stranded at Longniddry in 1869 (Plate 6) and the pod of pilot whales that came ashore near Dunbar in 1946 (Fig. 12.1). Now they are being seen more regularly offshore and alive, especially minke

Fig. 12.1 – A large pod of pilot whales stranded at Thorntonloch Bay near Dunbar in 1946. Such strandings have occurred from time to time throughout the environmental history of the Firth.

whales and orcas. This may reflect the decline of persecution in the wider seas as well as the cleaner Firth.[4]

So alongside the themes of unsustainability and environmental overuse, there are other currents that run against degradation and despoliation. They vary in different cases. As regards pollution, no one in the Victorian period doubted that raw sewage, or the foaming stink from factories and the film of oil in the rivers, were offensive – but they were seen as a necessary evil, a price for prosperity. Anyway, the price for cleaning them up was deemed too high. It needed the accretion of technical and bureaucratic expertise and commitment, combined with a decline in the kinds of Victorian industry that produced the pollution (textiles, paper, paraffin distillation and coal mining) before anything very effective could be done. Above all it needed a refreshing mid twentieth-century belief in big government and a willingness to spend a lot of taxpayers' money for a lot of public good.

In the case of the seabirds and seals, what changed was not bureaucratic and political, but a matter of sensibility. A carnage that to most Victorian gentlemen (or even to Victorian clerks) seemed a pleasurable manly recreation or the destruction of vermin came to seem to others, and perhaps especially to their wives, as an unspeakable cruelty towards other sensate life. Tentative legislation from 1869 in the case of seabirds, and from 1914 in the case of grey seals, was followed by progressively more effective protection, spurred on by a rising tide of popular interest in wildlife.

The problem today is less wilful destruction of mammals and birds than accidental disturbance by an increasingly mobile population with leisure at the weekends, seeking to enjoy the seaside (Fig. 12.2). Since the 1960s, breeding birds like terns and ringed plover have been driven off the beaches by people, and especially by their completely uncontrolled dogs. The indestructible plastic litter they leave (unknown in earlier generations) is compounded by what is washed up from fishing boats and commercial shipping, and dangerous to the occasional turtles that mistake it at sea for jellyfish.

In the nineteenth century, Edinburgh was already to the fore through its university and its museum in the scientific understanding of local wildlife below and above the surface of the sea, and in the amateur enjoyment of natural history.[5] In 1839, the Scottish Society for the Prevention of Cruelty to Animals (SSPCA) was founded in the capital to improve the lot of carthorses, and has grown to be a major player in lobbying for animal welfare and employing staff to assist animals in distress. In the twentieth century

CONCLUSION

Fig. 12.2 – A crowded beach at Kinghorn, July 1969: the post-war generation had more time and money than before, but 20 years later families like these might have preferred the Costa Brava.

birds became particularly popular, with the Scottish Ornithologists Club being set up in the capital in 1936, and the Royal Society for the Protection of Birds establishing a regional headquarters there after the Second World War with a lot of local support. In 1964, the Scottish Wildlife Trust was founded in Edinburgh, now with 36,000 members throughout Scotland. From the start it was active in public campaigning, notably over the grey seal cull. More recently, the Friends of the Earth and the Green Party have enjoyed comparative success. As long as protection did not materially affect the public's personal income, lifestyle or choice of diet, animal welfare and conservation became popular causes in the twentieth century.

Along with the voluntary bodies, the state became formally involved in conservation, both by passing legislation protecting species and habitats, and by setting up regulatory bodies, of which the first to be concerned with biodiversity was the Nature Conservancy established in 1949. It morphed over the years first into the Nature Conservancy Council and then in 1990 into Scottish Natural Heritage. With it came in due course a network of protected areas, Sites of Special Scientific Interest, Special Protection Areas, Special Areas of Conservation and now Marine Protection Areas, all backed by European as well as by British and Scottish law. Nature, so long just a material resource or a sporting diversion, also became heritage. After the Brundtland Report of 1987, governments formally agreed to worry about 'sustainability', and meant by that not only what to do when the oil ran out, (as the coal and iron ore had already run out in Scotland) but whether there would be the same heritage of biodiversity for future ages to enjoy as there had been in the past. Maybe, in realpolitik, they did not worry very much about biodiversity, especially about what happened at sea, where they mainly aimed to placate the fishermen, whose aim was always to catch more and blame others. But conservation was on the agenda and the public could at least try to hold them to democratic account, and sometimes did so, as in the controversies over seals. The story of pollution control was on the whole more successful, because the mess and public health risk affected more people more immediately. The establishment of a strong Scottish Environment Protection Agency (SEPA) in 1996 was the culmination of a long period of increasing powers to interfere with vested interest, and the progressive amalgamation of smaller bodies with formerly limited powers to do so, some of which went back to Victorian times. From the 1950s onwards these bodies, enlarged at every opportunity, became conscious that they had much to do, and from the 1970s, as their powers grew, they did it very well.

Fig. 12.3 – SSPCA officers at Musselburgh with oiled birds, many of them great-crested grebes, in February, 1978. This accident destroyed most of these grebes wintering in the Firth of Forth.

With so much big modern industry concentrated round the Firth of Forth, there must always be high potential risk to the natural

CONCLUSION

environment. So far, the Firth of Forth has avoided any major catastrophe, such as a serious oil spill from tankers making for Hound Point, a nuclear leak either from Torness power station or from the Polaris submarine base at Rosyth (where four nuclear submarines await decommissioning), a serious accident relating to the ethylene gas plant at Mossmorran and its terminal in Braefoot Bay, or a catastrophe at Grangemouth refinery and petrochemical works. There have been incidents. There were six oil spills in the Firth between 1970 and 1978, involving the death of many hundreds of birds, culminating in February 1978 of at least 700 sea duck and 241 great-crested grebes, which was most of the wintering population of the latter species in the Firth (Fig. 12.3).[6] No spills that have occurred since have been as serious. There have been fires at Grangemouth, notably in 1987 following an explosion in a hydrocracker (Fig. 12.4), and a serious a blaze in a fluidised catalytic cracker unit in June 2000, which took 60 firefighters several hours to extinguish and cost BP £1 million in fines. Torness experienced a failure in the gas circulator in 2002, blockages to the cooling water intakes by seaweed in 2006 and by several tons of jellyfish in June 2011, but in each case the problem was spotted in good time and no

Fig. 12.4 – The fire at the BP refinery in Grangemouth in 1987. It followed an explosion in the hydrocracker unit heard and felt up to 30 kilometres away.

leaks occurred. At Braefoot Bay, in January 1993, during a storm gusting to 93 mph, a gas tanker loading butane from Mossmorran broke loose from its moorings, grounded twice but remained intact. Relatively small quantities of radioactive material have been repeatedly found on the beach at Dalgety Bay since 1990, not from the submarines at Rosyth but from landfill relating to aircraft instrument panels scrapped after the Second World War. That the Firth has escaped a major disaster is due to the skills of the operators, the emergency services, and to high standards of safety and vigilance set by the regulatory authorities.

Although water pollution has been well solved compared to the past, a substantial local problem remains on the shores of the Firth from chronic air pollution, especially from the big coal-powered power stations. SEPA in 2008 publicised the fact that Longannet was making the local area the most contaminated in Scotland for a combination of nitrogen oxides, sulphur oxides, particulates and carbon dioxide. An independent report for the European Union in 2009 named Cockenzie as the worst emitter of nitrogen oxides in Europe, 50 per cent worse per unit of energy input than the next offender, Middlesbrough on Teesside. Action will now be taken, or these plants face being closed under European law. For all the talk of clean energy, Scotland's record in these respects is still poor, and clean energy itself may bring unappreciated environmental costs. Building and operating the marine windfarm projected off the Isle of May, in particular, might have serious consequences for seals, sandeels and certain seabirds – we still cannot be sure.

As well as constant vigilance by the regulators, the price of freedom from a major environmental disaster in the Firth of Forth is equally constant vigilance by the public. By and large, the citizens of Edinburgh and the rest of the Firth towns, aided by the Green Party, the Friends of the Earth and the RSPB, which keep their eyes on these matters for the common good, have proved adept at this. The biggest threat in recent years has been a proposal dating from 2005 to allow the ship-to-ship transfer of nearly 8 million tons of crude oil between Russian tankers in the Firth of Forth off Kirkcaldy. This proposal received the support of Forth Ports plc, the privatised successor of the Forth Ports Authority, which was in turn the successor, from 1968, of the Leith Docks Commissioners founded in 1838. In other words, though legally charged with controlling navigation in the Firth of Forth, it was no longer a public regulator but a profit-making company. Currently known as Forth Ports Limited, it was bought in 2011 by Arcus European Infrastructure Fund and now also owns the ports of Dundee and Tilbury, and Chatham terminal, having become one of the largest harbour operators in the world.

When the news of the proposals became known, the consequent outcry by the public was readily heard by politicians. Protests were backed by all the local authorities in the Firth of Forth and then by the entire Scottish Parliament: faced with the furore, in February 2008 Forth Ports abandoned their plans, and the Labour government in Westminster (which has jurisdiction over these matters, not the Scottish Parliament) agreed to bring in legislation to ban ship-to-ship transfers in British territorial waters. In 2010, however, the incoming Conservative and Liberal coalition blocked the new rules and announced a new consultation: their main justification was that the plans were (as framed) unworkable, as foreign tankers could still tranship oil just off the territorial water limit 12 miles offshore. Rumours abounded that Forth Ports were looking to reopen their proposals, and the Scottish National Party demanded the transfer of power of navigational regulation in Scottish waters to Edinburgh to prevent this possibility. The position is at present unresolved, but if ship-to-ship transfer were ever to be allowed, there would be substantial risk of an accident that would decimate the gannets of the Bass and the auks and seals of the May, and foul all the beaches and coves of Fife, Berwickshire and Lothian.

So the Firth of Forth remains in the twenty-first century as wonderful for wildlife and natural beauty as ever it was, albeit an ecosystem profoundly changed by human activity, as we have tried to show. Its environmental history has many aspects and has taken many twists. Its future in the medium and long term certainly depends upon global reactions to global threats, but no less on the vigilance and love of all the people who live around its shores. Perhaps if we understand and appreciate its past, we shall be the more empowered to protect its future.

NOTES

Introduction

1. See in particular C. Fleet, M. Wilkes and C.J. Withers, *Scotland: Mapping the Nation* (Edinburgh, 2011).
2. P.H. Brown, *Early Travellers in Scotland* (Edinburgh, 1891), p. 148; G. Baker, 'An unpublished account of an English catholic's Tour to Edinburgh in 1657', *Scottish Historical Review*, Vol. 90 (2011), p. 135.
3. N. Scarfe, *To the Highlands in 1786; the Inquisitive Journey of a Young French Aristocrat* (Woodbridge, 2001), p. 223.

1. Hunting and Gathering

1. C.R. Wickham-Jones and M. Dalland, 'A small mesolithic site at Craighead Golf Course, Fife Ness, Fife', *Tayside and Fife Archaeological Journal*, Vol. 4 (1998), pp. 1–19.
2. *British Archaeology* no. 60 (August 2001).
3. J.M. Coles, 'The early settlement of Scotland: excavation at Morton, Fife', *Proceedings of the Prehistoric Society*, Vol. 37 (1971), pp. 248–363.
4. H.F. James and P. Yeoman, *Excavations at St Ethernan's Monastry, Isle of May, Fife 1992–7* (Perth, 2008), p. 115.
5. J.E. Gordon and D.G. Sutherland (eds), *Quaternary of Scotland* (Joint Nature Conservation Committee, London, 1993); J.D. Hansom and D.J.A. Evans, 'The Carse of Stirling', *Scottish Geographical Journal*, Vol. 116 (2000), pp. 71–8.
6. G. Clark, 'Whales as an economic factor in prehistoric Europe', *Antiquity*, Vol. 21 (1947), pp. 84–104.
7. D. Sloan, 'A summary of some recent shell midden analyses', *Circaea*, Vol. 3 (1985), p. 145; D. Sloan, 'Sample Site and System: Shell Midden Economies in Scotland, 6000–4000 BP', unpublished University of Cambridge PhD thesis (1993), copy deposited in the National Monument Record Library, Royal Commission on the Ancient and Historic Monuments of Scotland, Edinburgh.
8. D. Grieve, 'On the discovery of a kitchen-midden on Inchkeith', *Proceedings of the Society of Antiquaries of Scotland* (henceforth *PSAS*), Vol. 9 (1870–2), pp. 452–5. See also *Discovery and Excavation in Scotland, 1973*, p. 26.
9. J.E. Cree, 'Notice of a prehistoric kitchen-midden and superimposed medieval stone floor at Tusculum, North Berwick', *PSAS*, Vol. 42 (1907–8), pp. 253–294; A.O. Curle, 'Notice of the examination of prehistoric kitchen-middens on the Archerfield Estate, near Gullane', *PSAS*, Vol. 42 (1907–8), pp. 308–319.
10. K.J. Edwards and I.B.M. Ralston, *Scotland, Environment and Archaeology, 8000 bc–ad 1000* (Chichester, 1997).
11. Much of the detail in the remainder of the chapter is cited (except where otherwise referenced) in A. Fenton, *The Shape of the Past*, II, chapter 3 (Edinburgh, 1986), see

also A. Fenton, 'Seaweed as fertiliser', in *Scottish Life and Society, IV: Boats, Fishing and the Sea* (Edinburgh, 2008), pp. 135–150. The 1479 reference is from *Records of the Parliaments of Scotland* (on-line), 7/10/1479; thanks to Richard Oram for this.
12. J.E.L. Murray, 'The agriculture of Crail, 1550–1600', *Scottish Studies*, Vol. 8 (1964), pp. 85–95; D. Beveridge, *Culross and Tulliallan* (Edinburgh, 1885).
13. K.A. Golding, 'The effect of waste disposal on soils in and around historic small towns', unpublished University of Stirling PhD. thesis, 2008.
14. R. Maxwell, *The Practical Husbandman* (Edinburgh, 1757), pp. 126–129.

2. Fishing in a Firth of Plenty

1. R. Sibbald, *The History Ancient and Modern of the Sheriffdoms of Fife and Kinross* (edn, London, 1803), pp. 115–118. See also his manuscript 'Treatise concerning the fisheing in Scotland, 1701', NLS: 33.5.16, and Sibbald, *Scotia Illustrata* (Edinburgh, 1699).
2. G. Markus, 'Gaelic under pressure; a 13th century charter from east Fife', *Journal of Scottish Name Studies*, Vol. 1 (2007), pp. 77–98.
3. H.D. Watson, *Kilrenny and Cellardyke, 800 Years of History* (Edinburgh, 1986), p. 15.
4. B. Poulsen, *Dutch Herring: an Environmental History c.1600–1860* (Amsterdam, 2008); M. Rorke, 'The Scottish herring trade, 1470–1600', *Scottish Historical Review*, Vol. 84 (2005), pp. 149–165.
5. This account is based on Rorke, 'Herring trade', and on unpublished research on the Crail archives kindly made available by Thomas Riis. The detail of the crears carrying yawls is in *Acts of the Parliament of Scotland*, V, 417–418.
6. *Records of the Privy Council of Scotland*, 1st series, IV, 123.
7. Rorke, 'Herring trade', p. 161.
8. W.C. MacKenzie, *The Book of the Lews* (Paisley, 1919).
9. See note 5 above.
10. *Records of the Privy Council of Scotland*, 1st series, VII, 85.
11. W. Mackay (ed.), *Chronicles of the Frasers* (Scottish History Society, 1903), pp. 239–241.
12. H. Scott, *Fasti Ecclesiae Scoticanae* (edn, Edinburgh, 1925), V, 182.
13. Watson, *Kilrenny*, pp. 36–37.
14. P. Lindsay, *The Interest of Scotland Considered* (Edinburgh, 1733), pp. 190–199.
15. T.C. Smout, *Scottish Trade on the Eve of Union, 1660–1707* (Edinburgh, 1963), p. 61.
16. Ibid., p. 223.
17. Sibbald, 'Treatise'; Sibbald, *Sheriffdoms*, pp. 337–346; Lindsay, *Interest*, p. 202.
18. *Third Report on the State of the British Fisheries, 1785*, in *Reports of the House of Commons, 1785–1801* (1803), X, 117.
19. J. Miller, *The History of Dunbar* (Dunbar, 1859), p. 264.
20. B. Harris, 'Scotland's herring fisheries and the prosperity of the nation', *Scottish Historical Review*, Vol. 79 (2000), pp. 39–60.
21. *The Chronicle of Fife, being the Diary of John Lamont of Newton* (edn. A. Constable, Edinburgh, 1810), pp. 136–137.
22. See 'Guide to the Records of the Scottish High Court of Admiralty, 1627-1750', CD-ROM available from the Early Scottish Maritime Exchange.
23. Admiralty Court Records, NRS: AC 7/2. ff. 6–8.
24. Admiralty Court Book of East Fife, 1686–1699, NRS: GD 1/357.
25. Cited in G. Gourley, *Anstruther, or Illustrations of Scottish Burgh Life* (Cupar, 1888), p. 130. See also St Andrews University Library MS B3/7/2.
26. D. Loch, *Essays on the Trade, Commerce, Manufactures and Fisheries of Scotland* (Edinburgh, 1778), II, 243.
27. Miller, *Dunbar*, p. 264.
28. Loch, *Essays*, II, pp. 232–241.
29. Ibid., pp. 244–245.
30. W. Daniell, *A Voyage round the Coast of Scotland and the Adjacent Isles 1815–1822* (edn, Edinburgh 2006, published as *Daniell's Scotland*), I, p. 322.

31. *Statistical Account of Scotland*, ed. J. Sinclair, 1791–1799. The most convenient edition is that edited by D.J. Withrington and I.R. Grant, where all the parish accounts are rearranged by county, *Lothians*, Vol. 2 (1975), ed. T.C. Smout, and *Fife*, Vol. 10, ed. R.G. Cant. All the following citations can readily be found by parish in this edition, though the reference to Carlyle's observation is, less obviously, under Inveresk, in Vol. 2, pp. 293–298.
32. The comment is from the anonymous editor of Sibbald, *Sheriffdoms* (see note 1 above), p. 119.
33. *Committee Appointed to Inquire into the State of the British Herring Fisheries, 1798*, in *Report of the House of Commons, 1785–1801* (1803), X, 338–340.
34. M. Gray, *The Fishing Industries of Scotland, 1790–1914: a Study in Regional Adaptation* (Aberdeen, 1978), p. 27.
35. Quoted in G. Gourlay, *Fisher Life: or the Memorials of Cellardyke and the Fife Coast* (Cupar, 1879), p. 64.

3. Oyster Wars

1. *Royal Commission on Sea Fisheries*, P.P. 1866, Vol. 18, evidence, p. 603.
2. T.W. Fulton, 'The past and present condition of the oyster beds in the Firth of Forth', *Fourteenth Annual Report of the Fishery Board for Scotland* (1895), p. 272; C.M. Yonge, *Oysters* (London, 1960), pp. 92–93.
3. NRS: GD 265/15/5, 'Admiralty decreet relative to the oyster scalps off Newhaven', pp. 97, 144.
4. T. McGowran, *Newhaven-on-Forth, Port of Grace* (Edinburgh, 1985), pp. 133–134.
5. Ibid., p. 21.
6. *Records of the Privy Council of Scotland*, third series, I, 484.
7. *Extracts from the Records of the Burgh of Edinburgh, 1681–1689*, pp. 55, 197, 239, *1689–1701*, pp. 18, 165, *1702–1718*, pp. 29, 129. *Statistical Account* of Prestonpans, East Lothian.
8. NRS: E72/15 Customs Books, Leith, outwards, 1672.
9. *Records of Edinburgh, 1655–1665*, pp. 337–340.
10. *Records of Privy Council*, I, 472, 484.
11. Ibid., 482; III, 69; XIII, viii.
12. A. Dawson, *So Foul and Fair a Day: A History of Scotland's Weather and Climate* (Edinburgh, 2009), pp. 118–129; H.H. Lamb, *Climate, History and the Modern World* (London, 1982); R.A. Dodgshon, 'The Little Ice Age in the Scottish Highlands and Islands', *Scottish Geographical Journal*, Vol. 121 (2005), pp. 321–337.
13. Fulton, 'Past and present', pp. 272–273; Yonge, *Oysters*, pp. 91–92.
14. Fulton, 'Past and present', p. 246.
15. Ibid.
16. R.M. Black, *Society of Free Fishermen of Newhaven: a Short History* (Edinburgh, 1951); McGowran, *Newhaven-on-Forth*.
17. *Acts of the Parliaments of Scotland*, 9/5/1695.
18. NRS: AC9/1473.
19. NRS: AC10/285.
20. Fulton, 'Past and present', pp. 246–247.
21. For all that follows in the remainder of this section, see NRS: GD 265/15/5, *Admiralty decreet relative to the oyster scalps of Newhaven*. This is referred to in Fulton, 'Past and present' and McGowran *Newhaven-on-Forth*. For advice on Alexander Wight, we are most grateful to Prof. John Cairns.
22. NRS: AC10/1278.
23. Black, *Free Fishermen*, p. 62; NRS: GD 265/9/1, Minute Book of the Free Fishermen, for 15 Oct 1812.
24. The details in this section where not otherwise referenced are based on Fulton, 'Past and present', pp. 249–261. Fulton also made extensive and careful use of material from the Free Fishermen's records, now in NRS: GD 265.

25. For pressgang, see Black, *Free Fishermen*; for weather, see Dawson, *So Foul and Fair*.
26. Yonge, *Oysters*, p. 155; *Times* citation from McGowran, *Newport-on-Forth*, p. 137.
27. Fulton, 'Past and present', pp. 250–251, gives two totals – I have taken the larger.
28. NRS: GD 265/9/1, f. 73.
29. Black, *Free Fishermen*, p. 89; McGowran, *Newport-on-Forth*, follows Black.
30. Fulton, 'Past and present', p. 254.
31. Quoted ibid., p. 255.
32. Ibid., p. 256.
33. NRS: GD 265/9/1, ff. 125–6.
34. *Glasgow Herald*, 23 Oct 1868; NRS: GD 265/9/1, f. 183.
35. NRS: GD 265/9/1, f. 131.
36. *The Scotsman*, 10 June 1869.
37. Fulton, 'Past and present', pp. 258–60.
38. Ibid., p. 261.
39. G. Hardin, 'The tragedy of the commons', *Science* Vol. 162 (1968), pp. 1243–1248. See also T.C. Smout, 'Garrett Hardin, The Tragedy of the Commons and the Firth of Forth', *Environment and History*, Vol. 17 (2011), pp. 357-378.
40. E. Ostrom, *Governing the Commons: The Evolution of Institutions for Collective Action* (Cambridge, 1990).
41. NRS: GD 265/9/1, ff. 74, 88, 104.
42. Ibid., ff. 63–64.
43. Ibid., ff. 114, 138.
44. Ibid., ff. 184–186.
45. Ibid., f. 104.
46. *The Scotsman*, 15 Dec 1864, 18 Jan 1865.
47. Yonge, *Oysters*, pp. 123, 155–158.
48. F. Fraser Darling, *Natural History in the Highlands and Islands* (London, 1947), p. 59.
49. Yonge, *Oysters*, pp. 160–167.
50. G. Leslie and W.A. Herdman, 'Invertebrate fauna of the Firth of Forth', *Proceedings of the Royal Physiological Society of Edinburgh*, Vol. 6 (1881), pp. 277–313; A.J. Berry and S.M. Smith, 'Aspects of the molluscan fauna of the rocky shores of the Firth of Forth estuary, Scotland', *Proceedings of the Royal Society of Edinburgh*, Vol. 93B (1987), pp. 431–447.
51. Fulton, 'Past and present', pp. 262–263.
52. Ibid., pp. 261–262, For climate, Dawson, *Fair and Foul*, pp. 176–186.

4. Herring Boom and Herring Bust, 1820–1950

1. B. Poulson, *Dutch Herring: An Environmental History, c.1600–1860* (Amsterdam, 2008), pp. 77, 181; D.H. Cushing, *Climate and Fisheries* (Bury St Edmunds, 1982), pp. 58–63, 199–208.
2. W.C. Hodgson, *The Herring and its Fishing* (London, 1957), p. 16.
3. P. Smith, *The Lammas Drave and the Winter Herrin'* (Edinburgh, 1985) is the classic account of the herring fishing in East Fife.
4. A. McCallum, *Cambridge County Geographies: Midlothian* (Cambridge, 1912), p.83
5. Smith, *Lammas Drave*, p. 6.
6. J.R. Coull, *The Sea Fisheries of Scotland, A Historical Geography* (Edinburgh, 1996), esp. chapter 7; M. Gray, *The Fishing Industries of Scotland 1790–1914: a Study in Regional Adaptation* (Aberdeen, 1978).
7. Gray, *Fishing Industries*, especially chapters 2 and 3.
8. Smith, *Lammas Drave*, p. 7; J. Miller, 'Traditional fishing boats', in J.R. Coull, A. Fenton and K. Veitch (eds), *Boats, Fishing and the Sea*, Vol. 4 of *A Compendium of Scottish Ethnology* (Edinburgh, 2008), p. 106; J.R. Coull, *Sea Fisheries*, pp. 110–111.
9. Gray, *Fishing Industries*, p. 58.
10. Smith, *Lammas Drave*, pp. 24–36.

11. Quoted ibid., p. 145.
12. *Report of the Royal Commission on Trawl Net and Beam Trawling Fishing*, P.P. 1884–5, Vol. 16, p. 84.
13. Ibid., pp. 89–90; Smith, *Lammas Drave*, p. 39.
14. *Report by the Commissioners for the British Fisheries*, 1861, p. 10; *Report on the Herring Fisheries of Scotland*, P.P. 1878, Vol. 21, p. xxviii; *Scottish Fisheries Board Report for 1883*, pp. xiv–xvi.
15. *Report of the Royal Commission*, 1884–5, p. 90.
16. *Report of the Royal Commission on the Sea Fisheries of the United Kingdom*, P.P. 1866, Vols. 17–18.
17. *Report of the Royal Commission*, 1884–5, p. xii.
18. *Scottish Fisheries Board Reports for 1883*, p. xxxvii, and for *1890*, pp. 179–90; Smith, *Lammas Drave*, pp. 84–90.
19. B. Patrick, *Recollections of East Fife Fisher Folk* (Edinburgh, 2003), pp. 52–3.
20. Coull, *Sea Fisheries*, pp. 105, 111.
21. http://www.scotfishmuseum.org/reaper/facts.html.
22. Coull, *Sea Fisheries*, pp. 111–112; *Scottish Fisheries Board Report for 1891*, p. 183.
23. S. Dick, *The Pageant of the Forth* (Edinburgh, 1911), pp. 175–176.
24. *Report of a Committee of the Fishery Board for Scotland as to the Regulation of Trawling*, P.P. 1888, Vol. 28, p. 60.
25. Gray, *Fishing Industries*, p. 75; Smith, *Lammas Drave*, pp. 43–55.
26. G. Gourlay, *Fisher Life or the Memorials of Cellardyke and Fife Coast* (Cupar, 1879), p. 97.
27. *Scottish Fisheries Board Report for 1884*, p. xviii; *for 1885*, p. xxxii.
28. B. Patrick, *Recollections*, p. 68.
29. *Scottish Fisheries Board Report for 1930*, pp. 16–17; Smith, *Lammas Drave*, pp. 102–109.
30. Angus Martin, *The Ring-Net Fishermen* (Edinburgh, 1981).
31. Smith, *Lammas Drave*, pp. 106, 114–116, 128.
32. *Scottish Fisheries Board Report for 1933*, p. 16; *for 1934*, p. 15.
33. Ibid., *for 1930*, p. 7.
34. Ibid., *for 1936*, p. 15; *for 1937*, p. 12; *for 1938*, pp. 15–16.
35. *Fishery Reports* for the years cited.
36. H.J. Thomas and A. Saville, 'The fisheries of the Forth-Tay estuaries', *Proceedings of the Royal Society of Edinburgh, Section B*, Vol. 71, pp. 171–188, discussion and quotation on p. 187.
37. Iain Sutherland, *From Herring to Seine Net Fishing on the East Coast of Scotland* (Golspie, 1986), p. 229.
38. J.M. Baxter, 'The offshore waters', in N. Maclean (ed.) *Silent Summer: the State of Wildlife in Britain and Ireland* (Cambridge, 2010), pp. 615, 632.
39. C. Clover, *The End of the Line* (London, 2004), p. 55.
40. Sutherland, *From Herring*, p. 10.

5. Lines and Trawls

1. J. Mitchell, *The Herring, its Natural History and National Importance* (Edinburgh, 1864), p. 347.
2. Quoted in P. Smith, *From the Sma' Lines and the Creels to the Seine Net and the Prawns* (Leven, 2001), p. 87.
3. B. Lenman, *From Esk to Tweed* (London, 1975), p. 2008.
4. P. Aitchison, *Black Friday: the Eyemouth Fishing Disaster of 1881* (Edinburgh, 2001), esp. ch. 2.
5. Smith, *Sma' Lines*, pp. 32–33; *The Scotsman*, 18 and 25 Nov, 1 Dec 1873.
6. A. Fenton, 'Shellfish as bait: the interface between domestic and commercial fishing', in T.C. Smout (ed.), *Scotland and the Sea* (Edinburgh, 1992), p. 152.
7. P. Smith, *The History of Steam and the East Fife Fishing Fleet* (Leven, 1998), pp. 16–43, 80–81.

NOTES

8. B. Patrick, *Recollections of East Fife Fisher Folk* (Edinburgh, 2003), p. 50.
9. Ibid.
10. Smith, *Sma' Lines*, p. 33.
11. Patrick, *Recollections*, p. 51.
12. Fenton, 'Shellfish', p. 141, citing *Report of the Committee on Scottish Bait Beds*, 1889.
13. NRS: AC9/1278.
14. Fenton, 'Shellfish', pp. 143, 146–147.
15. *Scottish Fisheries Board Report for 1894*, p. xvii.
16. Ibid., p. 142.
17. Fenton, 'Shellfish', pp. 146–147, 152.
18. Ibid., p. 147.
19. *Scottish Fisheries Board Report for 1885*, p. lxi; *for 1886*, p. lxiv; *for 1892*, p. xxiii.
20. Ibid. *for 1900*, p. xxiv.
21. E.V. Baxter and L.J. Rintoul, *The Birds of Scotland* (London, 1953), II, p. 431.
22. J.C. Wilcocks, 'Seine-trawling and beam-trawling in estuaries and sea-lochs in Scotland', in D. Herbert (ed.), *Fish and Fisheries, a Selection from the Prize Essays of the International Fisheries Exhibition, Edinburgh, 1882* (Edinburgh, 1883), p. 285.
23. *Hansard Parliamentary Debates*, 3rd ser., Vol. 171, pp. 261–262.
24. *Royal Commission on Sea Fisheries*, P.P. 1866, Vol.18.
25. http://www.tyne-wear-tees.co.uk/steam-trawlers.htm
26. C. Roberts, *The Unnatural History of the Sea: the Past and Future of Humanity and Fishing* (London, 2007), p. 163.
27. T.D. Smith, *Scaling the Fisheries: the Science of Measuring the Effects of Fishing, 1855–1955* (Cambridge, 1994), pp. 59–62.
28. *Report of the Commissioners on Trawl Net and Beam Trawling Fishing* P.P. 1884–5, pp. 63–67.
29. T.D. Smith, *Scaling the Fisheries*, pp. 86–94.
30. J.E. Gunter, *William Carmichael McIntosh, M.D., F.R.S.* (Edinburgh, 1977), pp. 86–88.
31. St Andrews University Library Muniments: MS 37113/1f.417.
32. *Report on Beam Trawling*, p. xii.
33. Ibid., pp. 90–91.
34. Ibid., p. 101.
35. Ibid., p. 102.
36. Roberts, *Unnatural History*, p. 163.
37. *Scottish Fisheries Board Report for 1889*, part 3, pp. 11, 175.
38. St Andrews U.L.: MS 37113/1f.432a.
39. *Report of a Committee of the Fishery Board for Scotland as to the Regulation of Trawling*, P.P. 1888, Vol. 27, pp. 57–58.
40. *Scottish Fisheries Board Reports for 1893*, p. 136; *for 1903*, scientific report, p. 13; *for 1905*, scientific report, p. 6.
41. Ibid. *for 1892*, scientific report, p. 29.
42. Ibid., p. 12.
43. J. Johnstone, *British Fisheries, their Administration and their Problems* (London, 1905), p. 61.
44. T.D. Smith, *Scaling the Fisheries*, pp. 88–94.
45. W. C. M'Intosh, *The Resources of the Sea* (2nd ed. Cambridge 1921), p. xii.
46. Roberts, *Unnatural History*, pp. 163–164.
47. *Scottish Fisheries Board Reports* for the years cited.
48. Ibid. *for 1920*, p. 33; Ibid. *for 1909*, p. 199; Ibid. *for 1923*, pp. 7–8; Ibid. *for 1933*, p. 43.
49. Ibid. for the years cited.
50. Ibid. *for 1921*, p. 7.
51. Ibid. for the years cited.
52. *Fisheries Scotland Report for 1966*, p. 11.
53. Roberts, *Unnatural History*, pp. 195–197.
54. *Fisheries Scotland Report for 1978*, p. 1.
55. Roberts, *Unnatural History*, p. 201.

56. J.M. Baxter, 'The offshore waters', in N. Maclean (ed.), *Silent Summer: the State of Wildlife in Britain and Ireland* (Cambridge, 2010), pp. 615–632.
57. R.S. Thurston, S. Brockington and C. Roberts, 'The effects of 118 years of industrial fishing on UK bottom trawl fisheries', *Nature Communications*, 1: 15, DOI: 10: 1038/ncomms 1013, 2010, p. 4.
58. Ibid., pp. 1–4.
59. P. Smith, *From the Sma' Lines*, appendix.
60. http://www.seafish.org/media/publications/SeafishResponsibleSourcingGuide_nephrops_201005.pdf.
61. http://www.snh.gov.uk/docs/B447222.pdf.
62. Personal communication from Stephen Bastiman, citing a colleague.

6. Traps and Nets in the Estuary

1. R. Parnell, *Prize Essay on the Fishes of the District of Firth of Forth* (Edinburgh, 1838); P.S. Maitland, 'The freshwater fish fauna of the Forth area', *Forth Naturalist and Historian*, Vol. 4 (1979), pp. 33–48.
2. J.G. Harrison, 'Fisher Row: planned housing and the declining fishing industry in late seventeenth-century Stirling', *Forth Naturalist and Historian*, Vol. 9 (1984–5), pp. 113–124.
3. For a discussion of fish traps in the Firth of Forth, see Tom Dawson, 'Locating fish-traps on the Moray and the Forth', http://www.scapetrust.org/html/fishtraps.html (2004).
4. J. Geddie. *The Fringes of Fife* (Edinburgh, 1894), p. 20.
5. H.M. Cadell, *The Story of the Forth* (Glasgow, 1913), ch. 14; J.C. Wilcocks, 'The best means of increasing the supply of mussels for bait', in D. Herbert (ed.), *Fish and Fisheries: a Selection from the Prize Essays of the International Fisheries Exhibition Edinburgh, 1882* (Edinburgh, 1883), p. 168.
6. W. Cobbett, *Tour in Scotland* (London, 1833), p. 134.
7. *Report of the Royal Commission on the Sea Fisheries of the United Kingdom*, P.P. 1866, Vol. 18, p. 605.
8. *Scottish Fisheries Board Report for 1890*, pp. 178–179; T.W. Fulton, 'The past and present condition of the oyster beds in the Firth of Forth', *Scottish Fisheries Board Report for 1895*, p. 263.
9. *Report by the Commissioners for the British Fisheries*, 1861, pp. 4, 10.
10. *Report of Royal Commission on the Sea Fisheries*, 1866, p. 596.
11. *Scottish Fisheries Board Report for 1908*, p. 208; J.G. Bertram, *The Harvest of the Sea* (London, 1865), pp. 238–240; *Scottish Fisheries Board Report for 1883*, pp. 48–60.
12. *Report by the Commissioners of the British Fisheries*, 1861, p. 7; *Report of the Royal Commission on the Herring Fisheries of Scotland* P.P. 1878, pp. 14–16.
13. J. Johnstone, *British Fisheries: their Administrations and their Problems* (London, 1905), p. 277; *Scottish Fisheries Board Report for 1908*, p. 247; *Fisheries of Scotland Report for 1965*, p. 65.
14. http://www.iainsmith.org/en/article/2009/129384/ . . .
15. *Scottish Fisheries Board Report for 1883*, p. xiii.
16. *Report by the Commissioners for British Fisheries*, 1861.
17. *The Scotsman*, 18 January 1843.
18. *Report by the Commissioners for British Fisheries*, 1861, p. 4.
19. Ibid., pp. 4–7; see also *Report of the Royal Commission on Sea Fisheries*, 1866, pp. 580, 589–591; *Report of the Royal Commission on the Herring Fisheries*, 1878, Appendix, p. 1.
20. NAS: GD 265/9/2, f. 45; *Report of the Royal Commission on the Herring Fisheries*, 1878, p. 14.
21. *Scottish Fisheries Board Report for 1908*, pp. 239–244.
22. *Report by the Commissioners for British Fisheries*, 1866, p. 5.
23. W.C. M'Intosh, *The Resources of the Sea* (2nd edn, Cambridge, 1921), p. 225.

24. Parnell, *Prize Essay*, p. 324; *Report by the Commissioners for British Fisheries*, 1867.
25. *Scottish Fisheries Board Report for 1909*, p. 200; *for 1911*, pp. xxviii, 193.
26. *Reports on the Fisheries of Scotland* for the years cited.
27. BBC News, 22 February 2001; http://nanovapor.org/Image/ToxicWaste/Rosyth%20article.pdf.
28. P. Hume Brown, *Early Travellers in Scotland* (Edinburgh, 1891), pp. 193–194, 227.
29. *Scottish Fisheries Board Report for 1904*, Part II, p. 116.
30. *A Fisheries Management for the Forth Catchment* (Forth District Salmon Board and River Forth Fisheries Trust, 2009), p. 44.
31. *Scottish Fisheries Board Report for 1898*, Part II, pp. 57–58; *for 1906*, Part II, p. 8.
32. R. Sibbald, *History of Stirlingshire* (Edinburgh, 1707), p. 36; *Scots Magazine* Jan 1746.
33. For an account of the variety of salmon traps on the Tay, see I.A. Robertson, *The Tay Salmon Fisheries since the Eighteenth Century* (Glasgow, 1998).
34. R. Williamson, *Salmon Fisheries in Scotland* (Pitlochry, 1991), pp. 9–10; A. Netboy, *The Atlantic Salmon, a Vanishing Species* (London, 1967), p. 260; R. Morris and M. Ramage, *Kincraig and Shell Bay* (Levenmouth, 2009), p.18.
35. Parnell, *Prize Essay*, pp. 382, 405–406.
36. *Scottish Fisheries Board Report for 1899*, Part II, Appendix, p. 17; *for 1906*, Part II, p. 8.
37. Ibid. *for 1907*, Part II, p. 81; W.L. Calderwood, *The Salmon Rivers and Lochs of Scotland* (2nd edn, London, 1921), p. 40.
38. P.S. Maitland, 'The sparling *osmerus eperlanus* in the Forth', *The Forth Naturalist and Historian*, Vol. 33 (2010), pp. 79–89.
39. Parnell, *Prize Essay*, p. 313.
40. *Scottish Fisheries Board Reports*, for the years cited.
41. Ibid.; Maitland, 'The sparling', p. 83.
42. Maitland, 'The sparling'.

7. Pollution

1. R.J. Morris, 'In search of twentieth-century Edinburgh', *Book of the Old Edinburgh Club*, n.s. Vol. 8 (2010), pp. 13–25.
2. *Report of the Royal Commission on River Pollution Fourth Report, Scotland*, P.P. 1872, Vol. 2, pp. 99, 123; *Fifth Report of the Scottish Advisory Committee on River Pollution Prevention: Rivers Almond and Avon and the Grange Burn* (HMSO, Edinburgh, 1935), p. 40.
3. M.W. Flinn (ed.), *Scottish Population History* (Cambridge, 1977), pp. 17–20, 368–397.
4. NRS: GD 335, Records of the Scottish Rights of Way Society, 1844.
5. P.J. Smith, 'The foul burns of Edinburgh', *Scottish Geographical Magazine*, Vol. 91 (1975), pp. 23–36.
6. W. Baird, *Annals of Duddingston and Portobello* (Edinburgh, 1898), pp. 421–423.
7. N. Goddard. ' "A mine of wealth", the Victorians and the agricultural value of sewage', *Journal of Historical Geography*, Vol. 22 (1966), pp. 27–90; *British Medical Journal*, 27 Sep. 1873.
8. *Report of the Royal Commissioners on Trawl Net and Beam Trawling Fishing*, P.P. 1884–5, Minutes of Evidence, p. 438.
9. *Report of the Royal Commission on River Pollution, Fourth Report, Scotland*, P.P. 1872, Vol. 1, p. 19.
10. Quoted in J. Glaister, 'The pollution of Scottish rivers', *Transactions of the Philosophical Society of Glasgow*, 1897, p. 86. A running commentary on the Esk Case may be found in the columns of *The Scotsman*.
11. *Report . . . River Pollution*, Vol. 1, pp. 20, 65; Vol. 2, p. 313.
12. Ibid., Vol. 1, pp. 22–24, 44–45, 69–70, 316.
13. Ibid., Vol. 2, p. 313.
14. R. Shand, 'The rise and fall of seaduck at Levenmouth', *Fife Bird Report 2006*, pp. 150–151.

15. *Report . . . River Pollution*, Vol. 2, p. 312.
16. Ibid., Vol. 2, pp. 112, 121; *Glasgow Herald*, 2 April 1869.
17. Glaister, 'Pollution of Scottish rivers', pp. 63–70.
18. *Hansard Commons Debates*, 4 June 1885.
19. *Report to the Secretary for Scotland, by the Inspector for Scotland under the Rivers Pollution Act, 1876*, P.P 1898.
20. *Report of the Commissioners on Salmon Fisheries Part I*, P.P. 1902, pp. 41, 408–414.
21. *Glasgow Corporation Water Works Commemorative Volume* (Glasgow, 1877), pp. 4–19.
22. *Municipal Glasgow: its Evolution and Enterprises* (Glasgow, 1914); For unpublished data on Stirling, thanks to Dr Jan Oosthoek.
23. J. Sheail, *An Environmental History of Twentieth-Century Britain* (Basingstoke, 2002), pp. 48–56.
24. *Scottish Fisheries Board Report for 1933*, p. 73.
25. H.M. Cadell, 'Land reclamation in the Forth Valley', *Scottish Geographical Magazine*, Vol. 45 (1929), p. 10.
26. *Fifth Report of the Scottish Advisory Committee on Rivers Pollution Prevention – Rivers Almond and Avon and the Grange Burn* (HMSO, 1935), pp. 5–32 (quotation on p. 32).
27. W.L. Calderwood, *The Salmon Rivers and Lochs of Scotland* (2nd edn, London, 1921), p. 39.
28. *Scottish Fishery Board Reports* for the years cited.
29. *Report of the Rivers Pollution Prevention Sub-Committee of the Scottish Waters Advisory Committee on the Prevention of Pollution of Rivers and Other Waters*, P.P. 1950.
30. W.F. Collett. 'The quality of the Forth Estuary (1)', *Proceedings of the Royal Society of Edinburgh*, section B, Vol. 71 (1971–2), pp. 138–141.
31. D.S. McLusky, 'Ecology of the Forth estuary', *Forth Naturalist and Historian*, Vol. 3 (1978), pp. 10–13.
32. *Scotland's Marine Atlas* (Edinburgh 2011), p.65; *SEPA View*, 40 (2008), p. 4.
33. See, in particular, the papers in the *Proceedings of the Royal Society of Edinburgh*, section B, Vol. 93 (1987) by M. Elliott and A.H. Griffiths, C.G. Moore, and D.S. McLusky; and also D.S. McLusky, D.B. Bryant, M.E. Elliot, M. Tear and G. Moffat, 'Intertidal fauna of the industrialised Forth estuary', *Marine Pollution Bulletin*, Vol. 7 (1976), pp. 48–51.
34. R.W. Covill, 'The quality of the Forth estuary (2)', *Proc. RSE*, section B, Vol. 71 (1971–2), pp. 143–170; R.W. Covill, 'Progress in pollution control in the Lothians area, Scotland', *Journal of the Water Pollution Control Federation*, Vol. 38 (1966), 1634–1644; *Hansard*, 28 July 1959.
35. Covill, 'Forth estuary'. Many of the papers in *Proc. RSE*: section B, Vol. 93 (1987) are directed to the problem, especially those of A.H. Griffiths, J.S. Buchanan, J.S.B. Buchanan, P. Read, D.C. Moore and I.M. Davies.
36. R. Forrester and I. Andrews, et al. (eds), *The Birds of Scotland* (Aberlady, 2007), Vol. 1, pp. 226, 238, 272; L.H. Campbell. 'The impact of changes in sewage treatment on seaducks wintering in the Firth of Forth, Scotland', *Biological Conservation*, Vol. 28 (1984), pp. 173–186.
37. Shand, 'Seaduck'; the late William Halcrow, personal communication.
38. N.H.K. Burton et al., *Effects of Reduction in Organic and Nutrient Loading on Bird Populations in Estuaries and Coastal Waters of England and Wales*, English Nature Research Report 586 (Peterborough, 2003).
39. Stephen Welch, personal communication.
40. *Significant Water Management Issues in the Scotland River Basin Districts* (SEPA, 2007).
41. *Marine Conservation Newsletter*, autumn, 2007, p. 15.
42. Thanks to Tom Leatherland and the late William Halcrow, personal communication, for guidance here.
43. R. Levitt, *Implementing Public Policy* (London, 1980); R.K. Wurzel, *Environmental Policy Making in Britain, Germany and the European Union* (Manchester, 2002).
44. O. MacDonagh, 'The nineteenth-century revolution in Government: a reappraisal', *Historical Journal*, Vol. 1 (1958), pp. 52–67; O. MacDonagh, *A Pattern of Government*

Growth, 1800–1860, Passenger Acts and Their Enforcement (London, 1961); O. MacDonagh, *Early Victorian Government, 1830–1870* (London, 1977).
45. *Report of the Rivers Pollution Prevention Sub-Committee*, pp. 16–17.
46. Wurzel, *Environmental Policy Making*, p. 212.
47. R.J. Morris, 'Externalities, the market, power structure and the urban agenda', *Urban History*, Vol. 17 (1990), pp. 99–109.
48. R.J. Morris and R.H. Trainor (eds), *Urban Governance, Britain and Beyond since 1750* (Aldershot, 2000); J. Garrard, *Leadership and Power in Victorian Industrial Towns* (Manchester, 1983).
49. R. Inglehart, *Silent Social Revolution* (Princeton, 1977).
50. Wurzel, *Environmental Policy Making*, pp. 227, 245.

8. Land Claim from the Sea

1. These terms are interchangeable. Geowise Ltd and Coastal Research Group, Glasgow University, 'Use of GIS to map land claim and identify potential areas for coastal managed realignment in the Forth Estuary', *Scottish Natural Heritage Research Report*, 1999, pp. 17–24.
2. *Third Statistical Account of Scotland*, parish of Grangemouth, revised 1961, http://tompaterson.co.uk/places/stat3_gran1.htm (accessed 9 March 2012).
3. G.W.S. Barrow (ed.), 1999, *The Charters of David I* (Woodbridge, 1999), p.114, doc. 98, quoted in R. Oram, 'Estuarine environments and resource exploitation in eastern Scotland, c.1125 to c.1400: a comparative study of the Forth and Tay estuaries', paper given at a conference on 'Landscapes or seascapes? The history of the coastal environment in the North Sea area reconsidered', Ghent University, April 2010; J. Reid, *The Place Names of Falkirk and East Stirlingshire* (Falkirk Local History Society, 2009), p. 206.
4. John Harrison, personal communication.
5. Quoted in Oram, 'Estuarine environments', p. 12.
6. A local rhyme, quoted in D. Hall, *Scottish Monastic Landscapes* (Stroud, 2006), p. 7.
7. R. Sibbald, *The History of the Sheriffdom of Stirling* (Edinburgh, 1710), p. 61.
8. D.S. McLusky, D.M. Bryant, and M. Elliot, 'The impact of land-claim on macrobenthos, fish and shorebirds on the Forth estuary, eastern Scotland', *Aquatic Conservation: Marine and Freshwater Ecosystems*, Vol. 2 (1992), p. 211; M.G. Healy and K.R. Hickey, 'Historic land reclamation in the intertidal wetlands of the Shannon estuary, western Ireland', *Journal of Coastal Research*, Special Issue 36 (2002), p. 365.
9. H.M. Cadell, *The Story of the Forth* (Glasgow, 1913), ch. 12. See also, Cadell, 'Land claim in the Forth Valley: I. Reclamation before 1840, II. Later reclamation schemes and the work of the Forth Conservancy Board'. *Scottish Geographical Magazine*, Vol. 45 (1929), pp. 7–22, 81–100.
10. J.P. Doody, 'Coastal Squeeze' – an historical perspective', *Journal of Coastal Conservation*, Vol. 10 (2004), pp. 129–138.
11. L.A. Boorman, 'Saltmarsh Review. An overview of coastal saltmarshes, their dynamic and sensitivity characteristics for conservation and management', *Joint Nature Conservation Committee Report, No. 334* (2003), p. 16; The Geowise *Report*, section 4.2, pp. 20–24, explains the intricacies involved in calculating sea-level and intertidal changes on the Forth.
12. T.P. Gostelow and M.A.E. Browne, 'Engineering geology of the upper Forth Estuary', *British Geological Survey Report*, Vol. 16, No. 8 (London, 1986) pp. 22, 24; Geowise *Report* (pp. 20–24) discusses the problem with trying to quantify the intertidal zone since 1600.
13. G. Bailey, 'Along and across the Carron', *Calatria*, Vol. 2 (1992); R. Sexton and E. Stewart, 'Alloa Inch: the mud bank in the Forth that became an inhabited island', *Forth Naturalist and Historian*, Vol. 28 (2005), pp. 79–101.
14. See for example, NRS: RHP5505, 'Reduced plan of lands in Haughs of Airth 1790'.

Unfortunately, for copyright reasons, this highly illustrative plan cannot be reproduced here.

15. 'The agriculture of the County of Stirling', *Transactions of the Highland and Agricultural Society*, Fourth Series, Vol. 15 (1883), p. 57.
16. A.J. Webb and A.P. Metcalfe, 'Physical aspects, water movements and modelling studies of the Forth estuary, Scotland', *Proceedings of the Royal Society of Edinburgh*, Series B, Vol. 93 (1987), pp. 259–272.
17. Cited in Babtie Group in conjunction with Northern Ecological Services and Coastal Research Group, University of Glasgow, *Feasibility and implications of managed realignment at Skinflats*, a report to the Forth Estuary Forum (2001), p. 21.
18. Scottish Environment Protection Agency consultation document, 'Significant water management issues in Scotland river basin district', p. 65.
19. R. Oram, 'The sea-salt industry in medieval times', *Studies in Medieval and Renaissance History*, forthcoming, 2012.
20. I. Bowman, 'Coal mining at Culross: 16–17th centuries', *Forth Naturalist and Historian*, Vol. 7 (1983), pp. 84–125.
21. Reid, *Place Names*; S. Taylor with G. Markus, *Place-Names of Fife*, Vols. 1–4 (Donington, 2006–12).
22. Reid, *Place Names*, p. 109.
23. *New Statistical Account of Scotland*, Vol. 2 (1843), Borrowstouness, p. 148.
24. *Statistical Account of Scotland*, Stirlingshire, Vol. 9, parish of Bothkennar.
25. J. Harrison, 'Between the Carron and the Avon', *Forth Naturalist and Historian*, Vol. 20 (1996), pp. 71–91.
26. G. Bailey, 'Excavations at the burgh of Airth', *Forth Naturalist and Historian*, Vol. 14 (1990), p. 113; A. Graham, 'Archaeological notes on some harbours in eastern Scotland', *Proceedings of the Society of Antiquaries of Scotland*, Vol. 101 (1969), p.212.
27. J. Reid, 'The lands and baronies of the parish of Airth', *Calatria*, Vol. 13 (1999), p. 64.
28. Harrison, 'Carron and Avon', p. 76; Reid, 'The carselands of the Firth of Forth', *Calatria*, Vol. 4 (1993), p. 8.
29. Reid, *Place Names*, p. 63.
30. Harrison, 'Carron and Avon', p. 72.
31. Sibbald, *Stirling*, p. 50.
32. Cadell, 'Land claim', p. 19.
33. Reid, 'Carselands', p. 29.
34. Harrison, 'Carron and Avon', p. 76.
35. R. Sibbald, *The History of the Sheriffdom of Linlithgowshire* (Edinburgh, 1710), p. 19.
36. R. Callander, 'The History of Common Land in Scotland', *Caledonia Centre for Social Development Working Paper No 1*, Issue No 2 (2003), p. 9, http://www.scottishcommons.org/docs/commonweal_1.pdf.
37. NRS: RHP 80865/1, 'Plan of Lands of Haughs of Airth showing division of the commonty of Airth', by Robert Sconce, 1784. NRS: SC 67/83/10, 'Extract act of the Lords of Council and Session, 1789, remitting to the sheriff of Stirling the action brought by James Bruce of Kinnaird and others against James Bruce of Powfoulis and others over ownership of the salt or sea greens on the Haughs of Airth, with two instruments of sasine of 1762 and 1764 relating to the lands in question'.
38. Cadell, 'Land claim', p. 81.
39. Ibid., p. 18.
40. Ibid., p. 21.
41. The reclamation work undertaken by the Borstal boys in Lincolnshire is outlined at http://www.justice.gov.uk/contacts/prison-finder/north-sea-camp (accessed 1 March 2012); Healy and Hickey, 'Shannon', pp. 365–373.
42. http://www.bo-ness.org.uk/html/timeline.htm (accessed 6 Jan 2011).
43. Cadell, 'Land claim' p. 82.
44. http://canmore.rcahms.gov.uk/en/site/79567/details/rosyth+hm+dockyard
45. Cadell, ' Land claim', p. 85.
46. Cadell, *Story of the Forth*, p. 235.

47. Cadell, 'Land claim', p. 84.
48. http://hansard.millbanksystems.com/lords/1921/may/03/forth-conservancy-order-confirmation (accessed 26 Oct 2011).
49. http://jncc.defra.gov.uk/page-1979/ (accessed 2 March 2012).
50. McLusky et al., 'The impact of land-claim', pp. 211–220; M. Elliot, M., M.G. O'Reilly and C.J.L. Taylor, 'The Forth estuary: a nursery and overwintering area for North Sea', *Hydrobiologia*, 195 (1990), pp. 89–103.
51. Peter Fothringham, Director, Forth Salmon Fishery Board, personal communication.
52. J.M. Morris, 'Firth of Forth National Vegetation Classification Survey 2003', *Scottish Natural Heritage Commissioned Report*, No. 092 (2005).
53. See Scottish Environment Protection Agency, Indicative River and Coastal Flood map at http://go.mappoint.net/sepa.
54. The Royal Society for the Protection of Birds plans for habitat creation along the shores of the Forth estuary are contained in 'RSPB Firth of Forth Vision: an analysis of environmental pressures and opportunities for the habitat restoration and creation in the Firth of Forth', unpublished document, n.d.; Babtie Group in conjunction with Northern Ecological Services and Coastal Research Group, University of Glasgow, 'Feasibility and Implications of Managed Realignment at Skinflats', a report to the Forth Estuary Forum (2001).

9. The Bass and its Gannets

1. G.-L. Leclerc, Comte de Buffon, *Histoire Naturelle, Générale et Particulière* (Paris, 1749–1778), Vol. 23, p. 376.
2. J. Fisher, *The Shell Bird Book* (London, 1966), p. 43.
3. J.G. Gurney, *The Gannet: a Bird with a History* (London, 1913), p. 171.
4. Gurney, *Gannet*, pp. 173–194. Subsequent quotations from these authors are from the reliable citations in Gurney unless otherwise stated.
5. B. Nelson, *Living with Seabirds* (Edinburgh, 1986); B. Nelson, *The Gannet* (Berkhamsted, 1978).
6. http://www.rps.ac.uk/1592/4/167 provides the latest edition of this Act.
7. E.T. Booth, *Rough Notes on the Birds Observed during Twenty-five Years' Shooting and Collecting in the British Isles* (London, 1881–7), III, p. 7.
8. Gurney, *Gannet*, p. 455.
9. Ibid., pp. 240–241; J. Sheail, *Nature in Trust: the History of Nature Conservation in Britain* (Glasgow, 1976), pp. 4–5, 22–25.
10. NRS: GD 103/2/44.
11. Gurney, *Gannet*, pp. 178–179; *Dictionary of the Older Scottish Tongue*.
12. Gurney, *Gannet*, pp. 183, 190, 456–457; Nelson, *Gannet*, p. 280; 'Journey through England and Scotland made by Lupold von Wedel in the years 1584 and 1585', *Transactions of the Royal Historical Society*, ns, Vol. 9 (1895), p. 242.
13. J. Walker, *Essays on Natural History and Rural Economy* (Edinburgh, 1808), p. 287; J. Fleming, 'Zoology of the Bass', in T. McCrie (ed.), *The Bass Rock, its Civil and Ecclesiastic History, Geology, Martyrology, Zoology and Botany* (Edinburgh, 1848), p. 406; Gurney, *Gannet*, pp. 254, 467.
14. Booth, *Rough Notes*, III, p. 3.
15. Walker, *Essays*, p. 287, Gurney, *Gannet*, p. 470.
16. A. Fenton, *The Food of the Scots* (Edinburgh, 2007), pp. 48–49.
17. P.H. Brown, *Early Travellers in Scotland* (Edinburgh, 1891), p. 154.
18. NRS: GD 25/9/18, Ailsa muniments, solan goose book; Gurney, *Gannet*, p. 81; Fenton, *Food*, p. 269.
19. Brown, *Travellers*, pp. 126–127.
20. Gurney, *Gannet*, pp. 201, 458; Brown, *Travellers*, p. 233.
21. Gurney, *Gannet*, pp. 250; NRS: GD 113/4/165/723.
22. NRS: AC 9/816 – 1723.

23. NRS: GD 110/735; GD 110/859/1; GD 110/866; T. Pennant, *Tour in Scotland, 1769* (edn, Edinburgh, 2000), p. 34; R. Scott-Moncrieff (ed.), *The Household Book of Lady Grisell Baillie, 1792–1733* (Scottish History Society, Edinburgh, 1911); Fenton, *Food*, p. 312.
24. Gurney, *Gannet*, pp. 253–254, 461–462; Booth, *Rough Notes*, III, p. 3.
25. NRS: GD 113/5/81d/31. W. Harvey at his visit to the Bass in 1633 speaks of a 'traffic in cooked eggs' but does not say they were gannets' eggs: *Disputations touching the Generation of Animals*, translated by G. Whitteridge (London, 1981), p. 66.
26. Gurney, *Gannet*, pp. 247, 253; Booth, *Rough Notes*, III, p. 2.
27. N. Macdougall, *James IV* (Edinburgh, 1989), pp. 136, 198.
28. Pennant, *Tour*, p. 34; P.J. Selby, *Illustrations of British Ornithology* (London, 1833), II, pp. 456–457.
29. J. Colquhoun, *The Moor and Loch, Containing Minute Instructions in all Highland Sports* (4th edn, Edinburgh, 1878), I, pp. 212, 236–237.
30. W.M. Ferrier, *The North Berwick Story* (North Berwick, 1980), p. 68.
31. C. Innes (ed.), *Natural History and Sport in Moray Collected from the Journals and Letters of the late Charles St John*, p. 204; Booth, *Rough Notes*, III, pp. 1–3; Colquhoun, *Moor and Loch*, p. 236; Gurney, *Gannet*, pp. 240, 252.
32. Sheail, *Nature in Trust*, pp. 22–29.
33. E. Bowles, *A Trip to the Bass Rock* (n.p., n.d.), pp. 8–9.
34. 'A.W.', *The Bass: its Historical and Other Features* (Aberdeen, 1912), p. 41.
35. Fleming, 'Zoology of the Bass', p. 395; Nelson, *Gannet*, p. 68; Sarah Wanless, personal communication, from whom the quotation is drawn.
36. Fleming, 'Zoology of the Bass', p. 405.
37. W. MacGillivray, *History of British Birds* (London, 1852), V, p. 409; Wanless, personal communication.
38. Gurney, *Gannet*, pp. 246–247, 253.
39. Pennant, *Tour*, p. 34; Gurney, *Gannet*, p. 239.
40. Gurney, *Gannet*, p. 238.
41. W. Harvey, *Disputations Touching the Generation of Animals*, trans. G. Whitteridge (Oxford, 1981), p. 67.
42. In the possession of Sir Hew Hamilton-Dalrymple.
43. Brown, *Travellers*, pp. 135, 234 – Brereton also mentioned 'kine' and rabbits; Walker, *Essays*, p. 287; Fleming, 'Zoology of the Bass', p. 407; J. Sinclair (ed.), *Statistical Account of Scotland* (edn, Wakefield, 1975), II, p. 528; F. Grose, *The Antiquities of Scotland* (London, 1789), p. 80; *New Statistical Account of Scotland* (1834–5), II, p. 330; 'A.W.', *The Bass*, p. 41.
44. MacGillivray, *British Birds*, V, p. 409; Fleming, 'Zoology of the Bass', p. 396; 'A.W.', *The Bass*, p. 41.
45. Gurney, *Gannet*, p. 183.
46. *Daniell's Scotland: A Voyage round the Coast of Scotland and the Adjacent Isles 1815–1822* (Edinburgh, 2006), I, p. 318: *New Statistical Account*, II, pp. 321–322.
47. Brown, *Travellers*, p. 233.
48. Booth, *Rough Notes*, III, p. 5.
49. B. Nelson, personal communication.
50. J. Hunt and J.B. Nelson, 'Reduced breeding of gannets on Bass Rock in 2011', *Scottish Birds*, Vol. 32 (2012), p. 31.
51. The whole of this paragraph is based on R. Furness, personal communication.
52. Gurney, *Gannet*, p. 233; Booth, *Rough Notes*, III, p. 2.
53. L.J. Wilson et al., 'Modelling the spatial distribution of ammonia emissions from seabirds in the UK', *Environmental Pollution*, Vol. 131 (2004), pp. 173–185.

10. The Isle of May and the Other Seabird Colonies

1. Robert Gordon, in a commentary on the map of Fife, published in Blaeu's Atlas of

NOTES

1654: see *The Blaeu Atlas of Scotland* (Edinburgh, 2006), with translations from the Latin by I.C. Cunningham. Gordon gives a remarkable botanical list from Inchkeith, later apparently copied without acknowledgement by Robert Sibbald in 1710. See *Atlas*, pp. 85.

2. J. Dickson, *Emeralds Chased in Gold: the Islands of the Forth* (Edinburgh, 1899), quotation from Carlyle, p. 101; on Inchcolm, p. 39. He gives a good account particularly of the political histories of each island.
3. NRS: AC 8/321.
4. The standard account of the May is W.J. Eggeling, *The Isle of May* (2nd edn, Kirkmichael, 1985, but mainly written in 1960): for the lighthouses, pp. 33–44, 62. Also, R. Dickson, *Strangers: Memories of a Lighthouse Keeper's Daughter* (Crail Museum Trust, 2009), p. 11.
5. R. Morris, *The Wildlife of Inchcolm* (Leven, 2003), pp. 41–42; R. Morris, *The Wildlife of Inchkeith* (Leven, 2003), pp. 43.
6. Alan Drever, personal communication.
7. John Young, personal communication.
8. The following paragraphs are based without further citation on H.F. James and P. Yeoman (eds), *Excavations at St Ethernan's Monastery, Isle of May, Fife*, Tayside and Fife Archaeological Committee Monograph VI (Perth, 2008).
9. P.H. Winkleman, *Nederlandse Rekeningen inde Tolregisters van Koningsbergen 1588–1602* (The Hague, 1971). Thanks to Petra Van Dam for this reference; St Andrews University Library Ms B65/23/315.
10. Eggeling, *May*, pp. 59–63. Robert Gordon (on Blaeu's *Atlas*, published in 1654) was the author of the observations attributed to Sibbald in 1710, but Sibbald merely copied Gordon verbatim and unacknowledged.
11. R. Dickson, *Strangers*, p. 11.
12. The most useful addition was published in Cupar in 1803, with supplementary information of that date in the footnotes.
13. Eggeling, *May*, p. 207.
14. J. Dickson, *Emeralds*, p. 245.
15. R.J. Berry, *Islands* (London, 2009), p. 25; Eggeling, *May*, pp. 92–94.
16. Eggeling, *May*, esp. ch. 9 and ch. 17.
17. Quoted ibid., p. 140.
18. Ibid., ch. 17.
19. The species accounts are based, except where otherwise specified, on A.H. Evans, *A Fauna of the Tweed Area* (Edinburgh, 1911); L.J. Rintoul and E.V. Baxter, *A Vertebrate Fauna of Forth* (Edinburgh, 1935); E.V. Baxter and L. J. Rintoul, *The Birds of Scotland* (Edinburgh, 1953); W.J. Eggeling, *The Isle of May* (2nd edition, Kirkmichael, 1985); V.M. Thom, *Birds in Scotland* (Waterhouses, 1986); C. Lloyd, M.L. Tasker and K. Partridge, *The Status of Seabirds in Britain and Ireland* (London, 1991); R. Murray, M. Holling, H. Dott and P. Vandome, *The Breeding Birds of south-east Scotland: a Tetrad Atlas 1988–1994* (Edinburgh, 1998); P.I. Mitchell, S.F. Newton, N. Ratcliffe and T.E. Dunn (eds), *Seabird Populations of Britain and Ireland: Results of the Seabird 2000 Census, 1998–2002* (London, 2004); R. Forrester, I. Andrews et al., *The Birds of Scotland* (Aberlady, 2007).
20. J. Fisher, *The Northern Fulmar* (London, 1952).
21. The most recent discussion is in Mitchell et al., *Seabird Populations*, pp. 59–61.
22. Murray et al., *Tetrad Atlas*, p. 44.
23. Quoted in Rintoul and Baxter, *Vertebrate Fauna*, p. 44.
24. N. Elkins et al., *The Fife Bird Atlas* (Dunfermline, 2003), p. 177.
25. Forrester, Andrews et al., *Birds of Scotland*, I, p. 777.
26. W.R.P. Bourne, 'Herring and lesser black-backed gulls nesting in Rosyth dockyard', *Sea Swallow*, 37 (1988), p. 65.
27. SNH data.
28. Morris, *Inchkeith*, p. 7.
29. Tom Leatherland, personal communication.

30. F. Daunt, S. Wanless et al., 'The impact of sandeel fishery closure on seabird food consumption, distribution and productivity in the north-western North Sea', *Canadian Journal of Fisheries and Aquatic Science*, Vol. 65 (2008), pp. 362–381.
31. M. Fredericksen, S. Wanless, M.P. Harris, P. Rothery, L.J. Wilson, 'The role of industrial fisheries and oceanographic change in the decline of North Sea black-legged kittiwakes', *Journal of Applied Ecology*, Vol. 41 (2004), pp. 1129–1139.
32. M.P. Harris, *The Puffin* (Waterhouses, 1984), p. 181.
33. R. van der Wal, 'The invasion of tree mallow on Craigleith', in R. Morris and B. Bruce, *The East Lothian Emeralds* (Kirkcaldy, 2007), pp. 79–85.
34. C.D. Preston, D.A. Penman, and T.D. Dines, *New Atlas of the British and Irish Flora* (Oxford, 2002), p. 219.
35. F. Daunt, personal communication.
36. S. Wanless, P.J. Wright et al., 'Evidence for a decrease in the size of lesser sandeels (*Ammodytes marinus*) in a North Sea aggregation over a 30-year period', *Marine Ecology Progress Series*, Vol. 279 (2004), pp. 273–246.
37. M.P. Harris, D. Beave et al., 'A major increase in snake pipefish (*Entelurus aequoreus*) in northern European seas since 2003: potential implications for seabird breeding success', *Marine Biology*, Vol. 151 (2007), pp. 973–983; M.P. Harris, M. Newell, F. Daunt, J.R. Speakman and S. Wanless, 'Snake Pipefish *Entelurus aequoreus* are poor food for seabirds' *Ibis*, Vol. 150 (2008), pp. 413–415.
38. K. Ashbrook, S. Wanless, M.P. Harris and K.C. Hamer 'Hitting the buffers: conspecific aggression undermines the benefits of colonial breeding under adverse conditions', *Biology Letters*, Vol. 4 (2008), pp. 630–633.

11. Seals: The Bone of Contention

1. Ian Boyd, SMRU, personal communication.
2. P.S. Hammond and K. Grellier, 'Grey seal diet composition and prey consumption in the North Sea', Final Report to DEFRA, project MF0319.
3. John Harwood, SMRU, personal communication.
4. B.J. McConnell, M.A. Fedak, P. Lovell and P.S. Hammond, 'Movements and foraging areas of grey seals in the North Sea', *Journal of Applied Ecology*, Vol. 36 (1999), pp. 573–590.
5. J. Matthiopoulos, B. McConnell, C. Duck, M. Fedak, 'Using satellite telemetry and aerial counts to estimate space use by grey seals around the British Isles', *Journal of Applied Ecology*, Vol. 41 (2004), pp. 476–491.
6. Special Committee on Seals Report, at http://www.smru.st-andrews.ac.uk/pageset.aspx/psr==411.
7. T. Haug, M. Hammill and D. Olafsdottir, *Grey Seals of the North Atlantic and the Baltic*, NAMMCO Scientific Publications, 6 (Tromsö, 2007).
8. L.J. Rintoul and E.V. Baxter, *A Vertebrate Fauna of Forth* (Edinburgh, 1935), p. 17.
9. W.J. Eggeling, *The Isle of May* (Kirkmichael, 1985), p. 83.
10. R. Lambert, 'The grey seal in Britain: a twentieth century history of a nature conservation success', *Environment and History*, Vol. 8 (2002), pp. 449–74.
11. P.P. Pomeroy, S.D. Twiss and C.D. Duck, 'Expansion of a grey seal (Halichoerus grypus) breeding colony: changes in pupping site use at the Isle of May, Scotland', *Journal of the Zoological Society of London*, Vol. 250 (2000), pp. 1–12.
12. John Young, former warden, personal communication.
13. The Scottish Government, *Scotland's Marine Atlas* (Edinburgh, 2011), pp. 20–21.
14. A. Mitchell (ed.) *Geographical Collections* (Scottish History Society, Edinburgh, 1906) II, pp. 24, 181; III, p. 314.
15. J. Lister-Kaye, *Seal Cull: the Grey Seal Controversy* (Harmonsworth, 1979), pp. 30–31: C. St John, *Short Sketches of the Wild Sports and Natural History of the Highlands* (London, 1847), p. 25.

NOTES

16. J. Sheail, *Nature in Trust: the History of Nature Conservation in Britain* (Glasgow, 1977), pp. 37–38.
17. NRS: AF 56/734.
18. Sheail, *Nature in Trust*, pp. 38–39; NRS: AF 56/1445.
19. B. P. Rae, 'Seals and Scottish Fisheries', *Marine Research in Scotland*, 2 (1960): see also *Report of the Consultative Committee on Grey Seals* (1963), pp. 28–33; H.R. Hewer, *British Seals* (London, 1974), p. 94.
20. Lambert, 'Grey Seal', pp. 459–461.
21. Lister-Kaye, *Seal Cull*, p. 39, Lambert, 'Grey Seal', pp. 462–463.
22. Lister-Kaye, *Seal Cull*, pp. 112–114.
23. John Harwood, personal communication.
24. Lister-Kaye, *Seal Cull*, pp. 40–44; Lambert, 'Grey Seal', p 463.
25. Lister-Kaye, *Seal Cull*, pp. 40–44.
26. Ibid., p. 18.
27. NRS: AF 62/4958. Memo by Fred Holliday of NCC, 10 Dec 1979; Minutes of a grey seal management in Scotland meeting in Garrick House, London, 4 June 1980.
28. NRS: SNH 5/8 Minutes of the Scottish Advisory Committee of the NCC, 25 Jan. 1980; NRS: AF 62/5700, letter from I.S. Macphail 21/9/1982; NRS: AF 62/4959, memo of I.G.F. Gray to Mr Barbour, 12 Feb 1982.
29. John Harwood, personal communication.
30. John Lister-Kaye, personal communication.
31. F.F. Darling, *A Naturalist on Rona: Essays of a Biologist in Isolation* (Oxford, 1939), pp. 93–96.
32. Lambert, 'Grey Seal', pp. 464–465.
33. NRS: AF 62/5700.
34. *Hansard*, 13 July 1981.
35. Ibid.
36. B.B. Parrish and W.M. Shearer, 'Effects of seals on fisheries', International Council for the Exploration of the Sea, 1977; C.F. Summers, 'Trends in the size of British grey seal populations', *Journal of Applied Ecology*, Vol. 15 (1978), pp. 395–400.
37. 'The case against a grey seal cull in Scotland', *Oryx*, 1979, pp. 253–258.
38. NRS: AF 62/4959, Notes of a meeting at Garrick House, 11 June 1981.
39. NRS: SNH 5/8.
40. Ibid., Letter from Mr Beverton CBE, Swindon, 16 June 1981.
41. We are grateful to Ian Boyd for these points.
42. http://www.anglingresearch.org.uk/node/254.
43. *Scotland's Marine Atlas*, pp. 18–19, 122.

Conclusion

1. P.J. Crutzen and E. Stoermer, 'The anthropocene', *Global Change Newsletter*, 41 (2000), pp. 17–18; P.J. Crutzen, 'Geology of mankind', *Nature*, 415 (2002), p. 23; J. Zalasiewicz, M. Williams, W Steffen and P.J Crutzen, 'The new world of the anthropocene', *Environmental Science and Technology*, 44 (2010), pp. 2228–2231.
2. W.F. Ruddiman, 'The anthropogenic greenhouse era began thousands of years ago', *Climatic Change*, 61 (2003), pp. 261–293.
3. J. McNeill, *Something New Under the Sun: an Environmental History of the Twentieth-Century World* (New York, 2000).
4. I.G. Cumming, 'Whales in the Firth of Forth', *Fife Bird Report 2006*, pp. 157–158.
5. For an account of the significance of Edinburgh in marine biology, see J.A. Adams, 'The science of the sea', in J.R. Coull, A. Fenton and K. Veitch (eds), *Boats, Fishing and the Sea, Scottish Life and Society: a Compendium of Scottish Ethnology*, Vol. 4 (Edinburgh, 2008), pp. 30–84.
6. L.H. Campbell, K.T. Standring and C.J. Cadbury, 'Firth of Forth pollution incident, February 1978', *Marine Pollution Bulletin*, 9 (1978), pp. 335–339.

SOME FURTHER READING

There is plenty to read. The contemporary plight of the world's seas is eloquently described in Callum Roberts, *The Unnatural History of the Sea* (London, 2007) and Charles Clover, *The End of the Line* (London, 2004). In 2011, the Scottish Government published an admirably full and balanced account of the local maritime environment, *Scotland's Marine Atlas*.

Essential for understanding the story of change is the set of scientific papers on 'the natural environment of the estuary and Firth of Forth' presented in 1987 to a symposium of the Royal Society of Edinburgh and published in the society's *Proceedings Section B (Biological Sciences)*, Vol. 93. These cover a great range of subjects, from the underlying geology, fisheries science, the benthic flora and fauna, to the problems of pollution and water quality. This followed an earlier but less comprehensive investigation of 1972, published in the society's *Proceedings Section B (Biological Sciences)*, Vol. 71. The volumes of *The Forth Naturalist and Historian*, published annually since 1975, are valuable for the study of the estuary, as is *Calatria*, the journal of the Falkirk Local History Society, published since 1991.

The history of Scottish fishing is covered in J.R. Coull, *The Sea Fisheries of Scotland* (Edinburgh, 1996) and Malcolm Gray, *The Fishing Industries of Scotland 1790–1914* (Oxford, 1978). J.R. Coull et al. (eds), *Boats, Fishing and the Sea*, Vol. 4 of the *Compendium of Scottish Ethnography* (Edinburgh, 2008) is invaluable, not least for chapters on the history of marine science, and on the utilisation of seaweed. B. Poulsen, *Dutch Herring, an Environmental History 1600–1860* (Amsterdam, 2008) sets the international historical context of the earlier days, and is informative on the historical ecology of North Sea herring.

The Statistical Account of Scotland is a famous source for local history that invariably has a great deal to say about the late eighteenth-century working environment. It was originally compiled and edited by Sir John Sinclair in the 1790s. The most accessible print version is that under the general editorship of D.J. Withrington and I.R. Grant, of which Vol. 2 dealt with

Lothian (Wakefield, 1975), Vol. 10 with Fife (Wakefield, 1978) and Vol. 9 with Stirlingshire (Wakefield, 1978). Its successor, *The New Statistical Account of Scotland* (Edinburgh, 1844) was often equally valuable. In the post-war world, the *Third Statistical Account* (ed. David Keir, various dates), was generally less useful. East Lothian published its own *Fourth Statistical Account*, in seven volumes between 2003 and 2009, largely as an exercise in reminiscence. The study of place names can be enlightening, for which see John Reid, *The Place Names of Falkirk And East Stirlingshire* (Falkirk Local History Society 2009), and Simon Taylor with Gilbert Markus, *The Place-names of Fife* (four volumes, Donington, from 2006).

Sir Robert Sibbald, Scotland's first distinguished naturalist who knew the shores of the Firth of Forth well, early in the eighteenth century wrote *A History and Description of Stirlingshire* (Edinburgh, 1707), and *A History Ancient and Modern of the Sherrifdoms of Fife and Kinross* (Edinburgh, 1710), the latter is best read in the edition published in Cupar in 1803, which contains much extra detail in the footnotes.

Contemporary local history seldom has much to say about the environment. T. McGowran, *Newhaven-on-Forth, Port of Grace* (Edinburgh, 1985), however, considers the decline of the oyster and is more sympathetic to the fishermen than was Thomas Fulton in his classic account in the *Report of the Scottish Fishery Board for 1895* to which every subsequent account (including this one) is much indebted. The sea fisheries of East Fife were described by Peter Smith in local books, of which *The Lammas Drave and the Winter Herrin'* (Edinburgh, 1985) is particularly useful for the level of detail. Belle Patrick, *Recollections of East Fife Fisher Folk* (Edinburgh, 2003) is the more valuable for being a vivid contribution from a woman's perspective. Peter Aitchison, *Black Friday: the Eyemouth Fishing Disaster of 1881* (Edinburgh, 2006) reminds us of the dangerous realities of a life at sea in the late Victorian world.

The general subject of salmon fishing can be approached through Peter S. Maitland, *Scottish Freshwater Fish: Ecology, Conservation and Folklore* (Edinburgh, 2007) and Iain A. Robertson, *The Tay Salmon Fisheries since the Eighteenth Century* (Glasgow, 1998), though not all the detail is applicable to the Forth. It is also worth consulting W.L. Calderwood, *The Salmon Rivers and Lochs of Scotland* (London, 1921).

Two older books that dealt with different aspects of the environmental history of the area are indispensable; H.M. Cadell, *The Story of the Forth* (Glasgow, 1913), dealt particularly with the reclamation of land in the estuary, an undertaking in which he was himself involved. J.H. Gurney, *The Gannet, a Bird with a History* (London, 1913), was pioneering in use of historical sources to investigate natural history, and especially good on the Bass Rock and the other Scottish gannet colonies. The past and the present of the Isle of May was well described by W.J. Eggeling, *The Isle of May: a Scottish Nature Reserve* (Kirkmichael, 1985), which should be read alongside the more recent account of the island's environmental archaeology in Heather F. James and Peter Yeoman et al., *Excavations at St Ethernan's Monastery, Isle of May, Fife, 1992–7* (Tayside and Fife Archaeological Committee Monograph 6, 2008).

The changing fortunes of the ducks, gulls, auks and other seabirds of the Forth is set in context by Ron Forrester and Ian Andrews et al., *The Birds of Scotland* (Aberlady, 2007) and Ray Murray et al., *The Breeding Birds of South-east Scotland: a Tetrad Atlas, 1988–1994* (Edinburgh, 1998), along with studies of successive censuses of British seabirds of which the most recent is P.I. Mitchell et al., *Seabird Populations of Britain and Ireland* (Joint Nature Conservation Committee, 2004). The seal controversy of the 1970s is summarised by John Lister-Kaye in *Seal Cull* (Harmondsworth, 1979), though scientific opinion has changed since that was written: the most up-to-date opinion and situation regarding seal numbers can now be accessed on-line in the Special Committee on Seals Report via http://www.smru.st-andrews.ac.uk.

The endnotes are intended to give the reader a guide to where we accessed our information, and where to look for further reading in more detail. Please use them to help you access the raw materials from which history is made. Nothing can compare in immediacy to contemporary sources from travellers' accounts, or from the press, or from the annual *Reports of the Scottish Fishery Board,* or perhaps especially from the great nineteenth-century Parliamentary reports. These last hold verbatim accounts of exchanges in hotels and town halls up and down the country, between the investigating Royal Commissioners, including great scientists like Tomas Huxley, and a host of ordinary people who came to be witnesses – like the St Andrews fisherman who had been examining fish eggs kept in a tin box in a rock pool, with a microscope he had bought for 7s 6d, or the citizen who told the Commissioners how he had thrown a lighted rag into a burn polluted with paraffin waste in West Lothian, and the flames had shot up 20 feet. The historian's day is made when the past speaks so suddenly and clearly to the reader in the present.

INDEX

Abercorn 139
Abercrumbie, William 207
Aberdeen 29, 76–8, 88, 94, 108, 119
    University of Aberdeen 205
Aberdour 44, 47, 90, 117, 130, 225
Aberlady 33, 97
    Aberlady Bay 26, 107, 198
Adam, Robert M. 203
Adamson, Robert 49, 100
Addiewell 153
Admiralty, Admiralty Court
    Scottish 36–8, 49, 55, 57–8, 226
    British 128, 158, 194
Agriculture and Fisheries for Scotland,
    Department of 261, 265–6
Ailsa Craig 202, 207
air pollution 276
Airth 177, 183–7, 189, 191
Airthrey 14
albatross 215, 241
Alexander, Colonel Sir J.E. 155, 157
Allan Water 14
Allman, Prof. George 85, 106
Alloa 163, 175, 178, 181
    fishing and fishermen at 97, 126–8, 130, 135, 138, 140, 142
Alloa Inch 188, 199
Almond, River, pollution of 151, 153–4, 156, 159–62, 164
Andersons, Edinburgh fishmongers 47–8, 64–7, 70–1, 129, 141
annelids (*see also* marine worms) 163
Anstruther 28–30, 33, 42, 76, 81–2, 86, 92, 99, 115, 119, 150
    Admiral Depute Court in 37, 226
    attitudes towards ring netting 92–3
    birds at 167, 241
    fish landing described by Belle Patrick 86, 102
    Fishery District of 78–9, 84, 86, 90, 93, 120
    railway at 82, 90
    Reaper at 88
    sea-angling from 124
    seaweed at 20–1, 23
    Union harbour of 90–1
anthills 231
anthropocene 8, 269
Arbroath 101, 256
Arcachon 71

Ariel, first steam trawler 107
Arcus European Infrastructure Fund 277
Ashton, Dr Elizabeth 67
Atlantic Ocean 13, 98, 122, 129, 166, 170, 201, 238, 253
    North Atlantic weather systems 36, 249–50
Auchterlony, James 53
Auckland, Lord 47, 129
Audubon, James 203
auks (*see also* guillemot, razorbill, puffin) 11, 118, 201, 223, 243, 246, 271, 296
Auld Haikes, Hakes 82–4, 86
Auroch 11
Avon River 145, 156, 159, 161–2, 181, 184, 187, 195

bag nets 139–40
bait (*see also* mussels) 26, 31, 40, 77, 84, 102, 104, 107–8, 113, 116, 118, 122
    baiting lines by women 40, 99–101
    pike-bait 143
Balbirnie and Balfarg, standing stones at 12
Baltic 166, 253
    trade to 28–9, 33, 52, 230
Bank herring 74
Bardot, Bridget 260
Barking on Thames 105
Bartram, James 131
Bass Rock 6, 12, 107, ch. 9 *passim*, 225, 230, 233, 235, 239–40, 242–7, 270, 277
    impressions of visitors 201–3
    recent effects of gannets upon 223
    shooting seabirds on 212–5
Bateman, John la Trobe 158
Bathgate 153, 155, 158
Baxter, Evelyn 7, 232–4, 236, 241, 244, 253
Baxter, William 49
Beddington, J.R. 264
Berwick Bank 244
Berwickshire 6, 20, 24, 33, 97–8, 227, 235, 238–40, 244–5, 247, 254, 256, 277
Beverton, Ray 266
big government 172
Billingsgate market 60, 71, 102
Berlin 174
Birmingham 135
Black Devon, River 141, 154–5, 157, 161, 200
black guillemot 233, 246
black redstart 232

297

blackbird 10
Blackford Hill 224
Blaeu, Joan, Blaeu Atlas 2, 19, 184, 209, 231
Blairdrummond Moss 48, 127
Blairhall 183
blue whale 14, 25, 271
bluethroat 232
Blundell, William 2
Boece, Hector 202, 205, 221
Bo'ness 44, 72, 126, 129, 142, 178, 186
   archaeological midden finds 15
      mussel beds 102–3
      peat on shore 128
      reclamation around 189, 193–7, 199
      sparling fishers 142
      sprat fishers 132–3

Booth, E.T. 205
Bough, Sam 25, 107
Boswell, James 145
Bothkennar 184
Bower, Walter 202
Bowles, Emily 214–5
Bowsy, John 20
Boyd, Morton 254
Boyter, David 84
Braefoot Bay gas terminal 275–6
Braid, Councillor James 267
Brereton, Sir William 2, 207, 209, 221
Bright, John 85, 107
Bristol 2, 44, 144
British Fisheries Society 35
British Geological Survey 179
British Petroleum (BP) *see* Scottish Oils
Brixham 105
Bronze Age 12, 16–7, 228
Brown, Alexander Wallace 109, 112
brown trout (*see also* sea trout) 152, 161, 270
Broxburn 154
Broxmouth 17
Bruce, George of Carnock and Culross 127, 183
Bruce, John of Airth and Stenhouse 185
Brundtland Report (1987) 274
Buccleuch, Dukes of 55, 64–5, 69, 103, 148, 151–3
Buchan herring 74
Buckhaven 33, 42–4, 78, 81, 90, 118–9
   Defoe visits 24
   fishermen turn coal miners 117
   hardiness of fishermen 97
   trawls and trawling at 84, 105
   sea-angling from 124
Burke and Hare, trial of (1829) 131
Burnham-on-Crouch 61
Burntisland 3, 20, 29, 32, 90, 94, 117, 130, 132, 268
   oysters and oyster scalps 47, 55, 57–8, 62, 64, 67
   sea-angling from 123
   winter herring boom 43–4
bush fisheries 31–2

Caddell, H.M. 128, 179, 189, 190–1, 194–5, 197
Caithness (*see also* Wick) 44, 80, 110
Calderwood, William L. 141, 161, 163, 258

Callendar 191
Cambuskenneth 177
Canada 13, 253
Canty Bay 206–7, 210, 213
Carlyle, Rev. Dr Alexander 39–41
Carlyle, Thomas 225
Carnegie, Capt. 25
Carr Craig 240
Carron ironworks 3, 172, 193
Carron, River 156–7, 161–2, 176–7, 181, 184, 186–8, 190, 193
Carse of Gowrie 135
Carse of Stirling 13–4, 127, 129, 175–7, 181, 183, 185–8, 200
Carson, Rachel 173
cats 226–7
cattle 154, 156, 226, 228, 250, 266
Cellardyke (*see also* Kilrenny) 23, 26, 33, 42, 81, 83, 112, 119, 210
   attitudes towards trawling 84–5, 90, 105, 114
   attitudes towards ring netting 92–3
Centre for Ecology and Hydrology 236
Chadwick, Edwin 146–9, 171
Chambers, Robert 46
Charles II 208
Chatham 277
chickens 223, 228
cholera 146
Christison, Prof. 151
Clackmannan, Clackmannanshire 157, 178, 186, 191, 200
clams 11, 16, 67, 113
Clark, George 61
Clarke, Dr W. Eagle 232
climate change impacts
   modern 6, 168, 179, 245, 249
   oysters and 53, 60, 68, 72, 76
   prehistoric 14
coal mining, coal miners (*see also* colliery pollution) 2–3, 117, 133, 172, 178, 183, 186, 196–8, 272
Coastal Anti-Pollution League 173
Cobbett, William 128
coble, net-and-coble 25, 54, 138, 140
Cobden, Richard 85, 107
Cockburn, Henry 210
Cockburnspath, 20
Cockenzie (*see also* Port Seton) 33, 40, 77, 93, 103–4, 113, 117–9
   oyster fishing from 49, 56–7, 59–60, 64, 70
   power station 4, 198, 276
cockles 11, 16
cod and codling 7, 24–7, 78, 84, ch. 5 *passim*, 126–7, 129, 223, 235
   caught on lines 27, 31, 38, 40–3, 96–105, 107, 117, 142, 229
   caught by trawl (*see also* Granton trawlers) 77–8, 104–9, 111
   present scarcity of 121, 270
   in seal diet 251–2, 268
cod-worm 252, 264
Colchester 136

INDEX

Coldingham 99
Coles, Prof. John 11
Collen, George 115
collieries, pollution from 145, 152, 155–6, 160
Colonsay 257–8
Colquhoun, John 213–4
Colville, Lord 58, 103
Common Fisheries Policy (EU) 80, 95, 115, 120–1
commonties 188–9
conger eel 126
Conservation of Seals Act (1970) 258
Control of Pollution Act (1974) 162, 169, 243
Coopers 77
cormorant 11–2, 229, 231, 233, 236, 239–40
Cornhill Magazine 256
Council for Nature 265
Court of Session 61–2, 152
Cove 33, 97, 119
Cowan, Alex and Sons 152
Cowie chipboard factory 175
Cowdenbeath 150
crabs 8, 41, 43, 69, 78, 81, 96–7, 105, 117–8, 123, 204, 236
    in prehistoric middens 11, 16–7
Craig fishing ground 81, 84
Craigentinny irrigation meadows 148, 150
Craigforth above Stirling 1, 138, 140
Craigleith 225, 236, 239–40, 244–8
Crail 29, 32–3, 37, 41, 83, 97, 117, 119, 139, 169, 221, 230, 241
    'Crail capon' 101
    harbour 34, 81
    sea-angling from 124
    seaweed use 20–1, 23
    town crest 26–7, 32
Cramond, Cramond Island 10, 48–9, 55, 59, 164, 225–6
Cramond Brig 154
Crapper, Thomas 147
crears (*see also* bush fisheries) 28–9, 31–2
Crown Estates Commissioners 189
crows 11
Crutzen, Paul 269
Culross 2, 20–1, 127–8, 178, 183
cured haddock 101–2

dabs 26, 105
Dalgety Bay, radioactive materials at 276
Dalhousie, Earl of 109–10, 150
Dalrymple, Sir Hew 208, 211, 213–4
Dalyell, Tam 263
Daniell, William 221
Danish fisheries and Danish seine net (*see also* seines), 74, 80, 95, 116, 119, 122–3, 244, 247–8
Darling, Frank Fraser 71, 262–3
Darwin, Charles 126
David I 228
deer 11, 14, 17, 262, 266
Defoe, Daniel 22, 24–5, 54
deindustrialisation 172
Department of the Environment (Westminster) 174

Devon, River *see* Black Devon
Dick, Stewart 88
Dickens, Charles 60
Dickson, Ruth 231
discards *see* fishery discards
dogs 226–7, 234, 236, 254, 263, 272
Dogger Bank 75, 95
dolphins 24, 169, 267, 271
Donibristle 59
Downs herring 75
Drummond, Sir James of Hawthornden 151
Drummond, William of Hawthornden 230
Dryburgh Abbey, 27
ducks (*see also* various species of duck) 7, 11, 150, 167, 197, 210, 275
Duddingston, Duddingston Loch 145, 167
Dunbar 1, 13, 17, 29, 33, 35, 41, 43, 45, 53, 75–6, 79, 82–4, 96, 130, 137, 229
    Admiral Deputes Court in 37–8
    charter 27
    decline of 117, 119
    fish carried to Edinburgh from 38–9
    fish reared at 116
    gannets and 204, 209
    harbour 90, 244
    kippers smoked at 24, 38
    pilot whales ashore near 271
    seaweed use 20–1
Dunbar, William 141–2
Dundas of Dundas 19, 26
Dundas, Lord 190
Dundas, Robert of Arniston 57–8
Dundee 81, 136, 277
    University of Dundee 264
Dundonnell, Earl of 128
Dunfermline 1, 150
dunlin 167
Durie, Rev. Robert 30–1
Dutch fisheries 24, 28, 33, 35, 74–5, 116
dykes and sea defences 176–8, 186–7, 190, 195
Dysart 29, 32, 117, 124, 183

Earlsferry 33
East Anglia (*see also* Lowestoft, Yarmouth) 77, 89, 177
East Lothian, (*see also* Lothian, Haddington) 17, 22, 29, 37, 47, 78, 134, 164, 168, 183, 198, 205, 238
e. coli 165
ecosystems 14, 40, 78, 105, 123, 166, 198, 219, 235, 265, 270, 277
Eden, River and estuary, 11, 102–3, 268
Edinburgh 1–2, 4, 10, 113, 131, 133, 141, 175, 191, 224–5, 273, 277
    investment in fishing from merchants 29, 31–2, 107
    oysters ch. 3 *passim*
    pollution and 5, 25, ch. 7 *passim*, 129
    proposed fishing company (1825) 25–6
    selling fish in 24, 30, 38–40, 42–3, 77, 81, 90, 97, 121, 125, 135, 142
    selling and eating gannets in 208–210
    selling seal skins in 256

University of Edinburgh 85, 106, 131–2, 189, 258, 272
 voluntary bodies in 273, 276
eels 25, 126–7
eider ducks 104, 167–8, 231, 234, 236–8, 248
Elphinstone, George, Viscount Keith 190
English Channel 75, 95
Erskine, Colonel (local Culross landowner) 21
Esk Case in Scottish legal history 151–3
esparto grass 152
Essex 57, 60–1, 71
European Bathing Water Directive 169–71
European Commission, European law 174, 178, 207, 244, 274, 276
European Economic Community (EEC) 169, 260
European Free Trade Association (EFTA) 172
European Parliament 260
European Urban Waste Water Directive 171
European Union (EU) 166, 172, 276
Evans, William 232, 244, 253
Ewart, Adam 208
Ewart, Prof. James Cossar 132
Eyebroughty 240
Eyemouth 24, 33, 38–40, 79, 96, 102, 115, 119, 204
 disaster for fishing fleet 98–9
 Fishery District 79, 84

Faeroes 202, 229, 239
Falkirk 3, 15, 145, 150, 175, 183–4
Fall, Charles and Robert 38
Farne Islands 119, 231, 247–8, 252–5, 258–9
Fast Castle 227, 244, 254
Fenwick, Henry, MP 106
Fergusson, Robert 25, 43, 46–7
ferrets 226
Fidra 213, 225, 244–5, 247
Fife Adventurers 30–1
Fife County Council 157, 167
Fife Ness 1, 78, 82, 84, 90, 114
 archaeological finds at 9–10, 253
Fifies (fishing boats) 88
Figgate Burn 148
Firth of Clyde 13, 104, 202
Firth of Tay 11, 268
fish-and-chip trade, development of 118
fish species in firth (1838) 125
fish traps 64, 126–9, 138–40
Fisher, James 202–3, 238
Fisherrow 26, 33, 40, 77, 93, 118–9
 fishwives of 26, 38–40, 77
 oyster fishing 49, 56, 59–60
fishery discards 115–6, 118, 201, 223, 238, 243
Flanders Moss 14, 241
Fleming, Professor J. 218–9, 221
float drave 33, 36, 40, 81, 88
flounders 41–3, 97, 117, 124
Forth–Clyde Canal 44, 91–2, 193
Forth Conservancy Board 159, 194–5
Forth District Salmon Fisheries Board, Forth River Salmon Board 155, 157
Forth Ports Authority, Forth Ports Ltd 276

Forth Rail Bridge 4, 19, 159, 194, 225
Forth River Purification Board 162–3, 169, 171
Forth Steamship Fishing Company 107
Foul Burn 148
fox 227, 236, 241, 247
France 71, 170, 174, 202, 260
Franck, Richard 137
Fraserburgh 88
Friends of the Earth 173, 273, 276
Fruid Reservoir 147
Fucus (seaweed) 20–2
fulmars 11, 206, 227, 236, 238–9
Fulton, Thomas W.
 critic of Thomas Huxley 115–6
 on oysters 54, 57, 60, 62, 72, 73, 129
Furness, Prof. Robert 223

'gandanooks' 127
gannets 45, ch. 9 *passim*, 230, 233, 242–3, 247, 251, 271, 277
 archaeology of 11–2, 229
 different ways of cooking and eating 206–11
 drugs from 205–6
 emissions from 223
 persecution of 211–3
 population estimates of 6, 217–23, 235–6
Gardner, Bill 215
'gardyloo' cry 152
Gardyloo ship 165
garvies (*see also* sprats) 127, 129–30
Geikie, Alexander 189
General Steam Fishing Company 107–8
George IV 51
Germany 91, 120, 173–4
Gladhouse Reservoir 147
Glaister, John 156
 University of Glasgow 152, 205
Glasgow Herald 155
Glasgow Philosophical Society 156
goats 226, 228
goldeneye 7, 166–7
Gordon, Robert, of Straloch 19, 209, 230–1, 243, 256
Gosford 47
Gourlay, George 90
gout cure 205
Grangemouth 14–5, 159, 164, 178, 182, 199, 270
 reclamation and sea defences around 175–7, 188–90, 193, 200
 refinery at 4, 15, 145, 163, 175, 178, 196, 275
Grangepans 183
Grant, William 58–9
Granton 3, 77, 79, 103, 114, 130, 133, 150, 153, 164
 oysters 47, 55, 58, 62, 64–7, 72
 trawlers 77–8, 101, 107–8, 110, 114, 117–120
Gray, Malcolm 44, 82
Gray, Robert 204, 240
great auk 229
great crested grebe 274–5
great lines (*see* line fishing)
greater black-backed gull 11, 243
Green parties 173, 273, 276

Greenpeace 244, 260, 263
Greenwich, Lady 55, 57, 64
Greenwood, Jeremy 264
grey seals (*see also* seals) 6, 215, ch. 11 *passim*, 271–3
   in archaeological record, 17, 228–9, 253
   modern science and 259, 262, 264–7
   popular and official attitudes towards 255–266
Grey Seals (Protection) Act (1914) 256, 272
Grimsay in Iceland 238
Grose, Francis 221
ground drave 33–4, 36–7, 40, 82
guano 201, 220, 247
   import from Peru of 22, 146
guillemot 2, 231, 233, 236, 245–6, 249–50
   in archaeological record 11, 229
gulls (*see also* various species of gull) 7, 150, 201–2, 214, 220, 229–35, 238, 248, 255, 271
Gurney, J.H. 203, 205, 209, 219–20, 223

Haddington 1
Haddington County Council 204
haddock 7–8, 38, 47, 78, 84, ch. 5 *passim*, 127, 223, 244, 251, 268, 270
   in archaeological remains 11, 229
   caught on lines 27, 40–3, 96–9, 115, 117
   caught by trawl 104–6, 115, 118
   cured 101–2
halibut 26, 38, 41, 96–7, 102, 124
Hamilton, Duke of 187, 195
hang net 138, 141
Hardin, Garrett 68, 71
harbour seals 253, 258, 261, 268
Harrison, John 186
Harvey, William 203, 206, 220–1
Harvie-Brown, J.A. 209, 232, 234
Haskeir off North Uist 256, 258
Haystack, Fife 240
Heath, Edward 74
Hebrides (*see also* Lewis, Uist) 12, 202, 206–7, 213, 252, 255, 259
hedgehog 11, 227
Helmsdale 82
herring (*see also* Lammas Drave and Winter Herrin') 5–6, 102, 114, 219, 235, 268, 270
   archaeological remains 229
   confusion with sprats and conflict with sprat fishers 129–32, 134
   fishing in medieval times 27–8
   fishing in sixteenth and seventeenth centuries 28–33, 35–7, 53
   fishing in eighteenth century 24, 33, 35, 37–44, 126–7, 226
   fishing in nineteenth and twentieth centuries 44–5, 69, ch. 4 *passim*, 96–9, 104–6, 108, 117, 124, 126, 136–7, 223
   seabirds feeding on 45, 201, 204–5, 214, 223, 239
   spawn destroyed by trawling 84–6, 106, 108, 110–1, 129
herring gulls 230, 240–2
Heugh, George 112
Higgins Neuk 190–1

Holland *see* Netherlands, Dutch fisheries
Holyrood Abbey and Palace 148, 177
honey buzzard 232
horses 20, 38–9, 156, 169, 183–4, 226, 272
   archaeological remains 17, 228
Hound Point 4, 47, 275
Hunman, William 113
Hutchison, Robert 107
Huxley, Prof. Thomas 85–6, 106, 109–12, 115–6, 119, 296

Ice Age 9, 12–3, 181, 269
Iceland 101, 120, 166, 238, 249
Imperial Chemical Industries (ICI) 163
Inchcolm 21, 44, 202, 225–7, 240, 247
Inchgarvie 19, 44, 130, 225–6
Inchkeith 47, 90, 96, 113, 130, 133, 224–7, 231
   archaeological remains on 16–7
   seabird colonies on 237–8, 240, 242, 244–5, 247
   seal colony on 253–4
Inchmickery 47, 60, 64, 67, 225–6, 237, 247
Innergellie 210
Innes, Alexander 208
Innes, Cosmo 214
Innes, Jane 210
Institute of Oceanographic Science 264
Institute of Sewage Purification 171
internal combustion engine, *see* motor-driven boats
International Council for the Exploration of the Sea 121
International Fund for Animal Welfare 260–1
intertidal zone 10, 164, 175, 178–85, 189, 198–200
Inveresk 139
Inverkeithing 2, 44, 49, 117, 130, 132, 195, 204
Inverness 14, 30
Ireland (*see also* Ulster) 44, 104, 179, 193, 258
Isle of May (*see also* various seabird species, seals, rabbits) 5–7, 12, 81–2, 99, 114, 202, 215, ch 10 *passim*, 253–6, 270, 276–7
   archaeology of 12, 227–9, 253
   bird observatory 226, 232–3, 253
   fishing and spawning grounds off 5–7, 78, 84, 86, 90, 93–4, 98, 100, 115, 118, 120, 122
   monks and hermits associated with 27–8, 225, 211, 228–30
Isle of Wight 263

James III 19
James IV 19, 55, 209, 211, 229
James VII 211
Jardine, William 219, 232, 234
jellyfish 272, 275
Johnson, Dr Samuel 145
Jonston, John 208

Kames, Lord 127, 177
Kennetpans 183, 190
Kent 44, 60, 71
Kilrenny (*see also* Cellardyke) 41, 81
   crest 27–8
Kincardine 97, 128, 130, 135, 142–3, 177, 181–4,

187, 190
  bridge at 4, 181, 193
  power station at 4, 176, 191, 198
Kincraig 139
Kinghorn 16, 24, 29, 32, 55, 117, 140, 225, 268, 273
'King William's Ill Years' 53
Kinneil 177–8, 182, 184, 187, 195, 197,199
kippers 24, 91
Kirkcaldy 32, 117, 130, 150, 276
Kirkwall 260
kittiwake 211, 231–4, 243–5
  archaeological remains of 11, 229, 243
  increase in twentieth century 236, 244
  recent problems 244–5, 247, 249–50
  shooting for fun 234
klondyking 92
Knox, Robert 131
Königsberg 230
Krakatoa 73

Ladies Scaup 184
laissez-faire ideology 71, 85, 107
Lake of Menteith 13
The Lamb 225, 240, 244–5, 227
laminaria (seaweed) 19–20, 22–3
Lammas Drave
  before the nineteenth century 29, 32–4, 36, 38, 40–1, 43–5, 74
  during the nineteenth century 75, 77–83, 86–7, 90
Lamont, John 35
landfill 7, 23, 147, 197, 200, 235, 242–3, 276
land reclamation ch. 8 *passim*
  for agriculture 184–191
  for industry 191–8
  map summarising 186
  reversal of (renaturing) 198–200
Lang, Professor Scott 111
Largo 33, 42, 97, 117, 119
Largo Bay 42, 123, 139–40
Lauder, George of Bass 203–4, 211
Laurentide Ice Sheet 13
Learmonth, Patrick 230
Learmonth, Thomas Livingston 189, 194
Leeds
Leith (*see also* Newhaven) 1, 16, 21, 29, 96, 117, 130, 135, 270
  Fishery District of 77, 79, 84, 93, 119–20
  oysters 16, 47, 49, 52–3, 66, 69, 72
  sewers of 7, 146–8, 151, 153, 164–5, 167–8, 241
  trawlers at 100–1, 107, 117, 119
Leith Docks Commissioners 276
Lennox, Duke of 30
Lerwick 88
Lesley, Bishop John 202, 205
lesser black-backed gull 229, 235–6, 240–2, 248
lesser crested tern 198
Leven, Levenmouth 4, 7, 29, 36, 167, 240–1, 270
Leven, River 154, 156–7, 161–2, 167, 169
Lewes fishing, Lewis (*see also* Fife Adventurers) 28, 30–2, 76, 207
Liberton 148

limpets 16–17, 102, 229
Lindsay, Patrick 31–5, 43
line fishing 27–8, 31, 33, 65–6, 84, 96–105, 107, 117–8, 126
  conflicts with trawlers 108–115
ling 41–2, 96–8
Linlithgow 1, 154, 175
Lister-Kaye, Sir John 261–2
Little Saltcots 177
Liverpool 44
Liverpool, HMS 137
lobsters 8, 41, 78, 97, 117, 118, 123, 267
  exported to England 41–3, 81, 97
Loch, David 37–8
Loch Katrine 158
Loch Lomond 13
Loch Vennacher 158
Loch Voil 158
Lochrin stream 148
London 2, 26, 42, 77, 90, 111, 144, 174, 261
  gannets sent to 208, 210
  lobsters sent to 41–2,
  oysters sent to 56, 60, 63, 71
long-tailed ducks 168
Longannet 128, 190, 241
  power station at 4–5, 143, 175–6, 191, 193, 198, 276
Longniddry 25, 291
Lord Advocate 57, 133
Lothian (*see also* East Lothian, Midlothian, West Lothian) 1–3, 33, 48, 53, 89, 92, 164, 168, 172, 198, 202, 228, 277
Lothians River Purification Board 162–5, 169, 171
Lothian Regional Council 165
Louisa, HMS 213
Lowestoft 38, 76, 91, 264
lumpfish 140
Lyell, Charles 256
Lyndsay, Alexander 2

M'Bain, Dr James, 62–3
MacDonagh, Oliver 170–1
Macdonald, Capt. Samuel 130
Macdougall, Norman 204
MacGillivray, William 203, 209, 219, 221, 232, 234
M'Intosh, Prof William 109–114, 116, 135
  riot against 111–2
mackerel 41–3, 50, 95, 114, 123–4, 127, 201, 205, 208, 223
McLusky, Donald 163
McNeill, Prof. John 169
Madden, Dr 151
Major, John, sixteenth-century academic 137, 202, 205, 208
Major John, twentieth-century prime-minister 170
Manchester 90, 135, 210
Mar Bank, Marr Bank 94, 98, 244
marine worms (*see also* annelids) 163, 165–6, 168
Marine Laboratories
  Aberdeen 94, 258
  St Andrews (Gatty Marine) 110–2, 116

INDEX

Marine Protection Areas 274
Marine (Scotland) Act (2010) 258
Markinch 148
Marr, James 85
Maunder Minimum 53,
May, Isle of *see* Isle of May
medical officers 148, 156
Megget Water 147
Meik, Messrs Thomas and Sons 194
Meiklejohn, A.H. 233
Meiklewood 14
Melilot 199
Melville, Lord 151–2
Mesolithic people 9–17
Methil 4
Middens, 145–6, 150
    as sources for archaeological evidence 11, 15–8, 21–2, 47, 227–9, 246, 253
Midlothian 57, 150, 156, 160
Millan, Bruce 260–1
Miller, Hugh 203
mink 236
Mitchell, John 96–7
Mitchell, William 208
monks, monasteries 1–2, 27–8, 117, 127, 177, 185, 226, 228, 230, 253
Monypenny, Thomas, of Kinkell 31
motor-driven boats, 91–3, 101, 122, 174
Moray, Earl of 59, 150
Moray Firth 81, 88, 256, 271
Morris, Ron 242
Morton, Fife 11–12
Morton, Earl of 58, 64
Mossmorran ethylene gas plant 4, 275–6
Muir, John, skipper 99
Murray, John 113
Murrayfield rugby ground 148
Musselburgh 39–40, 58, 81, 102–3, 113, 139, 140, 164, 208, 274
Musselburgh pools 198
mussels 1, 16, 65, 67, 99, 101–4, 108, 118, 163, 165, 236, 270

Napoleonic Wars 60, 80, 254
National Nature Reserves 227, 254
National Trust 254
National Trust for Scotland 227
Natural Environment Research Council (NERC) Seals Advisory Committee 259
Seals Research Unit, Sea Mammal Research Unit 254, 264
Nature Conservancy, Nature Conservancy Council 241–2, 254, 258–9, 264, 266, 273
Nelson, Dr Bryan 203, 206, 215, 222
Neolithic people 10, 12, 16–7, 67, 25, 228, 255
net-and-coble fishing *see* coble
Netherlands (Holland) (*see also* Dutch fisheries) 28–9, 31, 104, 110, 117, 174, 202
    and oyster trade 52, 55–6, 60, 62–4, 71–2
nethrops 8, 78, 122–3
Newbattle Abbey 177
Newcastle 52, 54, 56, 201

Newhaven-on-Forth 26, 33, 96–7, 113, 117, 119, 125
    and herring 77, 79, 90, 92–4
    and oysters ch. 3 *passim*
    and sprats 126, 130, 132–6
    Society of Free Fishermen of Newhaven 26, 59, 62–72, 134
Newton, Prof. Alfred 214
North Atlantic Oscillation 249–50
North Berwick 17, 21, 33, 81, 84, 101, 103, 119, 150, 213, 215, 225, 240, 248
    nunnery at 204
    trawlers and 107, 110, 113, 117
North British Railway Company
Norway, Norwegians, 13, 24, 98–9
    sealers 258–61
    sprat dealers 135
Norwich 2, 144
Nowell, Laurence 2

oil pollution
    of rivers by paraffin 151, 153–6, 160, 272
    of sea 168, 173, 178, 274–7
oligochaetes 163
Oram, Richard 183
Ore, River 161, 167
Orkney 12, 17, 21, 28, 31
    and grey seals 252, 255–6, 258–61, 263
ortolan 232
Ostrom, Elinor 68–9
overfishing
    of cod, haddock and other fish ch. 5 *passim*
    of herring ch. 4 *passim*
    of oysters ch. 3 *passim*
    of sparling 143
    in relation to seabirds and seals 223, 235, 238, 244, 252, 268, 281
Oxcars lighthouse 6, 114
oxygen sag 143, 157, 163–4
oystercatcher 104, 232
oysters 15, 25, 43, ch. 3 *passim*, 96–7, 104–5, 124, 129, 270
    in archaeological record 5–6
    seventeenth and eighteenth century exploitation 46, 51–9
    nineteenth-century catastrophe 59–73

paper works 145, 151–7, 160, 164, 172, 195
paraffin oil works *see* oil pollution
parliaments
    European 260
    Scottish, pre-1707 1, 2, 35, 188–9, 203–4
    United Kingdom 108–9, 114, 116, 147, 155, 157, 159, 164, 170, 172, 174, 189, 194, 258, 277
Paris, Mathew 1
Parnell, Richard 125
Patrick, Belle 86, 91, 99, 102
peat 13–4, 177, 229
    exploitation of 2, 101–2, 177, 183, 188
    pollution of Forth estuary by 48, 90, 128–9, 142
Penicuik 2
Pennant, Thomas 203, 209–11, 219
Pentland Firth 32

Pentland Hills 147–8
periwinkles, winkles 16–7, 19, 229
Peterhead 78, 81, 88
Peterson, Roger Tory 203
Pettycur 44
Phinn, Kelty 57–9
pied flycatcher 232
pilot whales 24, 271
Pittenweem (*see also* Treath) 29, 33, 42, 78, 81, 99, 102, 117, 19, 122, 210, 223
   attitude to trawling 85, 112
   attitude to ring-netting 93
   fish market at 7–8, 39, 240
   priory 211, 228, 230
   saltpans 2
   sea-angling from 124
   seaweed use 20–1, 23–4
Pittenweem Register 82, 97
plaice 26, 84, 96–8, 105–6, 110, 116–7, 124
plankton 238, 248–9
plastic litter 22, 168, 272
Playfair, Lyon 85, 106
pochard 7, 166–7
pollution (see also air pollution, oil pollution, paper works, peat, sewage) 3, 5, 8, 68, 72, 90, 123, 126, 129, 137, 141–3, ch. 7 *passim*, 199, 243, 270, 272, 274, 276
Pont, Timothy 2, 184
population, growth of human 3, 5, 14, 146–7, 234, 269–70
porpoises 24, 45, 82, 257, 267
Port Seton (*see also* Cockenzie) 48, 51, 103
Portmore Loch 147
Portobello 147–8
prehistoric environment ch. 1 *passim*, 227–8
Prescott, Dr Robert 119
Preston Island 176, 198
Prestongrange 59, 198
Prestonpans 3, 40, 113, 119, 167
   fishermen turn coal miners 117
   oyster fishing 16, 47, 49, 53–4, 56, 58–60, 65, 104
   salt making at 2, 183
Princes of Wales 213, 215
Pritchard, Hesketh 256
Privy Council of Scotland 29–30, 53, 203
puffin 11, 206, 229, 231–3, 236, 244, 246–51, 253–4
Pumpherston paraffin oil works 156, 160
purple sandpipers 167

Queen Victoria 148, 158, 211, 257
Queensferry 47–8, 129, 163–4, 175, 181–2, 204, 239
   bridges and crossing 3–4, 24, 124, 225
   fishing 24, 43, 90, 117, 126, 130, 132–5, 139–40
Queenshaugh by Stirling 177–8

rabbits 218, 221, 226, 229, 230–1, 233–4, 247–8, 262
Rae, Bennet 258
railways 3, 77, 82, 90–1, 101–2, 104, 108, 130, 135, 148, 193, 195–7, 201
Rainbow Warrior 260
rats, including black rat and brown rat 227, 247

Ray, John 203, 208–9, 221–2, 247
razor-shells 16, 123, 168
razorbill 231, 233, 236, 245–6, 249–50
   in archaeological record 11, 22
reclamation of land *see* land reclamation
red-backed shrike 232
redshank 178
Reid, John, modern scholar, 184
Reid, John, seventeenth-century illustrator, 184
re-naturing the estuary, 179,198–200
ring-netting 92–3, 105, 130
Rintoul, Leonora, 7, 232–4, 236, 241, 244, 253
Ritchie, Prof. James 258
Rivers Pollution Prevention Act (1876), 155, 170
Rivers Purification (Scotland) Act (1951) 162
Roberts, Callum 120–1
Rocheford, Jorevin de 137
La Rochefoucauld, Alexandre de 3
Rockal 99
Ross, near Burnmouth 97
Rosyth 4, 6, 137, 194, 226, 241, 275–6
Roy map (General Roy's military survey) 184, 187, 193
Royal Commissions
   on the Herring Fishing (Buckland Commission, 1877) 131–2, 134, 214
   on River Pollution (1872) 150, 152–6, 158, 161
   on Sea Fisheries (1866) 47, 71, 85, 106, 134
   on Sewage Disposal (1898–1915) 158, 170
   on Trawl-net Fishing (Dalhousie Commission, 1883) 85, 108–10, 113,150
Royal Institute of Chemistry, 171
Royal Society for the Protection of Birds (RSPB) 7, 200, 273, 276
Royal Society of Edinburgh 164–166
Royal Society of London 203
royal tern 198

St Abb's Head 6, 118, 227, 238–9, 243–7
St Adrian or Ethernan 228, 254
St Andrews 11, 19, 27, 29–30, 41, 96, 99–100, 102, 114, 116–7, 119
   Gatty marine laboratory at 110–112, 116
   rabbits on site of Old Course 230
   riot in 111–2
   Royal Commission on Trawl-net Fishing meets in 109
   University of St Andrews (see also M'Intosh, Prof. William) 78, 264
St Andrews Bay 41, 97, 110–111, 114, 16, 123
St Baldred 202–3
St Kilda 11, 206, 229, 238, 246
St Margaret's Hope 194
St Monans 33, 42, 81, 102, 110, 276
   harbour, with largest fishing fleet in Firth 90, 117, 119
   sabbatarianism 93
   sailing trawlers at, and disputes over 81–2, 105
St Mungo's College 156
saithe 42, 124, 235, 239
salmon 6, 11, 114, 124, 137
   caught in River Forth catchment and estuary

INDEX

(*see also* damage from pollution, below) 5, 64, 126–9, 139–40, 155, 267, 270
    caught at sea or in Firth below Queensferry 97, 137, 139–40
    damage to fishery from pollution or abstraction 141,155, 157–9, 161–3, 270
    damage to fishery from seals or cetaceans 254, 256–9, 264–5, 267–8
    poaching, 140–2
Salmon Act (1868) 170
salt marsh, salt-greens and sea-greens, 178, 181–3, 186–9, 191, 199–200
salt as preservative 28, 31–2, 35, 87, 92, 206–8
salt pans and salt manufacture 2, 49, 154, 176–8, 183, 186, 196
Saltcoats 133
sandeels 98, 122–3, 244, 249, 276
    as food for seabirds 201, 223, 235, 239, 243–6, 248–52, 276
    as food for seals 265, 269, 276
sandwich terns 234–5, 242
Sars, Georg Ossian 106
Saury or skipperfish 127
scallops, 8, 16, 102, 104–5, 108, 113, 123–4, 168
scaup 7, 166–7, 184
The Scotsman 25, 82, 107–8, 115, 125, 134–5
Scott, Sir Walter 50, 208, 224
Scottish Advisory Committee on River Pollution 159, 170–1
Scottish Board of Health 159
Scottish Environment Protection Agency (SEPA) 123, 200, 243, 274, 276
Scottish Fisheries Board 6, 48, 62, 79, 90–1, 93, 99, 103–4, 109, 129, 131–5, 138, 140–1 157–9, 170
    restrictions imposed or proposed by 85, 114–6, 132–4, 158
Scottish National Party 277
Scottish Natural Heritage 254, 274
Scottish Office 159, 171, 174, 259–65
Scottish Oils Ltd (later BP), 163, 195–6, 275
Scottish Ornithologists Club 273
Scottish Society for the Prevention of Cruelty to Animals (SSPCA) 204, 214, 234
Scottish Seabird Centre 215
Scottish Water 169
Scottish Wildlife Trust 199, 273
sea aster 199
Sea Birds Preservation Act (1869) 204, 214, 234
sea defences and sea walls 179,181, 189, 198–200
sea lamprey 126
sea-levels 11–4, 177, 179, 181, 200
Sea Mammal Research Unit (SMRU) 254, 264–5
Sea Shepherd Fund 261
sea trout (*see also* brown trout) 11,126–7, 129, 137–40, 161, 163, 257, 267, 270
sea walls *see* sea defences
seabirds (*see also* gulls and auks of various species, cormorant, fulmar, gannet, shag) 6, 11, 122, chs 9 and 10 *passim*, 251, 253–4, 257, 270, 27, 276
seagreens *see* salt-marsh
seals (*see also* grey seals, harbour seals) 19, 45, 122, 225, 227, ch. 11 *passim*, 270–2, 274, 276

seaweed 18–23, 144, 183, 275
seines and seine-net trawling 80, 92–3, 95, 105, 119, 120, 122, 130
Selby, P.J. 209, 211
sewage and sewers 5, 72, 90, ch. 7 *passim*, 198–9, 242, 272
shag 11, 229, 231, 233, 236, 239–40, 249
Shanks of Barrhead 147
Shannon estuary 179
sheep and sheep grazing 16, 183, 221, 226, 228–31, 233–4, 247–8, 250
shellduck 178
Sheffield 210
shellfish (*see also* various species) 58, 72, 113, 183–4, 236
    archaeological record 10–11, 16–7, 47, 229
    as bait 102, 104–5, 270
Shetland 13, 17, 28, 74–5, 89, 221, 238
ship-to-ship oil transfers 276–7
Sibbald, David 55–6
Sibbald, Sir Robert 24–5, 33–4, 138, 141, 177, 184, 186, 188, 202, 208–9, 231–2
Sites of Special Scientific Interest (SSSI) 178, 197, 274
Skara Brae 16
skate 25, 38, 41–3, 96–7, 101, 105, 124, 127
Skinflats reserve 198, 200
skipper fish 127
'sleeching' 2, 183
Slezer, John 220
Sloan, Derek 15–16
sma' lines *see* line fishing
Smith, Eric J. 263
Smith, James 90, 114
Smith, Peter 79
Smith, William 83
snails 17, 227
Society of Free Fisherman of Newhaven *see* Newhaven
sol 26, 43, 65, 69, 97, 105
Solemn League and Covenant 208
sparling 97, 124, 127, 129, 141–3
Special Areas of Conservation 274
Special Protection Areas (SPA) 178, 197, 274
sprats 69, 92, 94, 96–7, 105, 117, 124, 126–7, 129, 134–7, 142–3, 235
    disputes about 130–4
    as seabird food 235, 245–6, 250
    pollute Rosyth dockyard 137
stake nets (or 'fly nets') 139–40
Starleyburn harbour 62–3
steam drifters 88, 91, 93
steam liners 99, 101
steam oyster dredges 65, 70
steam trawlers 86, 99, 107–8, 110, 114, 117
    first ever built 107
Stevenson, R.L. 203
Stevenson family of engineers 90
Stewart, Colonel (Equerry to Princess Beatrice) 257
Stirling 1–2, 6, 13–4, 42, 126, 130, 159, 175, 181, 194, 241

305

Carse of 90, 127–8, 176–8, 186
    pollution around 150, 154–6, 163
    salmon fishing near 5, 64, 137–8, 158, 161
    sparling near 141–2
    University of Stirling 67, 163
Stirlingshire 3, 146, 160, 184
Strathcona, Lord 258
sturgeon 11, 127, 140
Sula Sgeir 207
Sutherland 32
Swedish sprat processors 135
'sweep net' 138

Tantallon Castle 114, 206
Tarbert, Viscount 55, 58
Taylor, John 207–9
Taylor, Simon 184
Tees estuary, Teesside 178, 276
Teith, River 157–161
telegrams 90
Tennant, William 45,101
Tentsmuir 11, 252, 268
textile mills, pollution by 145, 154, 172, 272
Thames, River 57, 105
Thatcher, Margaret 169–70, 261
thrushes 11
tide marks defined 179, 181–2
The Times 60
Tinbergen, Nikko 203
Tobermory 35
Torness nuclear power station 4, 275
Torrey Canyon 173–4
Torry Bay 127–8, 182, 198–9
Torryburn mussel beds 20, 58, 103
trawlers, trawling (*see also* Granton, motor-driven boats, seines, steam trawlers) 6, 84, 95, 98, 100, 104–24, 136, 150, 223, 238, 243, 270
    damage by overfishing 109, 111, 116, 120, 122, 132, 219
    damage to sea floor 84–6, 106, 108–9, 112–5
    The Treath or Fluke Hole 81, 83–6, 90, 93, 97, 106
Trotter, Marion 210
trout *see* brown trout, sea trout
tsunami in Scotland 13–14
tuberculosis 150
Tulla Water
Tulliallan 128, 190–1, 193
tuna 114
turbot 11, 26, 41–3, 84–5, 96–7, 102, 105, 124, 127
    first becomes appreciated 42
Turner, J.M.W. 203
Turner, William 205, 221
turnstone 167
turtle 272
turtle dove 232
twaite shad 126
Tweed, River and catchment 137, 147, 254
Tynninghame 97, 103, 198
typhoid and typhus 146

Ullapool 35
Ulster 31
United Nations (UN) 120
Urban Waste Water Treatment Directive 170

Valleyfield Lagoons 199
Veere 55
voles 11

wading birds (*see also* various species) 11, 164, 178
Walker, Rev. Prof. John 203, 206, 209
Wanless, Prof. Sarah 219
The Wash 193, 261
Water of Leith 147, 151, 153, 172
water temperature 54, 238, 244, 248–9
Watson, Margaret and John 208
Webb, A.J. and Metcalfe, A.P. 182
Wedel, Lupold von 205
Wee Bankie 98, 243–4
Wemyss (East Wemyss and West Wemyss) 2, 21, 33, 42
Wernerian Natural History Society 123
West Lothian 47, 145, 151, 172, 175
West Water 147
Wester Ross 28, 32
western sandpiper 198
whales (*see also* blue whales, pilot whales) 14, 16, 24–5, 44–5, 82, 169, 207, 226, 264, 270, 271–2
whalers, whaling 117, 119, 238
whelks 16–17
Whitburn 150, 159–60
White, Robert 155
White Spot fishing ground 86
whitebait 125, 135–6
whiting 41, 43, 96–7, 105, 112, 127, 223, 239, 251, 268
    giant shoal of 35, 76, 81, 88
Wick 35, 76, 81, 88
Wight, Alexander, Solicitor General 57
Willughby, Francis 203
Wilson, George Washington 203
Wilson, James, fisherman 59
Wilson James, manufacturer 107
Wilson James, naturalist 131
Wilson, Mrs Harold 264
Winter Herrin' 74, 77–81, 87, 89–91, 96, 117
    distinct stock 75
    first noted off Queensferry 44, 74, 126
    ring-netting and its demise 80, 93–4
Witherby, H.F. 203
wolf fish 6, 232
Woolf, Alex 202
wryneck 232

yare (yair) *see* fish traps 127, 129, 138
Yarmouth 24, 38, 44, 76, 105, 136
Young, James 'Paraffin' 153
Younger, George 261

Zetland, Lord 189